Advances in Intelligent Systems and Computing

Volume 909

The series "Advances in Intelligent Systems and Computing" contains publications on theory, applications, and design methods of Intelligent Systems and Intelligent Computing. Virtually all disciplines such as engineering, natural sciences, computer and information science, ICT, economics, business, e-commerce, environment, healthcare, life science are covered. The list of topics spans all the areas of modern intelligent systems and computing such as: computational intelligence, soft computing including neural networks, fuzzy systems, evolutionary computing and the fusion of these paradigms, social intelligence, ambient intelligence, computational neuroscience, artificial life, virtual worlds and society, cognitive science and systems, Perception and Vision, DNA and immune based systems, self-organizing and adaptive systems, e-Learning and teaching, human-centered and human-centric computing, recommender systems, intelligent control, robotics and mechatronics including human-machine teaming, knowledge-based paradigms, learning paradigms, machine ethics, intelligent data analysis, knowledge management, intelligent agents, intelligent decision making and support, intelligent network security, trust management, interactive entertainment, Web intelligence and multimedia.

The publications within "Advances in Intelligent Systems and Computing" are primarily proceedings of important conferences, symposia and congresses. They cover significant recent developments in the field, both of a foundational and applicable character. An important characteristic feature of the series is the short publication time and world-wide distribution. This permits a rapid and broad dissemination of research results.

** Indexing: The books of this series are submitted to ISI Proceedings, EI-Compendex, DBLP, SCOPUS, Google Scholar and Springerlink **

More information about this series at http://www.springer.com/series/11156

Michael E. Auer · Thrasyvoulos Tsiatsos
Editors

Mobile Technologies and Applications for the Internet of Things

Proceedings of the 12th IMCL Conference

 Springer

Editors
Michael E. Auer
Carinthia University of Applied Sciences
Villach, Austria

Thrasyvoulos Tsiatsos
Department of Informatics
Aristotle University of Thessaloniki
Thessaloniki, Greece

ISSN 2194-5357 ISSN 2194-5365 (electronic)
Advances in Intelligent Systems and Computing
ISBN 978-3-030-11433-6 ISBN 978-3-030-11434-3 (eBook)
https://doi.org/10.1007/978-3-030-11434-3

Library of Congress Control Number: 2018967409

This Springer imprint is published by the registered company Springer Nature Switzerland AG
The registered company address is: Gewerbestrasse 11, 6330 Cham, Switzerland

Committees

Steering Committee Chair

Michael E. Auer, CTI, Villach, Austria

General Conference Chair

Thrasyvoulos Tsiatsos, Aristotle University of Thessaloniki, Greece

IMCL2018 Chair

Kostas Apostolou, McMaster University, Hamilton, Canada

International Chairs

Samir A. El-Seoud, The British University in Egypt, Africa
Neelakshi C. Premawardhena, University of Kelaniya, Sri Lanka, Asia
Alexander Kist, University of Southern Queensland, Australia, Australia/Oceania
Doru Ursutiu, University Transylvania Brasov, Romania, Europe
Tania Bueno, i3G Institute Florianopolis, Brazil, Latin America
Hamadou Saliah-Hassane, TELUQ University, Canada, North America

Technical Program Chairs

Dan Centea, McMaster University, Hamilton, Canada
Sebastian Schreiter, IAOE, France

IEEE Liaison

Russ Meier, IEEE Education Society Meetings Chair

Workshop, Tutorial and Special Sessions Chair

Tom Wanyama, McMaster University, Hamilton, Canada

Publication Chair

Sebastian Schreiter, IAOE, France

Local Organization Chair

Kostas Apostolou, McMaster University, Hamilton, Canada

Local Organization Committee Members

Seshasai Srinivashan, McMaster University, Hamilton, Canada
Ishwar Singh, McMaster University, Hamilton, Canada

Program Committee Members

Kostas Apostolou, McMaster University, Hamilton, Canada
Abul Azad, Northern Illinois University, USA
Christos Bouras, University of Patras, Greece
Santi Caballé, Open University of Catalonia, Spain
Manuel Castro, Universidad Nacional de Educación a Distancia, Spain
Maiga Chang, Athabasca University, Canada

Dan Centea, McMaster University, Hamilton, Canada
Monica Divitini, Norwegian University of Science and Technology, Norway
Christos Douligeris, University of Piraeus, Greece
Daphne Economou, University of Westminster, UK
Golberi S. Ferreira, CEFET/SC, Brazil
Apostolos Gkamas, University Ecclesiastical Academy of Vella of Ioannina, Greece
George Ioannidis, Patras University, Greece
Pedro Isaias, The University of Queensland, Australia
Helen Karatza, Aristotle University of Thessaloniki, Greece
Despo Ktoridou, University of Nicosia, Cyprus
Barbara Kerr, Ottawa University, Canada
George Magoulas, Birkbeck College, UK
Nektarios Moumoutzis, Technical University of Crete, Greece
Demetrios Sampson, University of Pireaus, Greece
Ishwar Singh, McMaster University, Hamilton, Canada
Khitam Shraim, Palestine Technical University, Palestine
Seshaisai Srinivasan, McMaster University, Hamilton, Canada
Doru Ursutiu, University Transylvania Brasov, Romania
Minjuan Wang, Shanghai International Studies University (Oriental Scholar); San Diego State University, USA
Tom Wanyama, McMaster University, Hamilton, Canada
Ting-Ting Wu, National Yunlin University of Science and Technology, Taiwan
Dieter Wuttke, Technical University Ilmenau, Germany

Special Session "IMCL2018 Doctoral Consortium" Program Committee

Chairs

Kostas Apostolou, McMaster University, Hamilton, Canada
Dan Centea, McMaster University, Hamilton, Canada

3rd IMCL Student International Competition for Mobile Apps

Chairs

Tom Wanyama, McMaster University, Hamilton, Canada
Seshasai Srinivasan, McMaster University, Hamilton, Canada

Preface

IMCL2018 was the 12th edition of the International Conference on Interactive Mobile Communication, Technologies and Learning.

This interdisciplinary conference is part of an international initiative to promote technology-enhanced learning and online engineering worldwide. The IMCL2018 covered all aspects of mobile learning as well as the emergence of mobile communication technologies, infrastructures and services and their implications for education, business, governments and society.

The IMCL conference series actually aims to promote the development of mobile learning, to provide a forum for education and knowledge transfer and to expose students to latest ICT technologies and encourage the study and implementation of mobile applications in teaching and learning. The conference was also platform for critical debates on theories, approaches, principles and applications of mobile learning among educators, developers, researchers, practitioners and policymakers.

IMCL2018 has been organized by McMaster University, Hamilton, Ontario, Canada from 11 to 12 October 2018.

This year's theme of the conference was "Mobile Technologies and Applications for the Internet of Things".

Again outstanding scientists accepted the invitation for keynote speeches:

- Rory McGreal, Athabasca University, Canada. Speech title: Why Open Educational Resources are Essential for Mobile Learning.
- David Guralnick, Kaleidoscope Learning & Columbia University, New York, NY, USA. Speech title: How Mobile Technology Can Change the Way We Think About Education and Training.

Furthermore, two very interesting workshops have been organized:

- LoRa Workshop: Tour of Canada's Learning Factory and Introduction to LoRa by Ishwar Singh, McMaster University, Canada.

- Neural Workshop: Tour of Canada's Learning Factory and Introduction to Neural Network Models on Mobile Devices by Ishwar Singh, McMaster University, Canada.

Since its beginning this conference is devoted to new approaches to learning with a focus on mobile learning, mobile communication, mobile technologies and engineering education.

We are currently witnessing a significant transformation in the development of working and learning environments with a focus on mobile online communication.

Therefore, the following main topics have been discussed during the conference in detail:

- Mobile Learning Issues:

  - Dynamic learning experiences
  - Large-scale adoption of mobile learning
  - Assessment, evaluation and research methods
  - Mobile learning models, theory and pedagogy
  - Life-long and informal learning using mobile devices
  - Open and distance mobile learning
  - Social implications of mobile learning
  - Design of adaptive mobile learning environments
  - Cost-effective management of mobile Learning processes
  - Quality in mobile learning
  - Ethical and legal issues
  - Location-based integration
  - Emerging mobile technologies and standards
  - Interactive and collaborative mobile learning environments
  - 5G Network Infrastructure
  - Tangible, embedded and embodied interaction

- Mobile Learning and the Internet of Things (IoT):

  - Design of Internet of Things devices and applications
  - Internet of Things and Artificial Intelligence
  - Challenges for Internet of Things implementation
  - Open-source resources for IoT Development (hardware/software)
  - Industrial Internet of Things
  - Wearables & Internet of things
  - Mobile/online courses for Internet of Things

- Mobile Applications:

  - Smart cities
  - Online laboratories
  - Game-based learning
  - Mobile health care and training
  - E-health technologies

- Learning analytics
- Mobile learning in cultural institutions and open spaces
- Mobile systems and services for opening up education
- Social networking applications
- Mobile Learning Management Systems (mLMS)

The following submission types have been accepted:

- Full Paper, Short Paper, Distant/Pre-recorded Presentation
- Work in Progress, Poster
- Special Sessions
- Round Table Discussions, Workshops, Tutorials

All contributions were subject to a double-blind review. The review process was very competitive. We had to review about 145 submissions. A team of about 50 reviewers did this terrific job. Our special thanks go to all of them.

Due to the time and conference schedule restrictions, we could finally accept only the best 36 submissions for presentation.

The conference had again about 60 participants from 19 countries.

IMCL2019 will be held at Aristotle University of Thessaloniki, Greece.

Villach, Austria   Michael E. Auer
                                                          IMCL Steering Committee Chair
Thessaloniki, Greece   Thrasyvoulos Tsiatsos
                                                                     IMCL General Chair

Contents

Mobile Health Care and Training

Game Based Learning

Interactive Collaborative Mobile Learning Environments

Improving Grasping of Bionic Hand by Using Finger Compliance Design and Rapid Prototyping

Carin Yeghiazarian and Lucian Balan[✉]

Automotive and Vehicle Technology Program, Faculty of Engineering, W Booth
School of Engineering Practice and Technology, McMaster University, 200
Longwood Rd.S. - MARC 271, Hamilton, ON L8P 0A6, Canada
balanl@mcmaster.ca

Abstract. The purpose of the research work outlined in this paper is to use engineering design to improve the grasping capability of a robotic hand with six degrees-of-freedom within a restricted budget. The work presented in this paper is the result of a capstone project work with a group of students at School of Engineering Practice and Technology at McMaster University, Ontario. This paper aims to demonstrate that dexterity of such a hand improves when a compliance side motion is added to the robotic hand at each knuckle joint, in order to replicate the abduction/adduction motion of the four fingers. This proposed design can improve the grasping capability of the bionic hand, without adding more degrees-of-freedom or by increasing the complexity of the operating controller. By means of CAD software applications and 3D printing technologies, this method helps students bring to life their novel ideas. It can also help students study the grasping capabilities of a robotic hand through remote control by a mobile device or over an Internet connection.

Keywords: Bionic robotic hand · Engineering design · Compliance motion

1 Introduction

1.1 Context

The ultimate goal of bio-robotics is to produce a cost-effective bionic hand that can accurately mimic the functionality and dexterity of a human hand. Applications of such robotic hands include prosthetic limbs or telecontrol devices for human-related applications.

In general terms, the extraordinary dexterity of a human hand is the result of three factors: a large number of degrees-of-freedom (DOF), a significant number of sub-components (27 bones), and a sophisticated biological controller (human brain). A simplistic robotic hand typically has only 5 or 6 DOF [1–5]. This offers less dexterity than a biological hand, but it has the advantage of being cost-effective, and relatively easier to control.

The typical arrangement for a 6 DOF hand design consists of one revolute joint at the base of each finger and a combination of two revolute joints for the thumb motion.

© Springer Nature Switzerland AG 2019
M. E. Auer and T. Tsiatsos (eds.), *Mobile Technologies and Applications
for the Internet of Things*, Advances in Intelligent Systems and Computing 909,
https://doi.org/10.1007/978-3-030-11434-3_1

Typically, the joints are controlled by electric motors with some type of mechanisms, while the remaining articulations are not directly operated. Instead, their motion is limited or kinematically constrained to follow the motion of active joints [2–5].

1.2   Technical Background

The anatomical study of the biological hand is important when designing a robotic hand. Figure 1 shows the terminology of major bones in the human hand. The palm mainly consists of metacarpals and carpals bones. Each of the four fingers is composed of three phalanges, and the thumb is composed of two phalanges and a metacarpal bone.

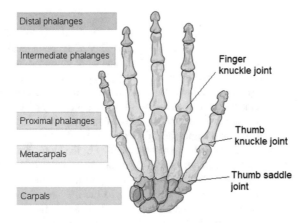

Fig. 1. Human hand skeleton consisting of 27 bones; main bones are labeled within figure (*Source* Wikipedia, edited)

The joints responsible for fingers motion are the condyloid joints, referred to as "knuckle joints" in this paper (Fig. 2a). These joints are spherical joints with two axes. The major axis is for flexion/extension of the fingers, while its second axis allows for a limited side motion of the fingers known as abduction/adduction movement.

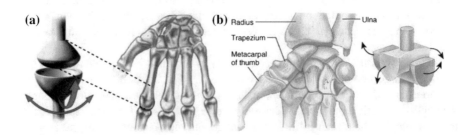

Fig. 2. **a** Schematic of the condyloid joint (knuckle joint) which enables the pivoting movement of the four fingers—(*Source* Wikipedia). **b** Schematic of the trapeziometacarpal used in thumb motion of a human hand—(*Source* web Proko-movable joints).

The thumb has two major joints. The first one, located between the proximal phalange and metacarpal bone, allows for flexion/extension, acting like a knuckle joint but without its side motion. The second joint is located at the base of the thumb, between metacarpal and carpal bones. It adds more flexion/extension and a side motion over a wide range. The latter is called trapeziometacarpal joint, or "saddle joint" (Fig. 2b).

In many simplified designs of robotic hands, the knuckle joint of fingers and thumb is not modeled [6], modeled as a fixed thumb [7], or modeled as simplified revolute joints with a single axis of rotation resulting in a hand with 5 DOF [1]. In other approaches [2–5], more dexterity is added to the thumb by using a secondary axis of rotation, thus producing a hand with 6 DOF.

Belter et al. [2] review various research and commercially available prosthetic hands. For commercial prosthetics, robustness and longevity of the product are of high importance. Due to this reason, several design aspects are different from that of research hands. The finger design for many of these hands only includes two joints, and the thumbs have actuated distal joints while the saddle joint must be positioned manually.

1.3 Problem Description

To replicate the complete dexterity of a human hand is not an easy task. Without accounting for wrist motion or rotation, a human hand has a total of 21 DOF: 4 in each finger (3 for extension/flexion and 1 for the side motion) and 5 for the thumb.

All these joints are active, meaning they need to be controlled by as many independent actuators (electrical motors, pneumatic cylinders, etc.). Aside from the complexity and cost of such a controller, incorporating the actuators in the restricted space of some articulations is a big technical challenge.

In order to produce a cost-effective hand, many engineers choose to keep only few of the most important DOF, sacrificing some dexterity of the hand at the benefit of a simpler controller.

2 Purpose

The purpose of the research work outlined in this paper is to identify simple ways to improve the engineering design of a robotic hand with 6 DOF such that it produces a more natural grasp without adding more DOF, nor by increasing the complexity of the operating controller.

3 Approach

The work presented is the result of capstone project work with a group of undergraduate students at School of Engineering Practice and Technology at McMaster University, Ontario. It starts with a legacy design of a simplified robotic hand with 6

DOF that was developed by a previous group of students [8]. This section of the paper discusses improvements to the legacy design of the robotic hand.

3.1   Legacy Design

The legacy design of the robotic hand with 6 DOF is shown in Fig. 3. Each finger has 1 DOF at knuckle joint, while the thumb has 2 DOF at saddle joint. Each DOF is operated by means of an electric motor concealed in the palm.

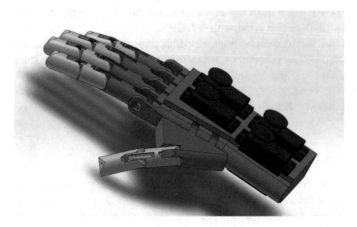

Fig. 3. Legacy design of robotic simplified hand with 6 DOF.

Each finger is operated by means of two cables and a linkage mechanism, as visible in Fig. 4a. When commanding a finger flexion motion, the corresponding electric motor pulls on the inner cable connected to the outermost finger segment and it releases the outer cable accordingly, and the finger begins pivoting.

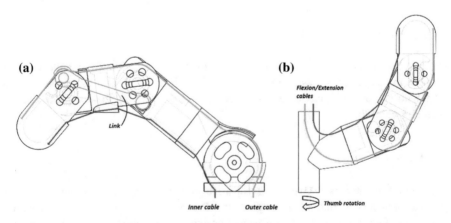

Fig. 4. a Detailed view of finger link and cable mechanism. **b** Detailed view of thumb cable mechanism with additional rotation

The linkage mechanism of the finger consists of a rigid link between the finger segment closer to the palm and the outermost segment. The link moves within a cavity in middle segment allowing the mechanism to be fully contained. The purpose of the linkage mechanism is to ensure that final finger segment moves in direct relation to second segment as the finger base rotates for grasping. This approach allows all three joints of one finger to be controlled by a single actuator while still providing some compliance to adapt to objects of various shapes.

The thumb has two DOF. One electric motor controls the flexion/extension movement by means of cables, while a separate motor rotates its base with respect to the palm, as shown in Fig. 4b.

3.2 Design Improvement 1—Finger Alignment

The first design improvement comes from observation that when a human hand is in a natural relaxed extension, the fingers tend to spread away, and when in flexion they tend to have a convergent posture as shown in Fig. 5. With the legacy design this is not possible, since all fingers move in parallel planes, and as the hand opens or closes, the side clearance between fingers stays the same.

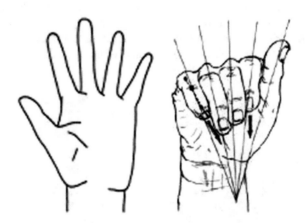

Fig. 5. Fingers in extension versus flexion posture

To address these issues, the upgraded design uses two curvatures. First, by placing the knuckle joints on an curved alignment with respect to the palm, a finger divergence is produced when hand is in extension (Fig. 6a). Second, by replacing the flat palm with a convex one, a finger convergence is obtained in flexion (Fig. 6b).

The updated design of robotic hand generates a more natural posture as compared to the legacy one. Furthermore, the proposed changes have very little impact over manufacturing cost considering that all parts are 3D printed with ABS plastic material.

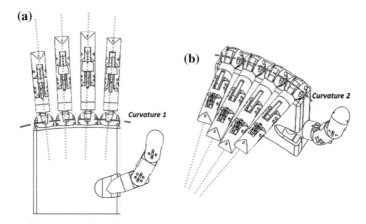

Fig. 6. **a** Curved alignment for knuckle joints to generate divergence in extension. **b** Curved palm to generate finger convergence in flexion.

3.3   Design Improvement 2—Fingers' Abduction/Adduction

The second design improvement is more significant, and it relates to the abduction/adduction motion of human hand which is depicted in Fig. 7. It is worth mentioning that abduction/adduction range of motion is considerably large when fingers are in extension, but almost entirely limited when in flexion.

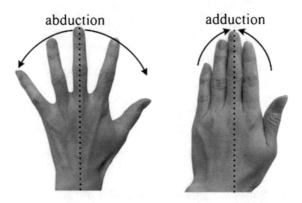

Fig. 7. Fingers abduction/adduction movement

The legacy design does not account for this type of motion, because each knuckle joint is modeled with a single axis of rotation. The authors believe that abduction/adduction motion of the fingers is important for generating a good quality grasp. When grasping different objects, it is this motion that allows the hand to self-adjust to the specific shape of the object. For example, when grasping a spherical object, the fingers will spread out around the convex shape of the object for a more stable grasp.

The upgraded design replaces the one-axis knuckle joints with two-axis spherical joints. The finger actuator still controls the first axis of rotation for extension/flexion of the finger. The second axis of rotation consists of a passive abduction/adduction motion controlled by a spring-like mechanism.

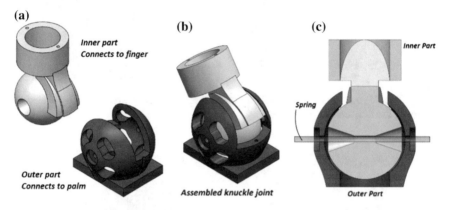

Fig. 8. Improved knuckle joint: **a** parts, **b** assembly, **c** section view

The upgraded knuckle joint design is shown in Fig. 8. The outer part of the knuckle is attached to palm, and the inner part connects with the base of the finger (Fig. 8a). The slot visible in outer part is larger at one end and smaller at the other. During grasping, the inner part protrusion slides into the slot (Fig. 8b). As grasping occurs, the object's contact surface pushes the fingers sideways as its shape dictates (Fig. 8c). This generates a knuckle compliance that is taken care by the deflection motion of the spring.

The finger can pivot sideways only as much as the clearance permits. The side clearance of the knuckle is variable such that more clearance is produced at extension than at flexion (see Fig. 9), similar with compliance of a real hand.

When abduction/adduction motion is not present, the elastic rod acts like a spring that brings the finger back into centered position of the slot. The stiffness of the elastic rod relates to the friction coefficient between fingers and object grasped. It has been determined in our trials that a lower friction coefficient requires a stiffer spring.

The effect of abduction/adduction compliance passively controlled by the knuckle spring is visible in Fig. 10. The CAD model of the robotic hand shows fingers in extension posture after exposed to the two different side forces.

The legacy design of the thumb was kept unchanged because the abduction/adduction motion is already incorporated into one of the two DOF controlled by a separate electric motor actuator.

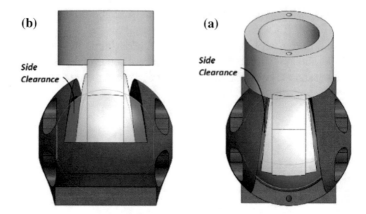

Fig. 9. Knuckle joint abduction/adduction compliance: **a** finger in extension, larger clearance. **b** Finger in flexion, smaller clearance

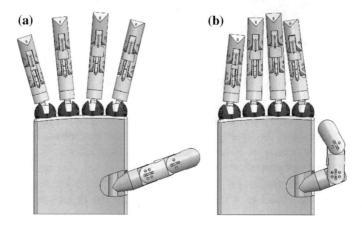

Fig. 10. Hand compliance: **a** extension with abduction, **b** extension with adduction

3.4   Design Improvement 3—Concealed Cables

Our last design improvement relates to the guidance of the cables used for finger and thumb operation of the legacy robotic hand. To avoid tangling, these cables should be contained inside the hand, similar to the tendons of a biological hand. For smooth operation and reduced friction, these cables must change direction gradually avoiding sharp cornering.

With traditional manufacturing, creating non-straight channels inside a part is a big challenge. Recommended solution takes advantage of 3D printing capabilities, which allows for the fabrication of parts with complex cavities inside. Figure 11 shows the upgraded finger design with curved inner channels for smooth cable operation.

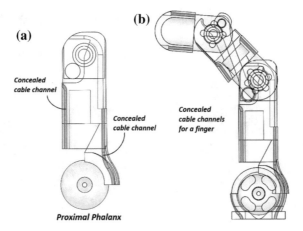

Fig. 11. Finger design with concealed cables: **a** detailed phalanx with cable channels, **b** full finger with cable channels

One benefit of this design is that almost the entire cable running the mechanism for closing and opening the finger would be traveling through enclosed channels within the finger parts themselves.

4 Results

After finalizing the hand design, a prototype of the robotic hand was created using 3D printing and off-the-shelf components. Small 3D printing tests were run to determine the adequate clearance for moving parts of the joints. The recommended optimal clearance is 0.25 mm.

The electric motors are housed within the palm. Each finger has one motor with the exception of the thumb, which has two. Each joint of the fingers is limited to approximately 90 degrees of movement. The printing of the entire hand took approximately 30 h to complete on a uPrint Plus Stratasys 3D printer with ABS plastic filament. The cost of the entire hand is within $250 CAD.

Due to a modular design and easy assembly, the parts of the hand can be re-printed if any are damaged or broken. An image with the hand prototype is shown in Fig. 12.

The hand prototype was tested for grasping several objects of various shapes. A separate prototype hand was made without the abduction/adduction compliance at finger knuckle joints, to test its effect on grasping quality. Figure 13 shows that a better grasp can be produced when the knuckle joint compliance of the fingers is considered.

By means of CAD software applications and 3D printing technologies, this design project helped students bring to life their novel ideas. It can also help students run more tests to investigate grasping capabilities of a robotic hand on objects of different shapes and surface smoothness.

Current development of this robotic hand includes the addition of a sensor-based glove with Arduino microcontroller for a wired control of the six actuators. As further

(a)   **(b)**

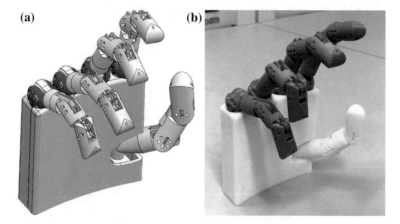

Fig. 12. CAD model and 3D printed prototype of robotic hand with upgraded design

(a)   **(b)**

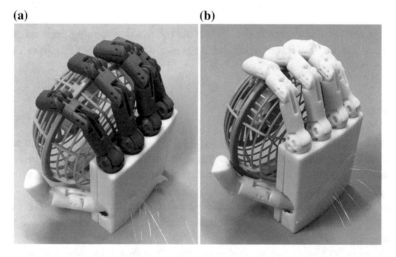

Fig. 13. Sample grasping: **a** with knuckle joint compliance, **b** without compliance

development, this robotic hand can be refined to allow for wireless remote control through a mobile device application over an Internet connection.

5   Conclusions

This paper presents a practical method for improving the grasping capability of a 6 DOF robotic hand, without significantly increasing the complexity of the hand controller. The major upgrade adds compliance for side motion of the fingers at the knuckle joint level. The spring-based compliance is passive and it mimics, to a certain degree, the functionality of a human hand. Design details were discussed, and a prototype was fabricated using 3D printing technology and off-the-shelf components. The upgraded design is applicable to more dexterous robotic hands with higher DOF. Further recommendations include the development of a mobility-based learning tool for undergraduate students.

References

1. Huang, H., et al. (2006). The mechanical design and experiments of HIT/DLR prosthetic hand. In *2006 IEEE International Conference on Robotics and Biomimetics*.
2. Belter, J. T., et al. (2013). Mechanical design and performance specifications of anthropomorphic prosthetic hands: A review. *Journal of Rehabilitation Research and Development, 50* (5), 599–618.
3. Zollo, L., et al. (2007). Biomechatronic design and control of an anthropomorphic artificial hand for prosthetic and robotic applications. *IEEE/ASME Transactions on Mechatronics, 12* (4), 418–429.
4. Pons, J. l., et al. (2004). The MANUS-HAND dextrous robotics upper limb prosthesis: Mechanical and manipulation aspects. *Autonomous Robots, 16*(2), 143–163.
5. Dechev, N., Cleghorn, W. L., & Naumann, S. (2001). Multiple finger, passive adaptive grasp prosthetic hand. *Mechanism and Machine Theory, 36*(10), 1157–1173.
6. O'Toole, K. (2007). Mechanical design and theoretical analysis of a four fingered prosthetic hand incorporating embedded SMA bundle actuators. *World Academy of Science, Engineering & Technology, International Science Index 7, International Journal of Medical, Health, Biomedical, Bioengineering & Pharmaceutical Engineering, 1*(7), 430–437.
7. Doshi, R. H., et al. (1998). The design and development of a gloveless endoskeletal prosthetic hand. *Journal of Rehabilitation Research and Development, 35*(4), 388–395.
8. Lau, R., Slawek, J., & Dancy, R. (2014, December). "Capstone project report: Remote operated electro-mechanical hand"—School of engineering practice and technology. Hamilton, Ontario: McMaster University.

Learning Analytics for Motivating Self-regulated Learning and Fostering the Improvement of Digital MOOC Resources

D. F. O. Onah[1]([⊠]), E. L. L. Pang[2], J. E. Sinclair[2], and J. Uhomoibhi[3]

[1] University College London, London, UK
d.onah@ucl.ac.uk
[2] The University of Warwick, Coventry, UK
[3] Ulster University, Coleraine, UK

Abstract. Nowadays, the digital learning environment has revolutionized the vision of distance learning course delivery and drastically transformed the online educational system. The emergence of Massive Open Online Courses (MOOCs) has exposed web technology used in education in a more advanced revolution ushering a new generation of learning environments. The digital learning environment is expected to augment the real-world conventional education setting. The educational pedagogy is tailored with the standard practice which has been noticed to increase student success in MOOCs and provide a revolutionary way of self-regulated learning. However, there are still unresolved questions relating to the understanding of learning analytics data and how this could be implemented in educational contexts to support individual learning. One of the major issues in MOOCs is the consistent high dropout rate which over time has seen courses recorded less than 20% completion rate. This paper explores learning analytics from different perspectives in a MOOC context. First, we review existing literature relating to learning analytics in MOOCs, bringing together findings and analyses from several courses. We explore meta-analysis of the basic factors that correlate to learning analytics and the significant in improving education. Second, using themes emerging from the previous study, we propose a preliminary model consisting of four factors of learning analytics. Finally, we provide a framework of learning analytics based on the following dimensions: descriptive, diagnostic, predictive and prescriptive, suggesting how the factors could be applied in a MOOC context. Our exploratory framework indicates the need for engaging learners and providing the understanding of how to support and help participants at risk of dropping out of the course.

Keywords: Learning analytics · MOOC · Self-regulated learning

The original version of this chapter was revised: Author name has been updated. The correction to this chapter can be found at https://doi.org/10.1007/978-3-030-11434-3_43

© Springer Nature Switzerland AG 2019
M. E. Auer and T. Tsiatsos (eds.), *Mobile Technologies and Applications for the Internet of Things*, Advances in Intelligent Systems and Computing 909, https://doi.org/10.1007/978-3-030-11434-3_3

1   Introduction

Online education has suffered the lack of learners' engagement which has serious issues generally over time within the educational system. Recent attempts to develop new emerging technologies to determine the future of the digital learning environment in a MOOC context has improved to some extent how contents are revised and tailored to participants. The flexible digital learning environment and design of innovative visualization of learning content have helped to improve time management among the learners [1]. However, the most dramatic effect on shaping the future of education comes from big data mining and learning analytics that could be related to educational improvement and learners' engagement. Although learning analytics and implementation is still in the early stages of experimentation in the digital learning environment, there are several controversial issues lingering around the implementation of learning analytics in educational settings. Undoubtedly, learning analytics has contributed significantly to the future of the digital learning environment. The growing implementation of new technologies and analysis techniques in education has shown the need for continuous research on current methods of engaging learners within digital learning environment.

MOOC has been claimed to solve many educational issues by the provision of free open access courses that enabling learners to explore independent learning [2]. This paper reviews issues relating to learning analytics in MOOC contexts, considering published data on MOOC learning analytics and discussing factors implicated in previous studies as being related to self-regulated learning [1, 3]. The free nature of MOOCs is said to be behind the reasons for profound risk of dropout [4, 5], and the students' ability to self-regulate their learning habits [6], while other studies point out personal reasons as a factor of learners' high dropout rate [2, 7].

2   Related Study

In the current literature, learning analytics research focuses on developing predictive models of learners' performance and identifies students at risk of dropping out in a digital learning environment [8, 9]. However, research related to learning analytics in MOOCs carries an intense and intrinsic possibility that might influence learning [10]. Learning analytics over the years faces challenging demands and difficulties when applied in MOOC contexts [11], but as Knox [12] mentioned, learning analytics has a strong potential for discovery when it is applied to MOOC datasets. There were few research studies that combine learning analytics practices in MOOCs [13–16].

2.1   Significance of Learning Analytics

Learning analytics platform monitors learners' activities in a digital learning environment [17]. The role of learning analytics in both online and traditional education settings could be found in (1) their role in reform of the learning activities; (2) how this could assist educators to improve educational content, teaching and learning; and (3) how lecture videos and audio could be revised for optimum engagement [7, 18]. Learning analytics has been successful as a mechanism that is essential for mitigating against the high dropout rate within online education.

Educational online institutions, academics and students require a standard foundation for which changes could be enacted. For academics, the need for real-time learning and tracking mechanism could facilitate the insight into the performance of learners, including those at risk of dropping out of the course. This tracking and observations can be of great significance in improving teaching, curriculum planning and learning activities. For the students, this could enable them to acquire better revised educational resources and could help them in effective retention and encourage continuing course engagement. They will also receive independent feedback and information relating to their individual performance and progress in relation to their personal goals and learning objectives. Learning analytics provides unprecedented level of feedback support to students in a digital learning environment. With learning analytics analysis, researchers could narrow their studies on the satisfaction of students in their studies by measuring how specific interactions affect their learning achievements [19, 20].

3   eLDa MOOC Platform

This study has introduced a novel platform to explore the four factors of the learning analytics described in this paper. eLDa is an online platform that supports a novel approach to MOOC development, which aims to actively involve participants in directing and regulating their own learning. It provides the necessary framework and support for participants to set their own learning goals and to access resources suitable for their needs. In order to support users' self-directed learning through informed choice, the system offered advice on (but not enforce) recommended prerequisites for each topic and provided a map for learners to visualize the elements they have studied so far in the learning environment. The course platform was designed for A-level teachers of Computer Science who participated in an online version of the platform. The platform was also applied in a blended-learning classroom to deliver undergraduate modules to student of Computer Science discipline from a top UK university. The online version of the course was also available for external participants worldwide. The platform has over 150 registered participants and about over 80 active participants engaging fully with the course elements.

3.1   Aims and Objective

The platform should support good data collection and analyse features to evaluate participants' self-regulated learning levels, the path followed and needed to integrate a variety of acknowledged MOOC 'good practice' features to support learners and mitigate participants' dropout.

3.2   Platform Design

The front end and back end of the design of eLDa platform was developed using WordPress content management system, phpMyAdmin, MySQL, PHP and Sensei plugins. The online module curriculum has 4 courses, 4 modules, 10 sessions and 50 lessons. The instructional course includes lecture content, PowerPoint narration and slides, audio and video lectures and transcripts of all audio and video lecture contents.

The transcript was provided to support non-speaker of English language to understand the lectures effectively by reading the transcribed scripts. Figure 1 illustrates the course interface.

eLDaMOOC – eLearning . Development . Adaptivity
Adapt Learn Platform
All Courses
My current course
Pre-course survey
Post course survey
Pre Course Self-Regulation Survey
Post Course Self-Regulation Survey
My Messages
My Profile
About
Programming Portfolio
My E-Portfolio
Computer Science: Computing concepts & Python programming
44 learners taking this course
Currently completed 2 lessons of 34 in total
6%
In Progress
COURSE DISCUSSION
This course will be teaching computing concepts and hands-on Python programming. There will
Daniel Onah
Log Out
WHO'S ONLINE
RECENTLY ACTIVE MEMBERS

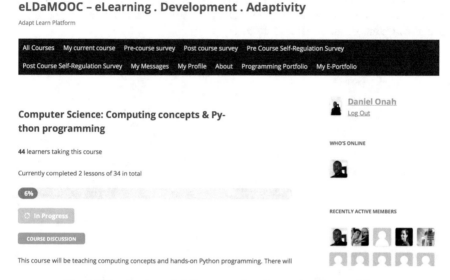

Fig. 1 Visualization of eLDa course interface and online participants

Figure 2 shows the visualization of lesson completed and lessons yet to be studied. This visualization of the whole course enabled learners to manage their study time effective and self-regulate their learning habits.

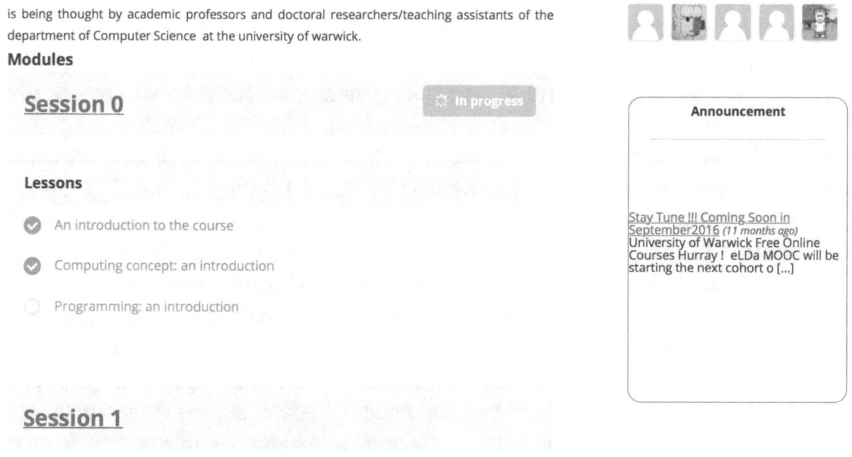

is being thought by academic professors and doctoral researchers/teaching assistants of the department of Computer Science at the university of warwick.
Modules
Session 0
In progress
Lessons
An introduction to the course
Computing concept: an introduction
Programming: an introduction
Session 1
Announcement
Stay Tune !!! Coming Soon in September2016 (11 months ago)
University of Warwick Free Online Courses Hurray ! eLDa MOOC will be starting the next cohort o [...]

Fig. 2 Visualization of course sessions and lessons

4 Proposed Framework

Learning analytics is the science of using data to build models that lead to better decisions that in turn add value to individual's self-regulated learning skills. The various activities and observations from previous studies could help in improving the present. The following factors help in making learning analytics to be effective in an educational context and as illustrated in Fig. 3.

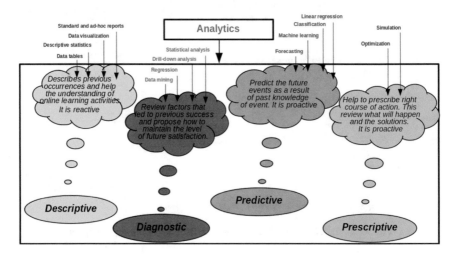

Fig. 3 Proposed framework of a learning analytics model

Descriptive Analytics: This is a reactive factor that helps in describing the previous occurrence and understanding of the individual learning activities and habits in a MOOC platform.

Diagnostic Analytics: This reviews all the factors that led to the previous success of an online course and proposed how to maintain and improve these factors for continuity. This explores the level of satisfaction among the learners with regard to the online course content.

Predictive Analytics: This is a proactive factor which helps to predict events as a result of previous observation and knowledge of past learning habits. The previous knowledge obtained as a result of the descriptive analytics and the diagnostic analytics help to support effective prediction of how the new learners could perform in the course based on the knowledge acquired from previous learners' activities and learning habits. This observatory knowledge could help in predicting students at risk of dropping out of the course and this will help in providing support to mitigate dropout among learners.

Prescriptive Analytics: This is a proactive factor which helps to prescribe a standard measure that could help to provide effective action among course instructors. This allows the course instructors and coordinators to review possible actions to take in order to resolve the effect of future dropout rate in an online course. The instructor acts based on the knowledge obtained from the predictive analytics and is able to identify areas of the online course that require urgent and adequate review in order to encourage and support self-regulated learning of the participants.

Figure 3 illustrates these four factors and the measure for the four analytic factors proposed in this paper.

5 Research Methodology

The overarching approach adopted to be used in this proposed work is derived from the design science research methodology (DSRM), which is a paradigm centred on the development and evaluation of an inventive artefact to investigate a precise problem domain.

5.1 Data Collection

Previous research conducted using the tool has seen data collected using newly established instrument known as MOOC online self-regulated learning questionnaire (MOSLQ) that was developed to measure and understand self-regulated learning within a novel MOOC platform. Mixed methods of qualitative and quantitative methods were used for the data collection. The data were analysed using statistical package for the social sciences (SPSS). Results based on the activities of students online were presented to illustrate resource management, self-regulated learning and course engagement. The same approach has been proposed in this study to explore learning analytics in the digital course environment.

6 Conclusion

This study introduced a proposed framework in applying learning analytics in a novel eLDa MOOC platform that could be used to penetrate uncertainty around MOOC and how resources are tailored or allocated. With this knowledge captured in a digital learning environment, it could tremendously improve the quality of teaching and learning experience in a MOOC environment. This study presents new methods and results relating to learners' activities in a novel MOOC platform developed for continuing professional development training organized for Computer Science teachers. This further demonstrates the various learning activities, independent learning habits and self-directed learning.

Thus, we explore the effect of learning analytics in mitigating MOOC dropout issues and how we could use the analytics models to support students while engaging with their studies in a self-regulated learning manner in a digital learning environment. This research was conducted on the basis of applying the proposed framework approach in a wider study implemented in eLDa MOOC platform. Several investigations have already been conducted to identify measures that could be used in the analyses of the learning analytics models described in this paper.

Acknowledgements. The first author wishes to acknowledge Mr. Adakole S. Onah's financial support in his research and family members and friends for their moral support.

References

1. Onah, D. F. (2017). *Investigating self-regulated learning in massive open online courses: a design science research approach* (Doctoral dissertation, University of Warwick).
2. Onah, D. F., Sinclair, J., & Boyatt, R. (2014). Dropout rates of massive open online courses: Behavioural patterns. In *EDULEARN14 Proceedings*, 5825–5834.
3. Onah, D. F. O., & Sinclair, J. E. (2017). Assessing self-regulation of learning dimensions in a stand-alone MOOC platform. *International Journal of Engineering Pedagogy (iJEP)*, 7(2), 4–21.
4. Alario-Hoyos, C., Pérez-Sanagustín, M., Delgado-Kloos, C., Muñoz-Organero, M., & Rodríguez-de-las-Heras, A. (2013, September). Analysing the impact of built-in and external social tools in a MOOC on educational technologies. In *European Conference on Technology Enhanced Learning* (pp. 5–18). Berlin, Heidelberg: Springer.
5. Alario-Hoyos, C., Muñoz-Merino, P. J., Pérez-Sanagustín, M., Delgado Kloos, C., & Parada G, H. A. (2016). Who are the top contributors in a MOOC? Relating participants' performance and contributions. *Journal of Computer Assisted Learning*, 32(3), 232–243.
6. Zimmerman, B. J. (2000). Attaining self-regulation: A social cognitive perspective. In *Handbook of self-regulation* (pp. 13–39).
7. Onah, D. F. O., Sinclair, J., Boyatt, R., & Foss, J. (2014). Massive open online courses: learner participation. In *Proceeding of the 7th International Conference of Education, Research and Innovation* (pp. 2348–2356).
8. Dawson, S., & Siemens, G. (2014). Analytics to literacies: The development of a learning analytics framework for multiliteracies assessment. *The International Review of Research in Open and Distributed Learning*, 15(4).
9. Dawson, S., Gašević, D., Siemens, G., & Joksimovic, S. (2014, March). Current state and future trends: A citation network analysis of the learning analytics field. In *Proceedings of the Fourth International Conference on Learning Analytics and Knowledge* (pp. 231–240). ACM.
10. Reich, J. (2015). Rebooting MOOC research. *Science, 347*(6217), 34–35.
11. Clow, D. (2013, April). MOOCs and the funnel of participation. In *Proceedings of the Third International Conference on Learning Analytics and Knowledge* (pp. 185–189). ACM.
12. Knox, J. (2014). From MOOCs to learning analytics: Scratching the surface of the'visual'. *ELearn, 2014*(11), 3.
13. Moissa, B., Gasparini, I., & Kemczinski, A. (2015). A systematic mapping on the learning analytics field and its analysis in the massive open online courses context. *International Journal of Distance Education Technologies (IJDET)*, 13(3), 1–24.
14. Vogelsang, T., & Ruppertz, L. (2015, March). On the validity of peer grading and a cloud teaching assistant system. In *Proceedings of the Fifth International Conference on Learning Analytics And Knowledge* (pp. 41–50). ACM.
15. Kloos, C. D., Alario-Hoyos, C., Fernández-Panadero, C., Estévez-Ayres, I., Muñoz-Merino, P. J., Cobos, R., & Chicaiza, J. (2016, September). eMadrid project: MOOCs and learning analytics. In *2016 International Symposium on Computers in Education (SIIE)* (pp. 1–5). IEEE.

16. Ruipérez-Valiente, J. A., Muñoz-Merino, P. J., Gascón-Pinedo, J. A., & Kloos, C. D. (2017). Scaling to massiveness with analyse: A learning analytics tool for open edX. *IEEE Transactions on Human-Machine Systems, 47*(6), 909–914.
17. Fournier, H., Kop, R., & Sitlia, H. (2011, February). The value of learning analytics to networked learning on a personal learning environment. In *Proceedings of the 1st International Conference on Learning Analytics and Knowledge* (pp. 104–109). ACM.
18. Koller, D., & Ng, A. (2012). The Online Revolution: Education at Scale. *L Educ.*
19. Kolowich, S. (2012). Who takes MOOCs. *Inside Higher Ed, 5,* 2012.
20. Vries, P. (2013). Online learning and higher engineering education the MOOC phenomenon', European Society for Engineering Education (SEFI), [Brussels]. In *Paper Presented at the 41st SEFI Conference.*

BYOD Collaborative Storytelling in Tangible Technology-Enhanced Language Learning Settings

Federico Gelsomini[1,2(✉)], Kamen Kanev[1], Reneta Barneva[3],
Paolo Bottoni[2], Donna Hurst Tatsuki[4], and Maria Roccaforte[2]

[1] Shizuoka University, Hamamatsu, Japan
`federico.gelsomini@uniromal.it, kanev@inf.shizuoka.`
`ac.jp`
[2] Sapienza University of Rome, Rome, Italy
`bottoni@di.uniromal.it, {federico.gelsomini,maria.`
`roccaforte}@uniromal.it`
[3] State University of New York at Fredonia, Fredonia, USA
`reneta.barneva@fredonia.edu`
[4] Kobe City University of Foreign Studies, Kobe, Japan
`dhtatsuki@gmail.com`

Abstract. Research on pedagogical and didactic advancements in the context of new mobile technologies is currently gaining importance. This paper presents a novel pedagogical framework, in which augmented interactions with the surrounding environment provide the students an enhanced learning experience through the gadgets they own. Our pedagogical model employs QR codes that are easily accessible through smartphone apps and stimulates the efficient development of essential communication skills. Hence, students can use their own devices to interact with the educational content, share information, and help each other while working collaboratively on different assignments. More specifically, we introduce and explore a new BYOD-based T-TEL setup that supports the creation and management of storytelling activities and fosters the development of communication skills through m-CSCL. This approach could be extended to environments specifically crafted for individuals with special needs.

Keywords: Bring Your Own Device (BYOD) · Mobile-Computer-Supported Collaborative Learning (m-CSCL) · Mobile-Assisted Language Learning (MALL) · Tangible Technology-Enhanced Learning (T-TEL)

1 Introduction

Recent advances in mobile computing and the ubiquity of mobile devices play an important role in supporting and integrating enhanced educative activities into the language learning process. Some scholars have analyzed how and to what degree collaborative language speaking and listening activities could be backed up or boosted by mobile devices [1]. The results of the study revealed that learning activities were

© Springer Nature Switzerland AG 2019
M. E. Auer and T. Tsiatsos (eds.), *Mobile Technologies and Applications for the Internet of Things*, Advances in Intelligent Systems and Computing 909,
https://doi.org/10.1007/978-3-030-11434-3_5

greatly facilitated by Mobile-Assisted Language Learning (MALL) modalities, with important influence of mobile devices on the language learning practices.

As mobile devices with appropriate applications are owned and used by a significant portion of the population, they have started affecting the learning settings and methodologies [2]. Context-aware applications are particularly suitable for execution on mobile devices, as they can be operated in various environments thus broadening the acquisition exposure and including authentic real-world experiences. Such applications often rely on device localization and tracking through embedded sensors and/or employ interactions with digitally encoded objects and surroundings [3, 4].

In this respect, employing digital codes, such as QR codes, for analysis and steering of the learning process could be highly instrumental, in particular, in the domain of MALL. The QR and other two-dimensional barcodes consist of high-contrast patterns of, e.g., black squares on a white background. Such patterns of square arrangements are used to uniquely encode different types of information such as plain texts, URL links, geolocations, e-mail addresses, and others. Through direct interactions with codes placed on various objects or embedded in the surrounding learning environment, users can be easily be directed to a variety of relevant sources of information for any particular subject [5]. Users can thus invoke actions linked to particular objects and access associated information sources in a seamless way, by just using a smartphone or a tablet equipped with a camera, a general QR code reading app, and possibly specialized software for surface-based interaction enhancements.

The development and support of different, unconventional interactions, both among and between students themselves and the surrounding learning environment, have a great potential in the context of the Bring Your Own Device (BYOD) model [6]. The BYOD approach is gaining popularity in various types of educational settings all around the world as technological advancements have made possible the creation of dynamic ad hoc Information and Communication Technology (ICT) environments operating at high speed. Current research on ICT-assisted learning, however, seems to be paying less attention to the technological issues per se, while dealing with pedagogical and didactic advancements. In this context, teaching activities toward socialization appear to be a good target for the employment of mobile devices in the learning process [7]. Findings, in fact, indicate that collaborative language learning through mobile devices could be further boosted by design-based learning approaches such as task-based language learning, situated language learning, seamless language learning, and communicative language learning. Both instructors and scholars will, therefore, benefit if a broader array of pedagogical strategies is made available to help identify and support the specific needs of students using technological advancements for more efficient learning of languages.

A properly designed pedagogical approach can establish and strengthen student interactions in such augmented learning settings, thus ensuring effective development of linguistic abilities through negotiation of meanings and adequate functional structures [8]. This paper, therefore, presents a novel pedagogical framework for enhanced learning experience in which augmented interactions with the surrounding environment are made possible by installing advanced computer vision software on the devices that students own and bring to class. Different aspects of this educational model will be investigated with the aim of fostering a more enhanced engagement of students involved in collaborative storytelling activities employing the BYOD concept. In order to validate the approach, we have carried out experiments with mobile devices, enabled for advanced image-based interactions with digitally augmented objects and surroundings.

2 Related Work

When speaking about mobile learning (m-learning), scholars refer to its three basic qualities that make it stand out, i.e., location independence, time independence, and meaningful content [9], which differentiate the potential of mobile learning from either e-learning or web-based learning [10]. Interest is growing regarding the possibilities offered by m-learning in the field of language education, allowing access to materials and activities without time or place restrictions, thanks to mobility, flexibility, and the seamless connectivity of the devices employed [11]. In this field, various ways to support collaboration-assisted language learning have been implemented through software and augmented reality (AR) applications via usage of QR codes.

In particular, m-learning is one of the fields in which the employment of QR codes in educational settings is most actively researched and analyzed, while mobile devices seem to be the best possible tool to implement such an approach. When it comes to QR codes in education, the potential of m-learning applications directly depends on the employed pedagogical frameworks where QR codes serve as an enabling technology rather than a target. Technology coupled with sound instructional design principles and guidelines assures learning enhancements that position the student in the center of the learning process [12].

We refer to the classification proposed by Rikala and Kankaanranta's review of QR code applications in educational settings [13], where QR code usages are divided into five main classes: trail activities or treasure hunts, outdoor or field activities, paper-based tasks, learner-generated content, and working instruction.

Trail activities or treasure hunts require students to solve problems related to findings based on exploring their own environment, either in a collaborative, competitive, or individual study modality [10]. *Outdoor or field activities* are centered on the exploration of particular life science subjects that can be found outdoor, such as flora or fauna topics. QR codes are used for hints related to the identification of species, deriving information about them, or accessing specific online resources for particular species. This could be practically done by sharing the retrieved information with the class through social network platforms or by creating, for example, ad hoc guidebooks for field studies, containing augmented information about the subject matter in the codes to scan. Studies have shown that such activities are both motivating and highly efficient for student learning [5, 10]. *Paper-based tasks* are useful in embedding multimedia resources such as audio or video materials into traditional paper sheets. They can be used in self-assessment processes as a guide for the students or in listening exercises. In the latter case, QR codes can provide a useful and flexible ubiquitous resource for students [10]. *Learner-generated content* activities are focused on letting students create written or oral materials such as reports or surveys and subsequently embed a link to their work into a QR code for sharing. Such an approach can very well support learner-centered education and stimulate interactive experiences. Finally, *working instructions*, details, and directives on how to proceed to fulfill a particular task are given by the teacher through QR codes in various forms. When instructions are given in a multimedia format, they also represent a valuable way to foster independent learning.

From the point of view of storytelling in language education settings, it is widely recognized that reading aloud directly or even listening to someone who is doing it, significantly improves the vocabulary, syntax, and sentence structure of the learner [14]. In a study of applied storytelling in English as a foreign language (EFL) classroom on a web-based multimedia system, it was found that student speaking skills improved significantly and were combined with an overall rise of learning achievements [15]. Notable language skills enhancement and, more importantly, increased motivation and engagement in studies employing digital storytelling activities and tools were registered by Figg and McCartney [16], while Abdullah [17] acknowledged significant vocabulary improvements for students using storytelling for learning in university settings. Other studies show that the addition of visual and aural components to storytelling can scaffold the story reading experience for young learners. In fact, improved oral comprehension was observed in English language students when reading was coupled with illustrated text instead of a traditional support [18]. These findings deal with the particular field of digital storytelling, that combines computer-based images, text, recorded audio narration, video clips, or music, trying to engage to a higher extent the students that use digital media in their daily life. Applications of the storytelling modality follow a socio-constructivist and inquiry-based pedagogical paradigm [19], aimed at providing students with the instruments that allow them to become self-confident and knowledgeable about their language learning productions, while exposing them to practical usages of the language itself in meaningful communicative contexts.

In previous research [20], we conducted experiments with collaborative activities in language learning classes, noting that interaction between students was not always effective. Thus, in this work, we implement a pedagogical approach that supports and promotes such interactions through Technology-Enhanced Learning (TEL) activities.

Dealing with language learning settings, we take into consideration the suggestion of the Common European Framework of Reference for Languages [21] to put the student in the center of the learning process—at the core of the learning environment itself. In our case, the learning environment incorporates various digitally enhanced physical objects and devices supporting surface-based interactions as a means to provide Tangible Technology-Enhanced Learning (T-TEL) support to the student. We have addressed the topic of T-TEL in our previous studies, analyzing and experimenting with digitally encoded physical objects and surfaces that help students attain augmented learning experiences through sound, images, and multimedia [6, 22]. From a technical point of view, the functionalities of vision integration needed to connect the real and the digital worlds, thus providing access to augmented learning experience, are based on sensors for tracking and registration, a processing unit for analyzing the sensor data and generating the augmentations, and finally, a display to show the integrated views [23]. As current tablets and smartphones are quite affordable and most of them possess the features listed above, they constitute a strong base for BYOD-based T-TEL enhancements [24].

Our current work focuses on T-TEL applications in the field of digitally assisted collaborative storytelling. Such stories are developed in a context linked to the environment surrounding the student participants in which digitally enhanced physical objects have been strategically placed and arranged in advance by the instructor. Our pedagogical model employs context QR codes that are easily accessible through standard smartphone apps and by doing so, students acquire and further develop essential communication skills, through reading and interacting with shared documents, listening to sample oral messages, recording their own voices, and help each other while working collaboratively on the shared assignments.

3   Method

To evaluate the functionality of our proposed approach, we carried out an experiment in an Italian language class for foreign students. It was followed by a questionnaire to measure learners' perception of the collaborative activity.

3.1   Participants and Setting

We ran our experiments in a class of Italian as a Second Language (ISL) offered to foreign students enrolled in Sapienza University of Rome, and organized by the University's Language Center. Typical ISL courses constitute of 40 contact hours, organized in 2-hour sessions, two or three times a week. Lessons are held at the Center facilities in classrooms equipped with standard board and projector.

The student population was very diverse, with 22 students from 11 different countries. With the exception of one doctoral student, all the other participants were enrolled in Bachelor or Master's courses in Sapienza. The class level was A2, as per the CEFR standard classification, and the students had a post-basic ability in Italian language. English was understood by all the participants and was used at times as a functional language. It was employed during the test to clarify some procedural instructions.

3.2   Activity

The standard classroom described above was used as a setting for the experimental activity. QR codes were prepared in advance by the teacher with the free online software QR Code Monkey[1] and placed in predetermined locations inside the classroom: three were posted as pictures hanging on the walls and two were posted as labels on objects.

As for the mobile device equipment, students were instructed to bring and use the devices they usually carry with them such as smartphones. Additional devices, such as

---

[1] https://www.qrcode-monkey.com/.

tablets or notebooks, could also be employed. In order to be able to read the codes, students were asked in advance, to download a QR code reader application of their choice, emphasizing that they should check different options and try to select the simplest and easiest for them to use. Students were also asked to check whether their devices were equipped with voice recorders and to install appropriate free software, if necessary. As a last requirement, all participants were asked to install Google Drive and Google Docs onto their mobile devices so that they could work collaboratively on shared files.

3.3   Collaborative Framework Design

An activity framework was devised to let students develop and practice the main four linguistic abilities: speaking, listening, writing, and reading. In the first phase, each student was provided with rudimentary information on how to create a story based on images and objects that could be found inside the classroom, or in the particular setting, the teacher considered adequate to be employed. This information was given to each student via a dedicated text message accessible through a QR code on the assignment sheets, dealing with one of the focal points of the story to be created and aimed at directing the student toward a particular location in the classroom. Indications were constructed in such a way that students could form pairs once the described spot was reached, and start their collaborative work by brainstorming on the story to be written.

As pairs were formed, the second phase began following the five-step story development of Freytag's pyramid as reported by Tatsuki [25]. For every story stage of *exposition*, *rising action*, *climax*, *falling action*, and *resolution* a different interaction

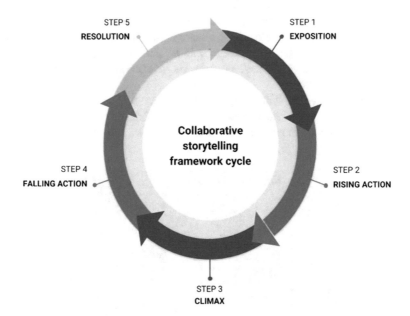

Fig. 1. The collaborative storytelling pedagogical framework cycle

location was specified and different indications were provided (Fig. 1). Such interactions and change of location were made possible by scanning the QR code attached to a specified location in the room through the mobile device camera, acquiring the needed information and moving on to the next story step. These processes were made up in such a way that every pair had to work on a different stage of the story, beginning with the exposition and then proceeding to the next step, each time in a different location, until a number of complete stories, one for each location, were created. At each step, except for the first one, the participants had to listen to the parts of the story developed by the other participants, and integrate collaboratively their follow up.

Once the entire story cycle was completed for all stations, the pairs were directed to go back to their original location and finalize the stories. At the third and last phase of the activity, each pair produced a written transcript of the story and presented it orally. A final debrief of all the participants followed.

3.4   Experiments

The experimental activity took an entire 2-hour class. Students had been exposed to the basic principles of storytelling and story creation in previous lessons. In particular, they were familiarized with the Freytag's five-step dramatic arc structural approach. All students were requested to bring their mobile devices to class.

The instructor introduced the activity explaining the basic technical details, paying specific attention to the interaction with QR codes and the function of locations. For the experiment, three images hung on the classroom walls and two objects placed inside the room were chosen. Two QR codes were attached to each image and object. One code was marked with the sign "i" at the center (Fig. 2, left), indicating that it was to be used as a source of information. It was connected to a text page giving directions about that particular image or object, possible words or sentences to be used with, and the five steps of the story creation. The second code, marked with "s" at the center (Fig. 2, right), was to be used for story creation, and was connected with a shared Google Drive folder. Five folders, one for each location/story, were created and managed by the instructor.

Each student received a white assignment sheet, with a QR code on it to begin the activity and form the teams. Participants scanned the codes, received indications about a picture or object and moved toward that location to find their partners and the

Fig. 2. QR codes marked with "i" and "s" signs

instructions for the exercise. As there were 22 participating students, the teacher decided to form two groups of three students and eight groups of two students, making a total of 10 teams. As there were five chosen locations to work with, the assignment was first given to 10 students and only after the first five teams were formed the remaining of the class received the assignment sheets. In this way, it was possible to have two pairs working in the same location, but with different shared files.

Once at a location, the teams interacted with the codes, acquired information, and created stories. At the first story step, the students interacted with a picture/object to collaboratively start the story. Then, they recorded the story opening with their devices and uploaded the recording into the shared folder. For each subsequent story step and location, they would listen to the previously recorded story parts and then continue the story according to both the content and the particular story step they were working on. After each recording, the file had to be labeled following the pattern "step 1, 2, 3, …, n_group1, 2, 3, …, n," to avoid overlapping with other teams.

After the fifth and last steps of the story were completed, the teams were asked to return to the activity starting point, listen to all the recordings in the shared folder related to that particular location, and connect all the parts to form a complete story. In doing so, they were advised to write down sentences on the assignment sheet to help them recall the story. Writing down the entire story was not requested but most of the students did it anyway. When all the teams finished writing, the instructor proceeded with the presentations and the debriefing.

3.5   Outcomes and Assessment

In order to measure students' activity perception and usability, at the end of the experiment the participants were asked to fill out a written questionnaire (See Table 1). The questionnaire was based on a ten-point Likert scale and was used for assessment.

Table 1. Survey questions

1. It was easy to use QR codes with the devices
2. It was easy to use the device's applications (QR code scan, voice recorder/player) during the activity
3. Mobile devices should be used more in such type of activities and learning environments
4. QR codes allow an easy access to the activity materials and the shared folder in Google Drive
5. Listening from mobile devices was easy and clear
6. Recording the stories and saving the audio file in the shared folder was difficult to do
7. Shared text documents connected to QR codes are cumbersome to work with
8. Using mobile devices and QR codes in the activity require too much attention to technology
9. It was difficult to collaborate with the activity partners
10. I have to learn too many things before I can perform this activity correctly

The percentages of students' answers of each question were calculated and analyzed. A summary of the results and their graphical visualization are given in the next section.

4   Results and Discussion

The experiment went smoothly without any reported disturbances. From teacher's perspective, the preliminary work for preparing the class took some extra time related to QR codes' generation and material preparation. However, the activity could be reused and modified with substantial ease, if a certain number of inputs are stored and reshuffled for the need of the particular class activity.

The teacher distributed a five-point Likert scale survey[2] in paper format at the end of the experiment, to have the students evaluate the experience (Table 1). The results were quite positive and promising for such type of learning modality, as all the participants reported that they liked the activity and were interested and motivated by it (Fig. 3). All the students reported that it was easy to use QR codes with the devices (27.3% agree, 72.7% strongly agree) and using the various devices applications of QR code scanner, voice recorder, and audio player during the activity (18.2% agree, 81.8% strongly agree). This strongly correlates with the fact that all the participants are in their early 20s and regularly use such applications and devices.

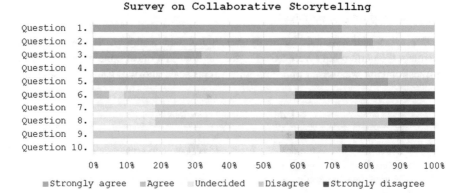

Fig. 3. Visualization of the questionnaire results

When asked about the easiness to access the materials contained in the shared folder through QR code scanning, the answer was very positive (45.5% agree, 54.5% strongly agree), probably given the fact that the instructor choose Google Drive, a free and commonly used cloud storage service, as a base for the repository of the shared materials. Different providers could be used with similar success, but given that

---

[2] Survey's questions are meant to be reshuffled.

Sapienza students receive a Google Mail account once they enroll in a university course, the access and usage of Google Drive in such an environment were seamless.

Regarding the listening experience, almost all students strongly agreed, stating that listening from mobile devices was easy and clear (13.6% agree 86.4% strongly agree), representing a great advantage from a language learning point of view, as the clarity of the input is of high value and importance. It has to be noted that no student used earphones in the experiment, indicating that the listening experience could be even improved if a headset is used. Also, as the recorded file could be played back several times, listening comprehension was better than having the teacher speak loudly in class.

While the passive action of listening to audio files was perceived by most of the participants as clear and easy to perform, the task of recording the stories and storing the audio files in the shared folders was found to also be very easy to complete. When asked if they found the recording and sharing difficult, a significant part of the participants disagreed with the statement (50% disagree, 40.9% strongly disagree). Nonetheless, it should be noted that one student was undecided about this and another one stated that he/she found some kind of complexity in performing the task.

Less than 20% of the students were uncertain when asked if it was cumbersome to work with shared text documents connected with QR codes, while the rest of the class did not find it inconvenient at all (18.2% undecided, 59.1% disagree 22.7% strongly disagree). This data could be linked to individual learning styles or to the abilities and self-regulation skills that are required to work collaboratively not only on the same subject but on the same document. It is also possible that the format of the shared folders displayed on mobile devices was not readable or sufficiently clear to be used seamlessly in a dynamic learning environment, so tablets or smartphones with bigger screens might perform better in such circumstances.

The actual collaboration during the work on the tasks was, however, quite good. When asked if the collaboration with activity partners presented some kind of difficulty, no student reported any problem (59.1% disagree, 40.9% strongly disagree).

Shifting the focus to mobile device usage, the survey results showed that more than 70% of the students were positive about using them in these types of activities and learning environments (27.3% undecided, 40.9% agree, 31.8% strongly agree) denoting a high level of awareness of their potential. This is also a proof of the value of the BYOD model itself in actual educational settings. The possibility of employing in the class the same device that one is normally carrying and using in everyday life could positively affect the instructional design practices and have an impact on the educational materials. Such employment can allow users to seamlessly access technologically enhanced pedagogical tools without lengthily advance preparation. The data of the survey confirmed this assumption as more than 80% of the participants in the experiment disagreed with the statement that using mobile devices and QR codes in the activity requires putting too much attention on the technology (18.2% undecided, 68.2% disagree, 13.6% strongly disagree). Also, 70% of the students stated that there was no need to learn too many things before they could perform the proposed activity correctly (31.8% undecided, 54.5% disagree, 13.6% strongly disagree), so that learners could concentrate their efforts on the linguistic practices and outputs of the task.

5 Conclusions and Future Work

We have proposed a BYOD-mediated Human–Computer Interaction (HCI) approach based on digital surface encoding and vision information processing that is instrumental in the creation of enhanced educational environments and the development of novel didactic methods. Our research targets language learning environments that take advantage of the BYOD approach and employ student-owned devices in the context of mobile-Computer Supported Collaborative Learning (m-CSCL), thus providing for more interactive and meaningful collaboration.

As a practical implementation, we designed and tested a new T-TEL setup supporting development and management of storytelling activities that foster students' language acquisition and communication skills development more effectively.

This approach could be extended to environments specifically crafted for persons with special needs, as for example hearing impaired or dyslexic students. We believe that enhancing the learning process through interactions with digitally encoded tangible objects and environments will help students with disabilities to take better advantage of it, while ensuring a more effective collaboration among all the learners in the class.

Acknowledgements. This work was partially supported by the 2018 Cooperative Research Projects at Research Center of Biomedical Engineering with RIE Shizuoka University.

References

1. Kukulska-Hulme, A., & Shield, L. (2008). An overview of mobile assisted language learning: From content delivery to supported collaboration and interaction. In *ReCALL* (Vol. 20, pp. 271–289). Cambridge University Press.
2. Traxler, J. (2009). Current state of mobile learning. In *Mobile learning: Transforming the delivery of education and training* (Vol. 1, pp. 9–24). books.google.com.
3. Kanev, K., Oido, I., Yoshioka, R., & Mirenkov, N. (2012). Employment of 3D printing for enhanced kanji learning. In *HCCE '12 Proceedings of the 2012 Joint International Conference on Human-Centered Computer Environments*. https://doi.org/10.1145/2160749. 2160784.
4. Kanev, K. (2012). Augmented tangible interface components and image based interactions. In *Proceedings of the 13th International Conference on Computer Systems and Technologies* (pp. 23–29). New York, NY, USA: ACM.
5. Lee, J.-K., Lee, I.-S., & Kwon, Y.-J. (2011). Scan & learn! Use of quick response codes & smartphones in a biology field study. *American Biology Teacher, 73,* 485–492.
6. Gelsomini, F., Kanev, K., Hung, P., Kapralos, B., Jenkin, M., Barneva, R. P., et al. (2017). BYOD collaborative kanji learning in tangible augmented reality settings. In D. Luca, L. Sirghi, & C. Costin (Eds.), *Recent advances in technology research and education* (pp. 315–325). Cham: Springer International Publishing.
7. Kukulska-Hulme, A., & Viberg, O. (2018). Mobile collaborative language learning: State of the art. *British Journal of Educational Technology, 49,* 207–218.
8. Burston J. (2014). MALL: The pedagogical challenges. *Computer Assisted Language Learning, 27,* 344–357 (Routledge).

9. So, S. (2008). A study on the acceptance of mobile phones for teaching and learning with a group of pre-service teachers in Hong Kong. *Journal of Educational Technology Development and Exchange (JETDE), 1*(1), 7.

10. Law, C.-Y., & So, S. (2010). QR codes in education. *Journal of Educational Technology Development and Exchange (JETDE), 3*(7). aquila.usm.edu.

11. Ogata, H., & Yano, Y. (2004). Knowledge awareness for a computer-assisted language learning using handhelds. *International Journal of Continuing Engineering Education & Lifelong Learning, 14*, 435–449.

12. Zhang, B., Looi, C.-K., Seow, P., Chia, G., Wong, L.-H., Chen, W., et al. (2010). Deconstructing and reconstructing: Transforming primary science learning via a mobilized curriculum. *Computers & Education, 55*, 1504–1523 (Elsevier).

13. Rikala, J., & Kankaanranta, M. (2012). The use of quick response codes in the classroom. In *mLearn* (pp. 148–155). academia.edu.

14. Paul, A. M. (2012). Why Kids Should Learn Cursive (and Math Facts and Word Roots). Retrieved July 4, 2015.

15. Hwang, W.-Y., Shadiev, R., Hsu, J.-L., Huang, Y.-M., Hsu, G.-L., Lin, Y.-C. (2016). Effects of storytelling to facilitate EFL speaking using web-based multimedia system. *Computer Assisted Language Learning, 29*, 215–241 (Routledge).

16. Figg, C., McCartney, R., & Gonsoulin, W. (2010). Impacting academic achievement with student learners teaching digital storytelling to others: The ATTTCSE digital video project. In *Contemporary Issues in Technology and Teacher Education. Association for the Advancement of Computing in Education (AACE)* (Vol. 10, pp. 38–79).

17. Abdulla, E. S. (2012, July). The effect of storytelling on vocabulary acquisition *4*(2015). iasj. net.

18. Huang, H.-L. (2006). The effects of storytelling on EFL young learners' reading comprehension and word recall. *English Teaching & Learning, 30*, 51–74.

19. Palmer, B. C., Harshbarger, S. J., & Koch, C. (2001) A. storytelling as a constructivist model for developing language and literacy. *Journal of Poetry Therapy, 14*, 199–212 (Springer).

20. Barneva, R. P., Gelsomin, F., Kanev, K., & Bottoni, P. (2017). Tangible technology-enhanced learning for improvement of student collaboration. *Journal of Educational Technology Systems*. 0047239517736875 (SAGE Publications Inc).

21. Council of Europe. (2001). *Common European framework of reference for languages: Learning, teaching, assessment*. Cambridge, UK: Press Syndicate of the University of Cambridge.

22. Kanev, K., Oido, I., Hung, P. C. K., Kapralos, B., & Jenkin, M. (2015). Case study: Approaching the learning of kanji through augmented toys in japan. In P. C. K. Hung (Ed.), *Mobile services for toy computing* (pp. 175–192). Cham: Springer International Publishing.

23. Turk, M., & Fragoso, V. (2015). Computer vision for mobile augmented reality. In G. Hua & X.-S. Hua (Eds.), *Mobile cloud visual media computing* (pp. 3–42). Cham: Springer International Publishing.

24. Burston, J. (2016). The future of foreign language instructional technology: BYOD MALL. *The EuroCALL Review, 24*, 3–9.

25. Tatsuki, D. H. (2016). How to Teach Narratives: A Survey of Approaches. kobe-cufs.repo. nii.ac.jp.

On the Usefulness of Animated Structograms in Teaching Algorithms and Programming

Ulf Döring[✉] and Benedikt Artelt

Technische Universität Ilmenau, Ilmenau, Germany
{ulf.doering,benedikt.artelt}@tu-ilmenau.de
https://www.tu-ilmenau.de/computer-graphics-group/mitarbeiter

Abstract. This document describes an online tool, which supports different synchronized views onto the same algorithm in parallel, e.g. structograms (Nassi–Shneiderman diagrams), Java code and the state of used variables. It can also animate the program and data flow. This is essential for self-controlled learning especially for novice programmers. Students can focus on the view they prefer and explore the relationship to the other views at their own speed, e.g. by stepping through the program and visualize the current state of the variables as well as the console output. The tool gives the possibility to modify the structure as well as the texts in the structograms. It supports exports into different file formats, i.e. for the creation of teaching material or solutions for exercises. The HTML5 implementation allows easy access to the tool on different platforms as well as its integration into web-based learning environments.

Keywords: Browser-based e-learning tool · Program visualization · Programming patterns · Multiple-views synchronization · Levels of abstraction · Java · HTML5

1 Introduction

Most engineering students at our technical university have to learn Java programming as well as data structures and algorithms in their first two semesters. Usually the previous knowledge of the approx. 500 students varies widely; motivation is often rather low and in the beginning it is quite hard for most of the students to understand the interplay of simple programming code, variables and output as well as the syntax of the programming language. The first weeks are especially critical, because if the students do not learn to understand those basics in time then it is nearly impossible for them to comprehend advanced topics like algorithm synthesis, object-oriented programming and the implementation of dynamic data structures.

The last years have shown that extra effort in classical ex-cathedra teaching had little effect. In general, it might make sense to introduce new topics and

© Springer Nature Switzerland AG 2019
M. E. Auer and T. Tsiatsos (eds.), *Mobile Technologies and Applications for the Internet of Things*, Advances in Intelligent Systems and Computing 909,
https://doi.org/10.1007/978-3-030-11434-3_8

basic ideas during a lecture, but the phases of self-controlled learning are much more important, especially if it can be supported where necessary. Accordingly, we achieved positive results by giving more time as well as individual support for solving tasks during exercises and tutorials. Nevertheless, this meant that we needed more teaching hours for supervisors and tutors.

Knowing that the personnel expenditure cannot be significantly increased, we started to develop e-learning applications, which are capable of giving general support and hints during the analysis of a given programming pattern or during the solution of a programming task [1]. One of these tools is based on structograms, which are also called Nassi–Shneiderman diagrams [2]. Reasons for using a structogram-based tool are the following:

- Structograms are a standardized [3] and very compact graphical representation of programs, especially suitable for small programming patterns and also for algorithms as they are handled in the taught subject.
- For novices, the visual parallelism of alternatives in if-else or switch-case control structures is more intuitive than the serialized notation of program code.
- Structograms are a better alternative to the often used pseudocode, because the strong visual structuring avoids that the students mix up syntactical elements with those from the programming language.
- They can be filled with native program code or with describing texts at different levels of abstraction.

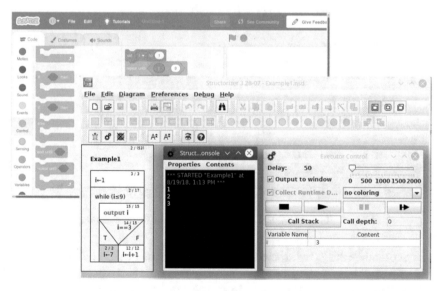

Fig. 1. Applications [4,5] dealing with structograms or a similar block-based design

There are several applications dealing with compact structogram-like visual representations (e.g. Scratch [4] and MOENAGADE [6]) or structograms (see

Fig. 1). The structogram-based applications are programmed in different languages like Python (e.g. [7]) or Delphi (e.g. [8]) but mostly in Java (e.g. [5,9]). Some of the applications are able to interpret program code written in different languages and to simulate the control and data flow. This and other features inspired our own software development to create a tool that combines visual programming and program visualization [10]. Even though it might be possible to use Java-applications in a web-based environment, we decided to develop an own HTML5-based application, because:

- The support of Java applets in web browsers decreases, mainly caused by security reasons.
- HTML5 became a powerful tool for implementing web-based applications, which are platform independent and thus easy to use for almost all students.
- An HTML5-based implementation can be seamlessly integrated into a web-based teaching and learning environment.
- It will be quite easy to automatically generate tailor-made versions of the HTML5 application, which are generated and delivered according to the use case in an online platform.
- It was not possible for us to find a relevant HTML5-based implementation, which deals with structograms. Nevertheless, at least Scratch 3.0 beta [4] shows, that parts of the community start to switch to HTML5.
- In a self-developed application it is easier to realize own didactical and methodical ideas immediately.
- We gathered programming experiences over many years in implementing digital libraries and according HTML5 applications (see e.g. [11]).

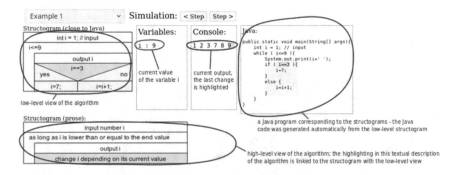

Fig. 2. Introductional example of the tool for the students

In September 2017, we started to use first versions of the tool during lessons for the 60 students of the Basic Engineering School [12]. Experiences and feedback accordingly helped to continuously improve features or to introduce new ones. At the beginning of the summer semester, the tool supported simple Java program simulations [13] as an alternative to debugging within Eclipse [14], which is usually used by our students to develop Java programs, see Fig. 2, which shows debugging features at this time. Since then the most obvious changes in the tool concerning program simulation are support of further control commands like `switch`, `case`, `break`, `continue`, and `return`, improved variable visualization, explicit control flow visualization as well as complete multidirectional synchronization of highlights between the *Java*-view and all the *Structogram*-views.

2   Currently Implemented Features and Their Utilization

2.1   Java as Target Language

Because of the educational objective the focus is currently definitely on Java, i.e. parsing and simulation follows the basic language rules of Java. But the coverage of the Java language features is by far not complete. Nevertheless, most examples from the first lessons can already be mapped into structograms as well as simulated using the tool. Currently supported are the following:

- definition of boolean and integer variables as well as a selected set of simple terms,
- definition and usage of one dimensional integer arrays,
- if-else, while and switch-case control structures,
- break, return and continue commands,
- console output of the defined variables.

Aside from these language features with simulation support, it is of course possible to use the tool as general structogram editor and enter arbitrary texts. The current limitations here are that texts cannot be wrapped within a structogram block and, compared with DIN 66261 [3], there are currently no blocks for post-tested iterations (e.g. do-while loops), continuous iteration and parallel processing.

From the students point of view, it is easier to implement a new algorithm based on an editable structogram because (similar to the applications shown in Fig. 1) the general structure and the specific commands can be handled separately. Furthermore, there is no danger to make syntactical mistakes during the structure definition (e.g. forgetting some bracket or using the wrong one).

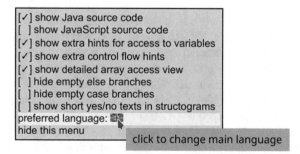

Fig. 3. Possibilities for interactive configuration

Each mistake bears of course the chance to learn the respective troubleshooting strategy and to memorize it permanently, but for novices without supervisor the incomprehensible cryptic error messages and respective unsuccessful trails to fix the error are usually very demotivating.

2.2 Interactive Configuration

Compared with applications like Structorizer [5] (which seems to be currently the most powerful and most actively developed application for handling structograms) there are only a few possibilities for the configuration, see Fig. 3. These are related to:

- *Language*: In general, the number of supported languages is not limited. Currently, we experiment with English, German and French. An important feature is that not only elements of the user interface (like labels, menus and tooltips) can be translated. It is also possible to define language-depended data (structogram texts). This especially helps foreign students, which have deficits in German (the current teaching language).
- *Source code output*: Aside from Java, it is also possible to show JavaScript code. From the methodical point of view, the checkboxes for hiding empty `else` and `case` branches are more important. The changes are shown immediately (in the generated source code) and allow to discuss or think about the consequences.
- *Detailedness of hints*: The presence of hints with detailed information for variable access, control flow or array access (see examples in Table 1) can be switched on and off. This allows the user to decide if the hints shall be shown or not. For instance, a teacher might wish to ask the students in each simulation step, what will happen with the variables (switching hints off) or he wants to give visual support to his explanations (switching hints on).
- *Structogram design*: Here, a lot of layout parameters like fonts, line widths, spacing, etc., could be managed, but currently it is only possible to force short names for the `if` and `else` branches.

2.3 Interactive Generation of Structograms

At the moment there are 3 ways to get structograms into the tool. Aside from selecting predefined sets in a selection box, it is possible to paste description code (which could be downloaded and saved as a text file in another session) into a textarea. The last way is to modify an existing or empty structogram with the help of special keyboard inputs. Using a pointer device or cursor keys the user can select (highlight) a block. This block can be edited, i.e. the text can be entered and changed. This applies to all 5 block types (pure text, if-else, while-loop, switch and case). The selected block can also be deleted. For each of the 5 block types exist 2 key codes, which allow to insert a new block before or after the selected block. The termination block symbol will be shown for text blocks with according contents (e.g. the keyword `break;`). For some block types there are constraints defined, for instance a case block can only be inserted directly into a switch block. Concerning switch-case blocks there is also a rule implemented, which ensures that during delete operations at least 3 cases are left in the switch block. Thus, the visual distinction between if-else and switch-case blocks is kept.

2.4   Work with Structogram Sets

It is quite easy to insert predefined sets of structograms into the HTML code. When the tool runs, these sets can be selected from a selection box. This way, tailor-made versions of the tool can be provided to the students by teaching persons. For instance, in lessons about sorting algorithms such algorithms, as well as more basic programming patterns like the switching of two variable values, may be added to the tool and used in a demonstration or for experiments performed by the students.

Discovering Relationships Structograms can be used to describe an algorithm at very different levels of abstraction and are, therefore, a useful didactical means. But the presentation of an algorithm at only one level (e.g. high or low) is not as expressive as to show both at the same time and visually link the corresponding elements together. Therefore, we provide in general a set of structograms for one algorithm. All structograms in this set as well as automatically generated Java code have synchronized highlighting. I.e. when pointer device or cursor key actions move a highlight in the currently active view then the highlights in all the other views are updated accordingly. This helps the students to understand the relationships between the different views. Thus, the transition between more abstract descriptions using natural language and low-level descriptions (which make use of the programming language itself) as well as transitions from structograms to the source code become more understandable for novice programmers. Figure 4 shows a set of 3 linked structograms.

The described synchronization works also during the simulation of the algorithm, i.e. the data and control flow. This way the students can decide at which level of abstraction they try to comprehend an algorithm, or they switch between the different levels, e.g. using a short textual description as general overview, a more detailed textual description as explanation and the Java-like view and/or the Java code to understand the low-level realization.

Modification The exploration process for a certain algorithm can also be supported by allowing modifications and by visualzing the consequences of these modifications. Instead of deleting or inserting blocks (as previously described) usually the slight changes of the program texts are the best eye-openers, for instance changing a < into a <=, using a wrong variable initialization value or changing the values in an array, which shall be sorted. As far as the set of supported command patterns is not left, the simulation of the data and control flow is performed immediately and the students can see if the output generated by the modified algorithm fits their expectation.

Export The export of structograms into different formats was one of the first implemented features. Aside from the PNG format, which can be retrieved sim-

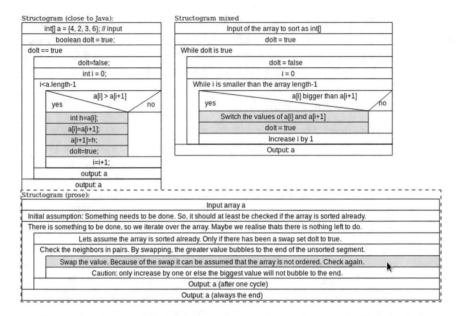

Fig. 4. Synchronized highlighting in a set of structograms for Bubble Sort

ply by using the `toDataURL`-method of HTML-canvas elements, different vector formats are supported as graphical output: EPS, SVG and LaTeX-pictures. Furthermore, the selected structogram can be saved as Java, JavaScript or the internal description format in order to load it later or to paste the code directly into the HTML code and make it permanently available via the selection box. For "classical" (static) teaching material like PDF-files or solutions of exercises the EPS- or PNG-export are mostly used; but for explanations in lectures or as experimental area in exercises and tutorials HTML-pages with the tool itself (adapted to the particular use case) are provided.

Simulation Simulation may be the strongest feature of the tool. Mandatory prerequisite for a successful low-level analysis of the algorithm is valid Java code in one of the structograms. Furthermore, the entered Java code must be quite simple, because there is only a restricted set of command and term patterns supported. These are detected using regular expressions. For each detected pattern like:

- define variable (`boolean` or integer types like `int`),
- change variable (`boolean` or integer types like `int`),
- control command (`break`, `continue` or `return`),
- term which evaluates to a Boolean expression

the needed parameters are extracted and used in the simulation. Thus, the detected meaning of the texts, which are included in the structogram, controls the control flow and the data flow during the simulation. In each simulation step data is appended to the simulation history, which contains the related structogram block, the changes of variable values, output actions, etc. A side effect of this simulation is that after the simulation the output console is filled with the output generated by the algorithm. Thus, users have immediately feedback if the structogram works as expected or not.

A further advantage of the history is, that it is very easy to trace steps backwards. Thus changes between one step and the other can be analysed by the students as long as needed simply by stepping forward and backward. The 'step back' debug feature is usually not supported by programming environments like Eclipse, because of the limited debug capabilities of the JVM (Java Virtual Machine).

Visualization For each step, the values of the used variables are known. This information is used to generate visual hints. Table 1 shows examples for different kinds of visual hints, which are shown as semitransparent overlay in the structogram. Aside from hints for the data flow there are also hints concerning the control flow.

Each teaching person may have his own ideas (which will probably change over time) about how a good explaining visualization should look like. Furthermore, the preferred kind of visualization may depend on the taught person(s). Thus, we want to make the tool configurable and extendable with respect to the graphical effects, which can be used for the visualization of the data and control flow. A first step into this direction is to build up a set of drawing functions, which have structogram text positions and values as well as graphical styles as parameters.

Table 1. Examples for hints concerning data and control flow

detected pattern	code example	generated hint
set a value	a = 1;	a=1;
change a value	a += 42;	a+=42; 43 = 1 + 42
select else branch	i==3	i==3 yes no i=7; i=i+1;
switch to a case	year-2000	year-2000 17 18 other limit-=23; limit-=27; limit-=30;

3 Selected Implementation Details

3.1 Generation of the Detailed Arrays View

In our teaching lessons, all explanations of array-based algorithms are usually supported by sketches on a sheet of paper or on the blackboard. We consider that as very important and supply corresponding automatically generated views in the tool. For all 1-dimensional arrays, the currently stored values, their indices and the relevant variables or terms that are used as indices are shown. See Fig. 5 as an example. Here t is a known integer variable (see *Variables*-view), but it is not shown in the *Detailed arrays view*. In the *var*-row of the *Detailed arrays view*, there are only those variables and terms listed, which are used as index in the according array a. For the analysed InsertionSort algorithm, aside from the variables i and m, only m-1 was detected as relevant term.

3.2 Generation of Vector Graphics

The code for generating vector graphics is quite compact. The main idea is to reuse the code for drawing in the HTML5-canvas. The called methods, which access the canvas context during the drawing process, are redefined for the supported vector graphic formats. The EPS-context for instance collects the generated EPS source code in its eps variable, i.e. general initializing EPS-code and all the keywords like gsave, moveto and lineto with the relevant parameters. Rounding values improves the readability of the generated EPS-code. The parameter transformation (this.h-y) is necessary because EPS is based on a so-called right-handed coordinate system (origin is in the lower left corner) whereas the HTML5-canvas uses a left-handed one (origin is in the upper left corner). In the following JavaScript code snippet r, h and rnd are helper to convert values for EPS. The lines after that show functions, which wrap the EPS-Code generation into the HTML5-context syntax.

```
r    : function(x){return this.rnd(x,1); },
h    : this.rel_h,
rnd  : function(val, dig){try { let fact=Math.pow(10, dig);
                           return Math.round(val*fact)/fact;
                         } catch (e){} return val;},
save      : function(     ){ this.eps+="gsave\n"; },
beginPath : function(     ){ this.eps+="newpath\n"; },
translate : function(x,y  ){ this.eps+=this.r(x)+" "+ this.r(this.h-y)+" translate\n";},
rotate    : function(a    ){ this.eps+=this.r(a*180/Math.PI)     +" rotate\n"  ;},
moveTo    : function(x,y  ){ this.eps+=this.r(x)+" "+ this.r(this.h-y)+" moveto\n"  ;},
lineTo    : function(x,y  ){ this.eps+=this.r(x)+" "+ this.r(this.h-y)+" lineto\n"  ;},
stroke    : function(     ){ this.eps+="stroke\n"; },
restore   : function(     ){ this.eps+="grestore\n"; },
fillText  : function(t,x,y){ this.eps+=this.r(x)+" "+ this.r(this.h-y)
                             +" moveto ("+t+") show\n"; },
```

3.3 Links Between the Structograms

The structograms of one set are linked by IDs. Currently, we use numbers as IDs. The simplest way is to enumerate the blocks in the most low-level diagram (usually the structogram with the Java code) in ascending order. The blocks in the more high-level structograms are linked to one or more of those low-level blocks by storing the respective numbers. During the interactive modification process some heuristics try to guess good IDs. But because the ambiguity of the possible assignments the IDs have to be adjusted where necessary. This currently has to be done by hand.

4 Planned Development, Risks and Chances

The relatively low effort for the development of a reasonable application encourages us to continue our work. Short and midterm aims include for example:

Fig. 5. InsertionSort simulation, display for 3 consecutive steps

- support of further data types (char, float, double, String),
- for-loop,
- method calls (at least recursion within one structogram),
- drag and drop of blocks,
- more and easier enrichments with hints and explanations,

- extended configuration capabilities,
- undo and redo of changes.

Thus, there are good chances to develop a more powerful tool. But we see also some risks:

- *Overload*: For freshmen, the use of an IDE (integrated development environment) like Eclipse is usually problematic not only because of the texts in the user interface or in the feedback to programming errors (those texts are written for advanced users). It is also the oversupply of menus, sub-windows, dialogs, etc. Each new feature (hint, view, animation) of our tool also bears the risk of leading to an oversupply of functionality, which may demotivate or prohibit effective usage as it often happens when freshmen use IDEs. Thus, configurability is very important for a successful application of the tool.

- *Durability of HTML5*: Choosing a programming language for a project carries the danger that the support of the language and according tools decreases over the time. Java applets may serve as an example. Because of security reasons, the use of Java within web browser applications decreases continuously. Our hope is, that JavaScript as immanent part of HTML5 is supported by web browsers for a long time and required changes, to keep the source code conform to the standards, do not cost to much effort.

- *Maintainability*: Until now we had good experiences with a quite agile style of development. Nevertheless this caused partly 'ugly' code, which is to be refactored in time. Else the maintainability of the continuously growing source code cannot be ensured.

- *Focus on own code*: Currently we use no third-party libraries. This has advantages as well as disadvantages. In general, at the beginning there is a larger development effort (especially with respect to possible browser dependencies). But the higher flexibility as well as possibly less maintenance effort (no changing library interfaces or trouble with license changes) let us still hold on this 'no external code' policy. This policy has also another advantage. It allows us to keep the source code as well as images, etc., in one file. That makes it easier for users to work offline or to store different versions of the tool in parallel.

- *Limitations of RegEx-parsing*: When terms become more complex the currently used approach of applying regular expressions to detect the meaning of texts entered into structogram blocks is not suitable. In general, the set of covered term syntaxes can be enlarged successively as needs rise up. But when students make modifications by their own it is not clear to them why the parsing of their entered text failed. It might be caused by a syntax error or a limitation of the tool. Thus, a more general syntax parser is needed. The tree it generates should be also worth to be visualized to the students to give them the possibility to observe the calculation process of terms step by step. This might be especially useful for the understanding of commands like i++ (vs. ++i).

5 Conclusions

Based on the described tool, structograms can be generated and used in a new way. The tool supports teachers during the preparation and presentation of their lectures but it especially supports the students in their own self-controlled learning phases. The integrated simulation of the program flow as well as of the data flow makes it much easier for novice programmers to analyse (and understand) a certain algorithm than it is possible for them in a professional programming environment. Special graphical hints, highlights and tooltips supplied by the teacher according to his didactical experience and approach improve the understanding. In general the tool might also be used, for example in primary, secondary or professional schools.

Acknowledgements. Parts of the presented work are supported by the BMBF funded BASIC Engineering School [12]. We would like to thank our colleagues and students, who have supported us in our work with tips and suggestions. Furthermore, we want to thank the developers of structogram editors or similar software for all the stimuli they gave us with their implementations.

References

1. Döring, U., & Fincke, S. (2017). Interaktive Ansätze zur Vermittlung von Programmierfähigkeiten im Rahmen des Ingenieurstudiums. In Tagungsband der 12. Ingenieurpädagogischen Regionaltagung 2017, 11–13 May 2017, Ilmenau, IPW 2017.
2. Nassi, I., & Shneiderman, B. (1973). Flowchart techniques for structured programming. *ACM SIGPLAN Notices, 8*, 12–26. https://doi.org/10.1145/953349.953350.
3. DIN 66261:1985-11. (1985). Sinnbilder für Struktogramme nach Nassi-Shneiderman.
4. Scratch 3.0 beta. (2018). https://beta.scratch.mit.edu.
5. Fisch, R., Gürtzig, K., et al. (2018). Structorizer. Retrieved August 16, 2018, from https://structorizer.fisch.lu.
6. Fisch, R. (2017). MOENAGADE—The mouse enabled game development tool. Retrieved August 16, 2018, from https://moenagade.fisch.lu/index.php?include=screenshot.
7. Linkweiler, I. (2018). PyNassi. Retrieved August 16, 2018, from http://www.ingo-linkweiler.de/diplom/doc/index.html.
8. Scheck, R. (2018). StruktEd. Retrieved August 16, 2018, from https://www.robert-scheck.de/tools/strukted.
9. Krummenauer, K. (2018). Struktogrammeditor. Retrieved August 16, 2018, from https://whiledo.de/index.php?p=struktogrammeditor.
10. Myers, B.A. (1990). Taxonomies of visual programming and program visualization. In *Taxonomies of visual programming and program visualization* (Vol. 1, No. 1, pp. 97–123). ISSN 1045-926X. https://doi.org/10.1016/S1045-926X(05)80036-9.
11. DMG-Lib. (2018). The Digital Mechanism and Gear Library. Retrieved August 20, 2018, from https://www.dmg-lib.org.
12. Basic Engineering School. (2018). What is the Basic Engineering School? Retrieved August 18, 2018, from https://www.tu-ilmenau.de/studieninteressierte/en/fields-of-study/basic-engineering-school/.

13. Döring, U. (2019) Aktuelle Möglichkeiten des Einsatzes von Struktogrammen als anschauliches Hilfsmittel beim Programmierenlernen. 13. Ingenieurpädagogischen Regionaltagung 2018, 7–9 June 2018, Ilmenau, IPW 2018 (to appear in spring 2019).

14. Wikipedia. (2018). Eclipse (Software). Retrieved August 19, 2018, from https:// en.wikipedia.org/w/index.php?title=Eclipse_(software)&oldid=852834401.

Problem-Oriented Learning Based on Use of Shared Experimental Results

Milan Matijević[1(✉)], Milos S. Nedeljković[2], Đorđe S. Čantrak[2], and Novica Janković[2]

[1] Faculty of Engineering, University of Kragujevac, Kragujevac, Serbia
matijevic@kg.ac.rs
[2] Faculty of Mechanical Engineering, University of Belgrade, Belgrade, Serbia
{mnedeljkovic,djcantrak,njankovic}@mas.bg.ac.rs

Abstract. This paper describes needs, possibilities, and implementation of open educational resources, which are supporting the problem-oriented engineering education based on realistic experimental results. Experimental results can partly substitute real experimentation which is necessary part of engineering education and research. Open educational resources and use of blended learning approach can improve relevancy and decrease costs of engineering education for the majority of engineering schools. These approaches include mobile learning concepts. Also, laboratories via the Internet are one way of sharing laboratory resources in order to increase the availability of experimental work. But there are many situations in which the use of Internet-mediated laboratories is nonrealistic. On the other hand, trained staff and hands-on laboratories and experimental installations are very limited resources at all engineering faculties in countries like Serbia. Because of that, the web portals for sharing experimental data (with defined experimental protocols and all relevant descriptions and tutorials) can be useful for mobile learning applications and engineering education and research.

Keywords: Experimental data · Data repository · Problem-based learning · Laboratory models · Sharing of experimental results · Open educational resources

1 Introduction

Experiments and problem-oriented education are necessary part of contemporary engineering education [1–3]. In this way, through the experience of solving realistic and open-ended problems, students can achieve better absorption of knowledge and skills. The student task is not focused on problem-solving with a defined solution but includes knowledge acquisition, enhanced critical appraisal, developing skills used for student's future practice, enhanced group collaboration and communication, and development of other desirable skills and attributes. There are a lot of publications dedicated to importance of problem-based learning in engineering education [1–6]. Problem-based learning in engineering education is dominantly based on the use of laboratory support, which is a very limited educational resource (regarding equipment, logistic support, and trained teaching staff). A solution of that problem can be sharing

© Springer Nature Switzerland AG 2019
M. E. Auer and T. Tsiatsos (eds.), *Mobile Technologies and Applications for the Internet of Things*, Advances in Intelligent Systems and Computing 909,
https://doi.org/10.1007/978-3-030-11434-3_9

of educational resources, i.e., building and using open educational resources, as well as the use of blended learning approach [7]. Blended learning refers to the combination of face-to-face and technology-supported learning activities [7]. From technological perspective, web technologies, including web laboratories, constitute the main technological context. According to [7], the technological context is supported by mobile apps with only 5% in reviewed papers.

Problem-oriented education is promoted in [8] by dissemination of inquiry learning spaces (ILS) as open educational resources integrating online labs, and complement science and technology knowledge with reflective and social abilities. Inquiry learning spaces creation, personalization, and exploitation are at the center of the Go-Lab (https://www.golabz.eu) and Next-lab European digital education initiatives. The golabz.eu sharing platform is a repository offering online labs, scaffolding apps, and inquiry learning spaces created by teachers for teachers [8]. The graasp.eu authoring platform is a social media enabling collaborative creation, agile personalization, and secure exploitation of open educational resources [8]. These platforms provide visibility and organization of online- or web-based laboratories, but could not provide their maintenance and sustainability. Because of that, education based on web laboratories can be very vulnerable [3–5].

Is it possible to organize a part of open educational resources and problem-oriented engineering education based on realistic experimental results without online experimental realization? Typical or/and the best illustrative experimental realizations by accompanying video clips, diagrams, and so on could be saved and organized within different web-based platforms like [9–11].

Data are the infrastructure of the science [12]. Nowadays, education is more data-intensive and collaborative than in the past, and there is very important issue concerning data accessibility, reuse, preservation and especially data sharing. Also, there are significant differences in needs of data management (collecting data, searching for, describing or cataloging, analyzing, short-term and long-term storage) which is based on subject discipline, funding, and so on [12]. In [13], an infrastructure for gathering, analyzing, and sharing data is discussed as well as some topics on the implementation and future of global experimental data repositories. It is not an easy task to offer mechanism for identification and archiving of datasets together with services for sharing them and making them citeable [14]. Many repositories support and implement minimal control on the data uploaded by users, and this problem is a drawback of new websites for sharing of experimental data but with weak institutional support. The study [14] has described the results of a study on the Zenodo repository.

Experimental datasets or repositories of experimental results for purpose of problem-oriented education can be implemented on websites, learning management systems (LMSs) like Moodle (https://moodle.com), multidisciplinary open access archive for the deposit and dissemination of scientific research documents like HAL (https://halshs.archives-ouvertes.fr), personalized learning environments (PLEs) like Graasp (http://graasp.eu), open research data repositories like Mendeley Data (https://data.mendeley.com), Zenodo (https://zenodo.org), 4TU (http://data.4tu.nl), Apollo (https://www.repository.cam.ac.uk), DRYAD (https://datadryad.org), ICPSR (https://www.icpsr.umich.edu/icpsrweb/), and so on [9, 10].

Of course, there is always the possibility to create the content of a web portal that should enable the integration of teaching resources via the Internet and the availability of experimental data for problem-oriented education. The web portal should provide descriptions of experimental installations, aims, and possible tasks of problem-oriented projects in engineering education, tutorials for basic theoretical concepts and use of software tools, illustrative examples of results, benchmark tests, students, and other papers. Proposed content of the web portal can be based on Go-Lab Tutoring Platform because of multiple goals including maintenance reasons [15–23]. The primary goal is to provide a basis for experimental results, in which the student will be able to use in various subjects in accordance with the instructions. The aim is that students can become familiar with actual physical phenomena, and the structure of the experimental system can understand theoretical concepts and can be able to apply them in the context of real experimental results.

This paper describes the needs, possibilities, and implementation of open educational resources which are supporting to problem-oriented engineering education based on realistic experimental results.

2 Laboratory Models and Problem-Based Learning

In [24], laboratory model PT400 is described along with its development, design, and fabrication as well as hands-on and web-laboratory use within problem-oriented education at Faculty of Engineering at University of Kragujevac. Application of problem-based learning concept has encompassed all phases of development of new laboratory model PT400, which was made in accordance with well-knowing laboratory models PT326 Feedback Ltd and The Quanser Heat Flow Experiment. Technical specification of these laboratory models is given in [11].

The laboratory model from Fig. 1 consists of a long tube (chamber) equipped with a few (three or four) temperature sensors located equidistantly. There are a coil-based heater and a blower at one end tube that is used to transfer heat conductively. The heater power is the input signal. Output signal is temperature in the tube. Disturbance signal can be changeable rate of airflow in the tube. This concept of the process trainer (PT) is common for a few producers of laboratory equipment (Fig. 2).

Fig. 1. Left—PT400 [11, 24], Right—The Quanser Heat Flow Experiment [11]

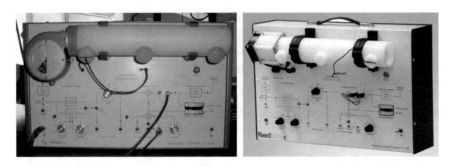

Fig. 2. Feedback Instruments Ltd. Process Trainers: Left—PT326 [11], Right—PT 37-100 [11]

Educational purpose of these laboratory models encompasses modeling and identification, measurement and control, sensors and actuators, signals and systems, as well as real-time programming via different computer supports (microcontrollers, PLCs, and PCs with AD/DA interface). Students can practice their skills, implement different theoretical concepts based on experimental results, and compare some theoretical and experimental results. Especially, problem-oriented project in design, fabrication, and testing of laboratory model PT 400 (Fig. 1) has a great educational potential, but this type of student project has to be supported by good laboratory logistics and very experienced and motivated teaching staff. Although overall fabrication cost of laboratory model PT 400 is more than 10 times cheaper than trade price of commercial laboratory models, this project task is unrealistic in Serbia for the needs of mass education. Moreover, because of very limited financial resources, hands-on laboratory exercises for 8 or 20 students per group (with 8 or 20 laboratory working places or half of ones) can also be unrealistic in Serbia. Because of that, this laboratory model was shared via Internet for web-based laboratory exercises [4, 24]. Some laboratory models, which are available for Internet-mediated experimentation at Faculty of Engineering at University of Kragujevac are integrated within repository of online laboratories [4]. Go-Lab Tutoring Platform integrates repository of online laboratories and additional e-services for contemporary education [7, 8, 15–23].

Also, in line of needs of problem-based engineering education, a complex laboratory model, depicted in Fig. 3, is designed and produced by teaching staff at Hydraulic Machinery and Energy System Department at Faculty of Mechanical Engineering, University of Belgrade [6]. Designing and implementation of this laboratory model is an excellent case study for problem-oriented education, and this setup is about four times cheaper than commercial ones. This laboratory model is unique at the faculty, and students in small group can do laboratory exercises. It is proved in practice that, although student groups for laboratory exercises are small, attendees do not learn all the elements planned in subject curriculum. Now, using Internet and remote control of the experimental setup, the students will still keep the "do-it-yourself" concept, but have more time to experiment and learn. They will be able to generate their reports via Internet, and teachers would have a possibility to follow their work and enable various access limitations [6].

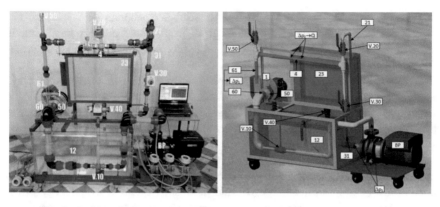

Fig. 3. Produced laboratory model: Pump system hydraulic performance [6]

When a laboratory model is unique at the Higher Education Institution (HEI), there is a huge risk that hands-on laboratory work by students can be substituted with demonstrative laboratory work by teacher. But remote experiments via Internet, with oral final exam where students would have a possibility to explain obtained results, can provide active role of students and better outcomes of educational process.

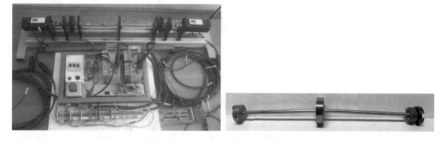

Fig. 4. Lab model of servo system with flexible coupling and hazard effect of its use [25]

In [25], remote experimentation is described with laboratory model in Fig. 4, which was designed and implemented by research staff at Faculty of Engineering, University of Kragujevac. Purpose of this laboratory model is to support the education and research. But Fig. 4 also illustrates the possible dangerous effects of this laboratory model use. This figure illustrates that remote experiments can protect students from different hazardous effects of their immediate hands-on use.

All of three mentioned examples of remote-controlled laboratory models are very vulnerable against to properly 24/7—maintenance. It is necessary to provide safe working regimes during long time for all components of laboratory models. Possible hazardous effects should be predicted and avoided by monitoring the range and dynamics of all relevant signals. Monitoring and maintenance of laboratory equipment is a serious institutional task. This institutional service is necessary for bought hands-on and Internet-mediated laboratory work. Because of sharing and integration of laboratory resources, maintenance of Internet-mediated laboratories should be much cheaper per time unit or regarding number of users. On the other hand, Internet-mediated laboratory can be much more restrictive for users regarding span of possible educational tasks and personal motivation.

A very common and general engineering and research task is the processing of measurement signals and experimental data, and their use for model verification, modeling, and/or identification of model parameters of physical objects and processes. Namely, for mentioned types of educational or research tasks, students need only experimental datasets.

There are a lot of educational aims in problem-oriented engineering education, where, without losing foreseen educational outcomes, we can use more modest laboratory infrastructure, but much cheaper and more safer one. Namely, it is possible to design a web portal with repository of experimental datasets. Overall, web content should encompass narrative description of the experimental plant, optional video clips of key phenomena in the system, recorded experimental data, possible tasks for problem-oriented learning, guides for learning of needed theoretical concepts (like signal filtering, or modeling of physical processes based on experimental data, etc.). Portal users can request new sets of experimental data based on concrete experimental protocol at laboratory model or experimental plant. In this way, the repository of experimental results and content of designed web platform can be updated.

This approach can be implemented via an independently designed web portal for the stated purpose, via Learning Management Systems like Moodle, or via Tutoring Platform like Go-Lab Portal (http://www.go-lab-project.eu/go-lab-portal). From the M-learning perspective, this approach can be tested by example of experimental dataset [10] and site http://graasp.eu/s/oy04aj which are related to a problem of identification of process model.

3 Experimental Data Sharing and Tutoring

In [10], experimental dataset entitled "I/O data for laboratory model PT326 Feedback Ltd" (see Fig. 2 and [11]) is published, with accompanying DOI, title, institution, authorship, categories for these data, description of these data, steps to reproduce these data, data files, and related links. A purpose is to publish reliable and sustainable open educational resource for problem-based learning. The same dataset is published on Go-Lab Tutorial Platform under http://graasp.eu/s/oy04aj [11] as well as on LMS Moodle under http://moodle.mfkg.rs/course/view.php?id=231. Previous web-based experiments from the faculty website http://fink.rs (E-učenje, Web Lab) were substituted by datasets which are published on different web platforms.

The problem was poor institutional ICT service and laboratory support. The core of this problem is the effects of economic crisis and long-term society transition in Serbia. The structure of institutional financial support in higher education produces poor reliability and sustainability educational services, and quality of education is highly dependent on enthusiasm, temporally projects, and personal creativity of teaching staff.

The second reason for publishing of representative experimental datasets instead of implementation of Internet-mediated experiments is the high complexity of experimental facilities and/or possible undesired consequences of test signals and control algorithms. For example, in [26], the development and implementation of automation, regulation, and measurement of heating energy consumption in the auditorium at the Faculty of Engineering is described (Fig. 5), which could be used for training and education in different engineering fields (civil engineering, control engineering, process engineering, and software engineering) [26]. Using a real remote-controlled system, educators are able to demonstrate the real-world principles of thermodynamics, fluid mechanics, and controls as well as to compare data from real systems and from models and software simulations [26]. It is clear that this kind of experimental plant could not have open access for control by any user, but it is possible to provide open access to monitoring and repository of experimental datasets.

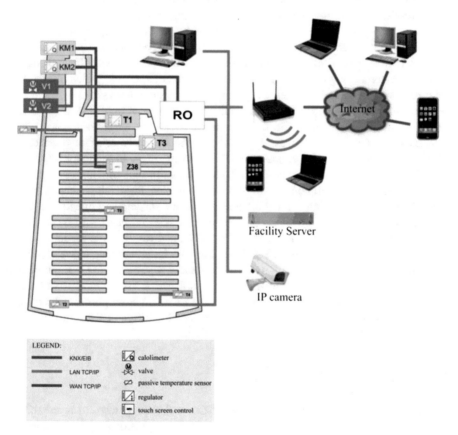

Fig. 5. Internet-controlled university amphitheater at Faculty of Engineering [26]

Contribution of this paper is promotion of a comprehensive and open infrastructure, based on repository of experimental datasets, for problem-oriented engineering education that exploits the possibility of experimental work on existing laboratory facilities. This approach offers integration of educational resources and efforts in order to reach an incentive educational environment with reliable support of developed open educational resources based on realistic experimental data. Teaching staff with several educational institutions can collaborate in order to achieve a higher level of quality of teaching/learning materials and communication with students concerning with the same projects, tasks, or case studies in problem-oriented engineering education. Because of that, developed Go-Lab Tutorial Platform is advanced compared to other web platforms. Web platforms for sharing experimental results, like https://data.mendeley.com, https://datadryad.org, https://zenodo.org, and https://www.repository.cam.ac.uk, do not offer possibility for development of tutorial's part of web support.

All mentioned experimental plants depicted at Figs. 1, 2, 3, 4, and 5 have been subjects of problem-oriented projects in last 10 years at the University of Belgrade and University of Kragujevac. Authors of this paper have a great experience with management of these experimental plants. Because of different limitations, degradation of capacity of experimental-based research/education work is necessary, beginning from a use of hands-on experimental realization, then a use of web-based experiments, up to a use of recorded sets of experimental data. On the other hand, publishing of experimental datasets has a great value for sharing experimental results, which are the base for further education and research as well as benchmarking analyses.

The web portal http://graasp.eu/s/oy04aj is an example of published experimental datasets with tutoring content for their use in the context of problem-oriented education. Experimental data are available in order to enable more relevant research and educational tasks. Designed educational environment offers Introduction, #1Experimental data (with described experimental plant, experimental protocol, and possible tasks for students), Preparation (Theory guide), Examples, Research papers (published research papers concerning with the same laboratory model/experimental datasets), Discussion, Conclusion, Official documentation (for relevant laboratory models), and Request for #Nth—Experiment's description, tasks, and results. A potential contribution of any a new user is planned. Published problem-based learning space at Go-Lab Tutoring Platform is linked with published Mendeley datasets [11]. Experimental datasets with filled standardized forms can be published at https://data.mendeley.com after review and accompanied DOI [10]. Because of that, we propose this practice of reviewing, description, and authorship of experimental datasets. Our example has used only part possibilities of Go-Lab Tutoring Platform.

4   Go-Lab Tutoring Platform

Go-Lab Tutoring Platform (https://www.golabz.eu and http://graasp.eu) is suitable for M-learning purpose as well as for problem-oriented educational tasks by adequate creation, personalization, and exploitation of so-called inquiry learning spaces.

Go-Lab is repository online labs, support applications, and inquiry learning spaces (ILS), which are available online 24 h a day and 7 days a week and open worldwide to any interested user [8]. ILS can be seen as interactive online educational content which can be delivered as a web application hosted by developers or in the repository. A single sign-on service is provided by the *graasp.eu* platform to enable seamless navigation between the platforms by the users (see Fig. 6). A backend learning analytics service is also available for support applications requiring advanced and secure processing of the activity traces [8]. Illustrations of additional services are *gateway.golabz.eu* and *composer.golabz.eu*. The architecture of the *graasp.eu* authoring platform will be extended to better handle ethical and privacy issues related to the data (learning analytics) and content (learning outcome) [8].

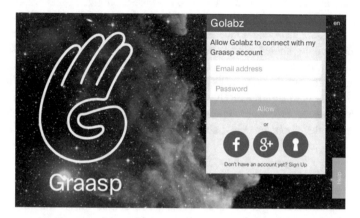

Fig. 6. The *graasp.eu* sign up or sign in dialog to import as a copy and personalize resources from *golabz.eu* [8]

With the *graasp.eu* account, the teacher can import as a copy of any interesting resource from *golabz.eu* and personalize it or create a new one like the one described in this scenario [7, 8]. Starting in 2017, the *golabz.eu* sharing platform will offer new apps and app categories supporting skills (collaboration and reflection), the creation of system models of the world, as well as apps for self and peer assessment [8] (Fig. 7).

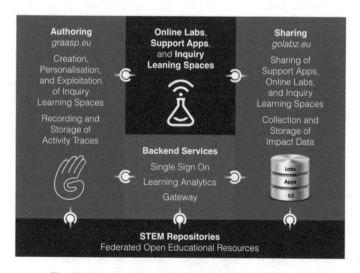

Fig. 7. The structure of Go-Lab Tutoring Platform [8]

There are five standard inquiry phases of the ILS, which can be freely populated with legacy or cloud resources by the teacher are Orientation, Conceptualization, Investigation, Conclusion, and Discussion. But teacher can change this concept and rename proposed phases or introduce some new phases. For example, ICL http://graasp.eu/s/oy04aj does not use standardly proposed phases. Applications which can be added in the various phases belong to a few categories, including general apps such as *File Drop* enabling students to submit a report or pictures taken during outdoor activities, or a *Quiz Tool*, etc. Once all phases are populated with the relevant content, ILS can be privately shared with users as a standalone web page using a secret URL displayed, when clicking the show standalone view button. The standalone page can also be set by the teacher and a QR code corresponding to the secret URL can be displayed to ease connection from mobile devices. Users (students) can freely navigate through the phases, which are represented by tabs [8]. Detail characteristics, possibilities, and guides about use Go-Lab Platform are given in [7, 8, 15–23, 27].

5   Conclusion

Developed open educational resources (OER) can enable the availability of experimental results for problem-oriented education. These datasets can be part of institutional designed web portals, LMSs, or different platforms for OER and sharing datasets. Mendeley's open research data repository (https://data.mendeley.com) after review process publishes datasets with accompanied descriptions, authorship data, and DOI. In this manner, storing, sharing, publishing, and finding datasets are free, effective, and reliable. This paper promotes using of published datasets at mentioned manner— directly or via other portals for educational or research purposes.

Go-Lab Ecosystem (http://support.golabz.eu/about) supporting M-learning, blended learning, and problem-based learning is proposed by this paper for using published datasets and relevant content for problem-oriented engineering education. Go-Lab Tutoring Platform offers better characteristics than LMSs and other OER platforms including own developed institutional/laboratory websites. Go-Lab platform provides a large collection of remote and virtual laboratories, where remote ones offer an opportunity with real equipment from remote locations.

Based on our 10 years experience with problem-oriented education and blended learning concepts with remote experiments, we propose in this paper a new option of using of Go-Lab platform with published datasets instead of remote labs. Advantages are decreasing of needed financial support and increasing of reliability of educational resources use. Also, Go-Lab platform has tools for simple implementation of different learning management contents. Very important, improving these tools and building of new tools and services are in progress [7, 8, 15–23, 26].

Further work will be motivated with our experience with laboratory model depicted in Fig. 3. Our intention is to record all typical and important situations (datasets, video clips) for potential education and research needs. Practically, our goal is to form a virtual lab which is composed of real datasets and illustrative video clips and diagrams. This will be the next step of the proposed approach in this paper.

Acknowledgements. This work has been partly funded by the SCOPES project IZ74Z0_160454/1 "Enabling Web-based Remote Laboratory Community and Infrastructure" of Swiss National Science Foundation and partly by projects TR 35046 and TR33047 Ministry of Education, Science and Technological Development Republic of Serbia which is gratefully acknowledged.

References

1. de Graaff, E., & Kolmos, A. (2007). History of problem based and project based learning. In *Management of change* (pp. 1–8). Sense Publisher.
2. Mills, J. E., & Treagust, D. F. (2003). Engineering education—Is problem-based or project-based learning the answer. *Australasian Journal of Engineering Education, 3*(2), 2–16.
3. Kalúz, M., et al. (2014). ArPi Lab: A low-cost remote laboratory for control education. In *19th World Congress the International Federation of Automatic Control* (pp. 9057–9062), August 24–29, Cape Town, South Africa.
4. Matijevic, M., Nedeljkovic, M., Cantrak, D. J., & Jovic, N. (2017). Remote labs and problem oriented engineering education. In *IEEE Global Engineering Education Conference (EDUCON)* (pp. 1390–1395), April 25–28, Athens, Greece.
5. Heradio, R., et al. (2016). Virtual and remote labs in control education: A survey. *Annual Reviews in Control, 42,* 1–10.
6. Nedeljkovic, M., et al. (2018). Engineering education lab setup ready for remote operation—Pump system hydraulic performance. In *IEEE Global Engineering Education Conference (EDUCON)* (pp. 1175–1182), April 17–20, Canary Islands, Spain.
7. Gillet, D., et al. (2017). Monitoring, awareness and reflection in blended technology enhanced learning: A systematic review. *International Journal of Technology Enhanced Learning, 9*(2/3), 126. https://doi.org/10.1504/IJTEL.2017.10005147.
8. Gillet, D., et al. (2017) Cloud ecosystem for supporting inquiry learning with online labs: Creation, personalization, and exploitation. In *Proceedings of 2017 4th Experiment at International Conference: Online Experimentation* (pp. 208–213). https://doi.org/10.1109/expat.2017.7984406.
9. Matijevic, M. (2018). I/O data for laboratory model PT326 Feedback Ltd., Mendeley Data, v2. http://dx.doi.org/10.17632/xw8sg9x8cm.2.
10. Mendeley Limited, Registered Office the Boulevard, Langford Lane, Kidlington, Oxford, OX5 1 GB, United Kingdom. https://data.mendeley.com/datasets.
11. Matijevic, M. (2018). Inquiry Learning Space (ILS): Heat flow experiment—Modeling and control—Sharing of experimental results, sharing and authoring platform. http://graasp.eu/s/oy04aj, http://graasp.eu and https://www.golabz.eu.
12. Tenopir, C., et al. (2011). Data sharing by scientists: practices and perceptions. *PLoS ONE, 6*(6), 1–21. https://doi.org/10.1371/journal.pone.0021101.
13. Almeida, R., Mendes, N., & Madeira, H. (2010), Sharing experimental and field data: The AMBER raw data repository experience. In *Proceedings—International Conference on Distributed Computing Systems* (pp. 313–320). https://doi.org/10.1109/icdcsw.2010.75.
14. Sicilia, M. A., García-Barriocanal, E., & Sánchez-Alonso, S. (2017). Community curation in open dataset repositories: Insights from Zenodo, *Procedia Computer Science, 106,* 54–60. https://doi.org/10.1016/j.procs.2017.03.009.
15. Govaerts, S., et al. (2015). Tutoring teachers—Building an online tutoring platform for the teacher community. In *Immersive education, communications in computer and information science* (Vol. 486). Springer. https://doi.org/10.1007/978-3-319-22017-8.

16. Manske, S., et al. (2015). Exploring deviation in inquiry learning: Degrees of freedom or source of problems?. In *Proceedings of the 23th International Conference on Computers in Education* (pp. 1–10).
17. Rodríguez-Triana, M. J., et al. (2015). *Orchestrating inquiry-based learning spaces: An analysis of teacher needs*. Lecture Notes in Computer Science (including subseries Lecture Notes in Artificial Intelligence and Lecture Notes in Bioinformatics) (Vol. 9412, pp. 131–142). https://doi.org/10.1007/978-3-319-25515-6_12.
18. Cao, Y., et al. (2014). Helping each other teach: Design and realisation of a social tutoring platform. In *Journal of Immersive Education (JiED)—Proceedings of 4th European Immersive Education Summit* (pp. 1–11).
19. Sergis, S., et al. (2017). Using educational data from teaching and learning to inform teachers' reflective educational design in inquiry-based STEM education. In *Computers in human behavior* (pp. 1–16). https://doi.org/10.1016/j.chb.2017.12.014.
20. Manske, S., & Cao, Y. (2015). Go-Lab global online science labs for inquiry learning at school specifications of the lab owner and cloud services (Deliverable D4.8 Grant Agreement no. 317601), Go-Lab consortium.
21. Li, N., et al. (2014). Enforcing privacy for teenagers in online inquiry learning spaces. In *Understanding Teen UX workshop at CHI Conference on Human Factors in Computing Systems*. http://infoscience.epfl.ch/record/197892.
22. Vozniuk, A., et al. (2015). Contextual learning analytics apps to create awareness in blended inquiry learning. In *2015 International Conference on Information Technology Based Higher Education and Training, ITHET 2015*. https://doi.org/10.1109/ithet.2015.7218029.
23. Gillet, D., & Bogdanov, E. (2013). Cloud-Savvy contextual spaces as agile personal learning environments or informal knowledge management solutions. In *12th International Conference on Information Technology Based Higher Education and Training, ITHET 2013* (pp. 2–7). https://doi.org/10.1109/ithet.2013.6671011.
24. Matijević, M., et al. (2014). The development and implementation of a thermal process trainer for control and measurement via the Internet. *Computer Applications in Engineering Education, 22*(1), 167–177. https://doi.org/10.1002/cae.20543.
25. Matijević, M., et al. (2017). Laboratory model of coupled electrical drives for supervision and control via internet. In *Proceedings of the REV2017—14th International Conference on Remote Engineering and Virtual Instrumentation* (p. 592–607), March 15–17. New York, USA: Columbia University. https://doi.org/10.1007/978-3-319-64352-6.
26. Stefanovic, M., Matijevic, M., & Lazic, D. (2013). Experimental plant for supervision and monitoring of an intermittent heating system for engineering training. *International Journal of Engineering Education, 29*(3), 799–807. ISSN 0949-149.
27. Go-Lab Support Area. (2018). Video-tutorials devoted to the Go-Lab Ecosystem, its interfaces, and particular applications. http://support.golabz.eu/videos?category=5&page=0.

Interactive Content Objects for Learning Digital Systems Design

Heinz-Dietrich Wuttke[(⊠)], Rene Hutschenreuter,
Daniel Sukiennik, and Karsten Henke

TU Ilmenau, Ilmenau, Germany
dieter.wuttke@tu-ilmenau.de

Abstract. Interactive content objects (ICOs) are immersive digital tools (e.g., simulations or interactive experiments) that students can use to generate responses, analyze data, etc. That way they can follow given examples or experiment with their own conceived variants and thus come to new insights interactively. The integration of such tools into a learning system, in which theory is usually taught by means of texts and graphics, opens up new possibilities to apply what has been read directly and thus to understand it better. To teach a systematic approach to methods of designing digital systems, we have realized a collection of such interactive tools, which show in particular the connections between different methodical approaches of design procedures. Based on the latest web technologies, the design of the tools takes into account the possibility of embedding them in different learning scenarios (e.g., mobile or stationary). In the paper, we present some of these tools and show how to use them to provide a deeper understanding of the interrelationships between different theoretical approaches.

Keywords: Mobile e-learning tool · Web technology · Problem-based learning

1 Introduction

Interactive content objects (ICOs) are immersive digital tools (e.g., simulations or interactive experiments) that students can use to generate responses, analyze data, etc. [1]. In the past, we have developed a series of Java applets in the form of interactive content objects (ICOs) with the following characteristics.

They explain a small content, are self-contained, calculate formulas in the background (which are the subject of investigation and always deliver correct results), allow to explore given or own created examples, are robust against wrong inputs, give immediate and meaningful feedback about correctness of inputs, visualize results in different forms, and can be integrated into websites. With these properties, they are a kind of "Learning Objects" because they are small, reusable in different instructional contexts and self-contained. In addition to common learning objects, great emphasis is placed on interactivity and the possibility to create own examples and to allow a free experimentation with the underlying simulation algorithms.

The applets were developed for basic courses in computer science to provide a systematic step-by-step approach to the design of digital systems with which the single

© Springer Nature Switzerland AG 2019
M. E. Auer and T. Tsiatsos (eds.), *Mobile Technologies and Applications*
for the Internet of Things, Advances in Intelligent Systems and Computing 909,
https://doi.org/10.1007/978-3-030-11434-3_11

design steps can be followed in close connection with theory. For each step of the design process (e.g., structure-oriented versus function-oriented design description using Boolean algebra, normal forms, minimization, circuit realization, and analysis, etc.), there is an interactive content object (ICO) that allows learners to explore the learned content with given or own created examples. The applets were developed independently of each other as self-contained units [2].

For teaching purposes, however, we also wanted to show how design decisions in the early stages of the design process influence the outcome of the entire design. This was more or less impossible with these applets. The only way to transfer the design from one ICO of a previous design step to another ICO to explain the next design step was to copy a line with a Boolean expression and paste it into the next ICO [2]. Changes in one ICO did not affect the others, so that an update in several other open applets always required copying and pasting into all ICOs. In order to run through the entire design process, a tool that enabled switching between the different ICOs was missing.

Therefore, we developed a complex ICO, called "SANE-Workstation" (SANE is a German abbreviation), where all the applets work over the same data structure (Fig. 1).

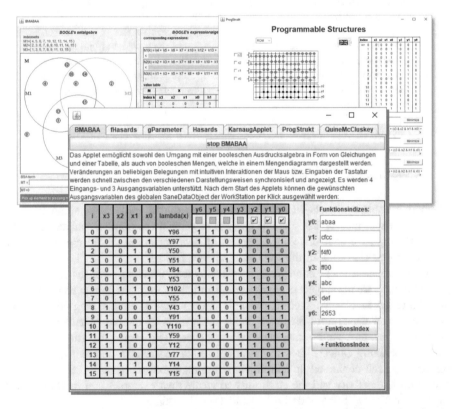

Fig. 1 SANE-workstation as central control panel and two corresponding applets in the background

By using this workstation, changing a value in one applet causes simultaneously changes in all other applets because of the central data structure (SANE-data, see Sect. 3). In the workstation, any number of exit functions can be entered either via their function index (a unique number, identifying a Boolean function) or by clicking on the values in a truth table. Since the applets only process up to four output variables, the functions to be examined must be selected by clicking on them. By default, the functions are assigned to the applets starting from the left in the truth table (y_0).

In Fig. 1, for example, the functions y_2, y_1, and y_0 are selected on the workstation panel to demonstrate the relationship between the description of Boolean functions using truth tables or Boolean expressions and the programming of programmable logic devices (PLDs). These functions are transferred simultaneously to the applet dealing with Boolean algebra and the applet dealing with programmable logic devices, where they can be examined.

Unfortunately, Java applets do not run on mobile platforms. That is why we now have realized a redesign based on HTML5 and JavaScript. In the paper, we describe the architecture of this new educational tool as well as some examples for using it in a problem-based teaching scenario.

The rest of the paper is organized as follows: First, we discuss some related work with a focus on our goals. We then give an overview of the entire design philosophy of the new SANE-tool set and its architecture. The next section shows some examples of the tools implemented so far, followed by a section describing their use in various learning scenarios. In the last section, we give a summary and an outlook on further tools and features to be developed.

2 Related Work

Numerous tools on the Internet support the design of digital systems. They can be divided into two main categories:

- Tools for industrial applications that support the design of complex systems and
- Tools for educational purposes that cover certain aspects of the design and are only suitable for "academic" examples.

Industrial tools are usually very complex and to master all their possibilities, you need long training periods and a lot of experience. They are tailored to specific product groups. Due to the NP-completeness of the problems to be solved, they usually work with heuristics that the user can only influence implicitly (e.g., via certain optimization goals such as time or resource requirements). The underlying algorithms that are of interest for teaching are hidden from the users of the tools. Further design tools see, e.g., [3].

Educational tools, on the other hand, are tailored for training purposes, limited to simple examples and are thus able to calculate all variants completely and thus contribute to an understanding of the underlying algorithms. They also illustrate intermediate results that remain hidden in industrial tools.

According to our previous research, there are training tools that deal with certain aspects of our teaching concept but none that cover all the focal points of our teaching concept.

Table 1 gives some examples of educational design tools and their teaching contents and properties.

Table 1. Educational materials supporting digital systems design

Refs.	BE	BF	ND	Par	Min	Haz	Cir	Int	Own	Web	Se	UD	QT
[4]	x				x			x		x			
[5]	x	x			x		x	x	x	x			
[6]						x	x	x		x			
[7]						x	x	x	x	x			
[8]	x	x		x	x		x	x	x	x	x		x
[9]	x	x		x	x	x	x	x	x	x	x		
[10]	x	x	x	x	x	x	x	x	x	x	x	x	x

Table 1 is organized as follows Ref—gives a reference to the specified tool, the further rows classify them by content and ICO related issues and the abbreviations have the following meaning:

- Content-related topics: BE—Boolean expression, BF—Boolean function, ND—non-determined function, Par—partial function, Min—minimization of expressions, Haz—hazard detection, Cir—circuit realization and analysis,
- ICO properties: Int—interactive, Own—create own examples, Web—can be integrated into websites, Se—self-contained, UD—work on unique data structure, QT—usable for questions and tests.

As Table 1 shows that most tools are tailored to different aspects of the design process. On the one hand, you find tools, dealing with Boolean functions and expressions [4, 5] and on the other hand tools that deal with circuit design [6, 7].

The tools, dealing with circuit design, are usually based on the graphical editing and simulation of circuits without showing the connection to the underlying Boolean functions.

Tools dealing with Boolean algebra use different syntax rules to display Boolean expressions and can only be coupled with each other through time-consuming renaming and copy/paste.

For example, in [8, 9], you can find various mini-tools dealing with truth tables and normal forms (disjunctive normal form (DNF) and conjunctive normal form (CNF)) as well as with Karnaugh-Veith (KV)-diagrams and the Quine-McCluskey algorithm to minimize Boolean functions. In [9], there is also a minimalistic logic simulator called Amilosim [11] available, where you can simulate given examples or design and simulate own creations of combinational and sequential circuits. However, even these are separate tools with non-transferable examples and it is not easy to combine them and transfer a result from one tool to another.

The focus of our lecture is teaching the entire design process, from truth tables and Boolean expressions to the verified hazard free circuit by using such tools. Therefore, different tools that work on the same data structure have to be coupled. That was the motivation for our work.

3 Architecture of the SANE-Tool Set

SANE's architecture follows strictly the View-Controller-Model (MVC). This results in a redux-like data flow [12] where the application state is stored in a central data object. Any change of the data object is only possible via its own inherent functions. The data flow is therefore strictly unidirectional. This central data object is the so-called SANE-data—object. Only inherent functions can change the state of the SANE-data—object. The different views communicate as submodules with the main module of the SANE-tool set (SANE-app). It accommodates all necessary event handlers, which are able to handle the events as required by calling the functions of the SANE-data object.

Following the MVC-method, SANE-date corresponds to the model, SANE-view to the views and the SANE-app to the event handlers as control. Figure 2 depicts this architecture concept.

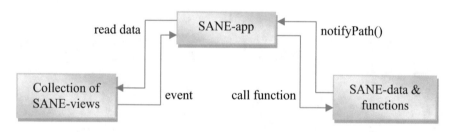

Fig. 2 MVC-based SANE architecture

The main module SANE-app is always assigned to exactly one SANE-data object and can be assigned to any number of SANE-views. For that reason, adding further views is very easy.

The communication between SANE-view, SANE-app and SANE-data works as follows: If an object of a SANE-view sends an event (e.g., caused by user interaction), it triggers the corresponding event listener in the SANE-app. The listener, in turn, is able to call a function of the SANE-data object. If the corresponding function changes data, the data object announces this and all objects that have a data binding to the changed data get a notification and receive the new data.

Figure 3 shows the communication process between the tree instances SANE-view, SANE-app, and SANE-data.

Since SANE is developed under the strong influence of polymer, the system is divided into two main groups. On the one hand, the polymer elements are embedded in HTML5 in connection with DOM (Document Object Model) elements, and on the

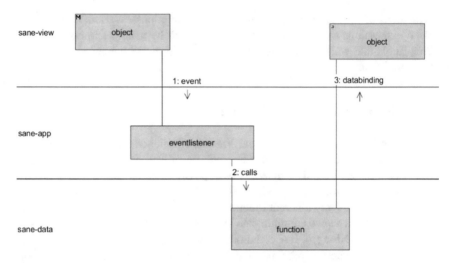

Fig. 3 SANE communication process

other hand, these embedded elements are extended by all necessary functions using JavaScript (Fig. 4).

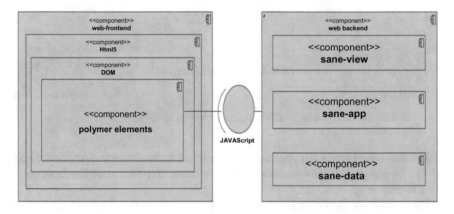

Fig. 4 Front-end <> Back-end cooperation and embedding

Based on this concept, a responsive design is inherent and tools of the SANE-tool set run on desktops as well as on tablets and other mobile devices. The toolset adapts to different screen sizes and is displayed in either a compact or a wide view in which two tools are displayed in one window. In the settings, it is also possible to choose between different languages.

4 Examples of SANE-Tools

The main advantage of the SANE-tool set is that it accesses a unified data structure for all tools and keeps it always up to date. This makes it possible to demonstrate connections between different representations of Boolean functions. For example, the transition from a function-oriented representation, as found in truth tables, to a structure-oriented representation as Boolean expressions in different normal forms (e.g., CNF, DNF and NAND normal forms (NANF)) can be demonstrated, see Fig. 5.

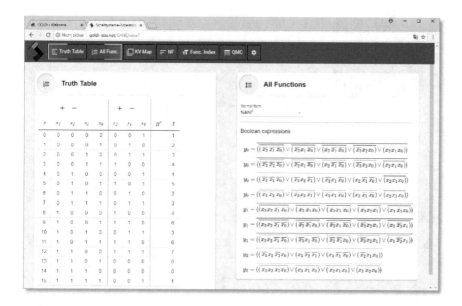

Fig. 5 Truth table and equivalent NANF of three boolean functions y_2, y_1, and y_0

Different minimization methods such as KV diagrams or the Quine-McCluskey algorithm can also be compared (see Figs. 6, 7).

In addition to these comparison options, the SANE-tool set can also be used to solve more complex problems, as the following example shows:

- Task: *Find a minimum, hazard-free circuit for the function y_0 given in the truth table of* Fig. 5.
- Hint: *To solve the problem, first set up the KV diagrams for the CNF and the DNF and then examine them for structural hazards.*

Figure 7 shows that the CNF realization is cheaper, since it gets by with fewer gates. In addition, compared to the DNF implementation, it is also free of structural hazards because of all neighboring blocks in the KV-diagram overlap.

The next step is to challenge the students to find out whether this is always the case, i.e., whether counterexamples can be found or whether a law has been found here. Such experiments can be carried out quickly and easily with the SANE-tool set for different

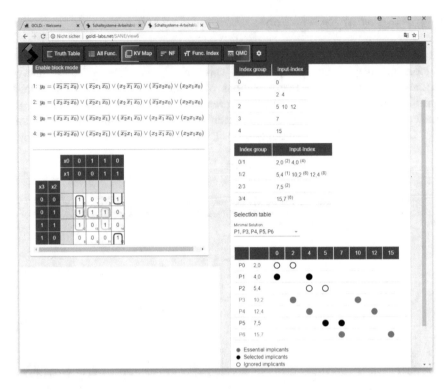

Fig. 6 KV-diagram and Quine-McCluskey minimization of the same boolean function

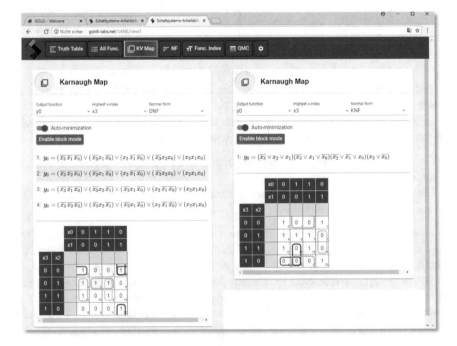

Fig. 7 Minimization of the same boolean function as DNF a CNF

Boolean functions and promote a better understanding of interrelationships. They also sensitize students to the diversity of the problems and thus prevent them from drawing thoughtlessly unthinking conclusions.

5 Learning Scenarios, Using the SANE-Tool Set

The set is designed so that it can be integrated into different learning scenarios. Some of these possibilities are listed below:

1. Demonstration during traditional lessons
2. Self-study to prepare for examinations and hands-on experiments
3. Integration in the circuit simulator "BEAST"
4. Integration in the remote lab "GOLDi"
5. Integration in online courses in learning management systems, e.g., LMS moodle
6. Task generation for flipped classroom.

In scenarios (1) and (2), the tool set can be used as it is. Figure 8 shows the KV-diagram, running on a mobile phone.

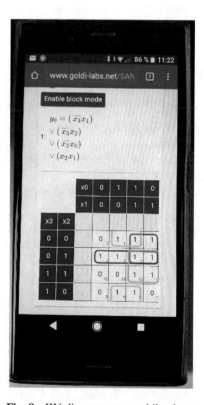

Fig. 8. KV-diagram on a mobile phone

For the scenarios (3) and (4), there exists an export function, which transforms the Boolean expressions in a readable file format for the remote lab and the simulator. Thus, real experiments with the developed own example are possible. The exported version is used without changes after exporting it. In case of detected errors in the simulator, the corrections have to be made in the SANE-tool.

For (5) and (6), individual tools of the tool set are integrated into an LMS for interactive practice of theory, so that the students have the textual material of the lecture as a script or PowerPoint presentation at their disposal and interactive exercises are available as part of the text. To do this, individual views of the toolset are configured so that the correct output is initially only determined in the background without being displayed. Instead, students are asked to enter their own solution. The tool with the correct solution then compares this and a feedback is generated from it [13].

The listed scenarios are still under development and only partly involved in our actual running lectures. Future work will examine the acceptance of the tools among students and possible effects on the improvement of learning outcomes.

6   Summary and Outlook

The paper describes a new developed tool set for teaching digital systems design. It can be used in problem-based teaching scenarios where students perform an entire design and verification of a digital circuit and examine the influence of design decisions in earlier phases of the design process on the finished circuit.

The implementation, based on current web technologies, allows running the web-based toolset as well on mobile devices.

A further development of the tool set goes in three main directions

1. Examine the acceptance of the tools among students and possible effects on the improvement of learning outcomes,
2. Closer integration into the GOLDi-Environment by import/export possibilities during the design process and
3. Extension to sequential logic functions and closer integration into the GIFT-tool set.

Acknowledgements. The work started in the frame of a student software project, supervised by David Sukiennik. We thank the students and David for their successful contribution to the new SANE-tool set.

References

1. Saul, C., & Wuttke, H.-D. (2013). An adaptation model for personalized e-assessments. *International Journal of Emerging Technologies in Learning (iJET), 8*(S2), 5.
2. Wuttke H. D., & Henke, K. (2002). *Teaching digital design with tool-oriented learning modules "living pictures"* (pp. S4G-25–S4G-30). Piscataway, N.J.: Champaign, Ill, IEEE.
3. Top Ten Online Circuit Simulators—Electronics-Lab| Rik. (2018). http://www.electronics-lab.com/top-ten-online-circuit-simulators/. Accessed August 21, 2018 (Online).

4. Boolean Logic Simplificator—Boole Calculator—Online Software Tool. https://www. dcode.fr/boolean-expressions-calculator#q1. Accessed August 20, 2018 (Online).
5. Online minimization of boolean functions. http://tma.main.jp/logic/index_en.html. Accessed August 25, 2018 (Online).
6. Donzellini. *Deeds downloads.* https://www.digitalelectronicsdeeds.com/downloads.html. Accessed August 20, 2018 (Online).
7. Hades Simulation Framework. https://tams-www.informatik.uni-hamburg.de/applets/hades/ webdemos/index.html. Accessed August 20, 2018 (Online).
8. Combinational Logic Circuit Design. http://electronics-course.com/combinational-logic-design. Accessed August 21, 2018 (Online).
9. Thormaeh. Technische Informatik - Philipps-Universität Marburg - AG Grafik und Multimedia. https://www.uni-marburg.de/fb12/arbeitsgruppen/grafikmultimedia/lehre/ti. Accessed August 24, 2018 (Online).
10. Schaltsysteme-Arbeitsblätter im Netz – SANE. http://goldi-labs.net/SANE/. Accessed August 26, 2018 (Online).
11. Amilosim: A minimalistic logic simulator. http://www.mathematik.uni-marburg.de/ ~thormae/lectures/ti1/code/LogicSimulator/simulator.html?filename=./examples/d-flipflop_ gated.sim. Accessed August 24, 2018 (online).
12. https://redux.js.org/. Accessed June 05 2018 (Online).
13. Wuttke, H. D., & Henke, K. (2008). *LMS-coupled simulations and assessments in a digital systems course* (pp. 726–731). IEEE.

Using Virtual Experiments in Teaching Control Theory

Mostafa M. Soliman[✉]

W Booth School of Engineering Practice and Technology, McMaster University,
Hamilton, Canada
Solimml2@mcmaster.ca

Abstract. A common criticism for theoretical and mathematically intense courses, including control theory, is that many of the concepts taught are abstract with limited practical relevance. An effective way to address this criticism is to increase the hands-on and experiential learning component, including laboratory activities. The paper proposes the use of virtual laboratories as a method for increasing student engagement and experiential learning in control theory courses. The experiments are designed to mimic as close as possible the behavior of different real dynamical systems such as a car suspension. This can be achieved using LabVIEW software and the 3D virtual reality tools.

Keywords: Virtual laboratories · Remote learning · Control systems education

1 Introduction

Hands-on experiential learning is deeply recognized as an effective method for teaching. It significantly improves the student understanding of theoretical concepts, and can offer an advantage for the student while working in the industry after graduation. By designing the labs to go hand in hand with the lecture, the student is able to immediately implement, verify, and see the concepts in action.

Typical challenges of engineering laboratory activities are limitations on the timetable, infrastructure, appropriate health, and safety training. One option to overcome these challenges is the use of remote laboratories. In remote experimentation, students operate with the real system, although they are not physically present in the laboratory. Another option that rose in recent years is the virtual laboratory (VL). VLs do not suffer from access restrictions and they can be performed at any time in any place at the student convenience. They can also be used with large class sizes [1, 2, 3].

The use of a DSP with Matlab and LabVIEW to provide remote control labs was proposed by Hercog et al. [4]. A virtual Web-based laboratory for control design experiments that uses MATLAB Web server (MWS) is described by Uran and Jezemik [5]. The use of Matlab to develop virtual control experiments is detailed in Rossiter [1], Rossiter et al. [2].

The paper proposes the use of VLs as a method for increasing student engagement and experiential learning in control theory courses. The VL experiments should be designed to mimic as close as possible the behavior of different real dynamical systems

M. E. Auer and T. Tsiatsos (eds.), *Mobile Technologies and Applications*
for the Internet of Things, Advances in Intelligent Systems and Computing 909,
https://doi.org/10.1007/978-3-030-11434-3_12

such as a car suspension, wind turbines, etc. This will be achieved using LabVIEW software and the 3D virtual reality tools.

The following assumptions are made:

- Each student has access to a computer that is running LabVIEW Run-Time Engine. This software can be downloaded and installed for free without any restriction.
- Students with no programming background will be using the VLs. This assumption dictates that the VL should be simple and easy to use by everyone.
- The use of 3D animations to mimic real dynamical systems helps students in understanding the theory.

This paper demonstrates an easy methodology to develop 3D virtual experiments that can be used in teaching control theory courses. The experiments are created using the following:

- Core Simulation using LabVIEW Control Design and Simulation Module: This module allows the designer to construct plant and control models using transfer function, state space, or zero-pole-gain.
- 3D Virtual Scene creation using 3D Picture Control Module: This module allows the user to design virtual worlds and draw/import 3D virtual objects.

Once the experiment is designed, LabVIEW is used to generate an executable file to be posted with a question sheet at the course shell. The students can then download the executable file and run it from their computers without needing any license or installing extremely large software.

The proposed approach offers several advantages compared to the existing approaches that were proposed in the literature. Most of the virtual labs proposed in the literature are restricted to 2D demonstrations. By using 3D animations and virtual reality, the VLs can be designed as close as possible to real physical systems. Furthermore, the proposed approach does not require the student to purchase expensive software licenses. It also does not require an internet connection. This allows the student to use the experiment with the least possible requirement in terms of hardware or software. The methodology proposed in this paper does not require high programming skills, thus it can easily be adopted and customized by the course instructor. Finally, the students can perform these experiments more than once on their own computer. They are not restricted to any hardware or time limitations that are typically found in physical labs. All of these factors help in achieving deep learning of complex theoretical concepts.

To illustrate the use of this methodology, a mass–spring–damper system experiment that was used in teaching the control theory course (PROCTECH 3CT3) in the fall of 2017 is described.

2   Motivating Example: Mass–Spring–Damper System

One standard topic in any control theory course is the study of second-order systems with a transfer function (1), where ω_n is the natural frequency and ζ is the damping ratio.

$$G(s) = \frac{\omega_n^2}{s^2 + 2\zeta\omega_n s + \omega_n^2} \tag{1}$$

The step response of the system can be one of the following cases, depending on the value of the damping ration. The step response for all these cases is shown in Fig. 1.

$$
\begin{aligned}
\zeta > 1 &\Rightarrow \text{overdamped} \\
\zeta = 1 &\Rightarrow \text{critically damped} \\
0 < \zeta < 1 &\Rightarrow \text{underdamped} \\
\zeta = 0 &\Rightarrow \text{Undamped}
\end{aligned} \tag{2}
$$

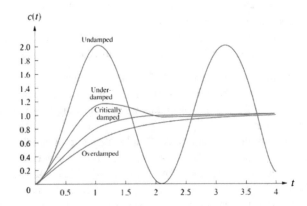

Fig. 1. Typical responses of a second-order system [6]

The students often get confused and intimidated by the mathematical quantities ζ, and ω_n, and they usually struggle to interpret their values and correlate them to the real physical world.

In this paper, the design of a virtual experiment that helps the student in understanding the types of second-order system responses and the physical interpretation of the different mathematical quantities is detailed. A mass–spring–damper system, shown in Fig. 2, will be used to develop a virtual experiment. The system model is given by (3)–(5), where M is the mass, K is the spring constant, B is the damper constant, f is an external force, and x is the displacement. This system can represent a car suspension system in real life.

$$G(s) = \frac{1}{Ms^2 + Bs + K} \tag{3}$$

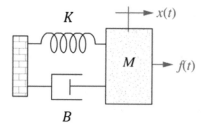

Fig. 2. A mass–spring–damper system [6]

$$\omega_n^2 = K/M \tag{4}$$

$$2\zeta\omega_n = B/M \tag{5}$$

3 Using LabVIEW to Develop Virtual Laboratories

LabVIEW (short for Laboratory Virtual Instrumentation Engineering Workbench) is a software and a programming environment that is commonly used for developing data acquisition, instrument and control application. The programs are developed using a graphical programming language, named as "G" language.

This section demonstrates how to use LabVIEW to build 3D virtual experiments that can be used in teaching control theory courses.

3.1 Dynamic Simulation Engine Using LabVIEW

One of the powerful LabVIEW modules is the Control Design and Simulation Module. This module allows the user to construct dynamic models using transfer functions and state-space models. This can be used to simulate the behavior of plants and controllers, analyze system performance with pole-zero maps and Bode plots.

The main Control Design and Simulation palette is shown in the figure (Fig. 3).

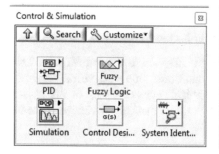
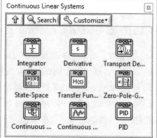

Fig. 3. Palettes of the control design and simulation module

The "**Continuous Linear Systems**" sub-palette contains many blocks such as integrator, transport delay, state space and transfer function that allow the user to simulate linear dynamical systems.

To build the virtual experiment, a dynamic simulation model must be built in LabVIEW. For the case of the mass–spring–damper system in (3), the LabVIEW simulation model is shown in Fig. 4. The model consists of a Control and Simulation Loop, where all the simulation functions, such as integrators, gains, and summing junctions, are placed. The user can specify the solver settings as required.

The differential equation of the mass–spring–damper system can be realized by two integrators, three gain blocks, and one summing junction. The value of the gains can be entered by the user through the front panel. This can allow the user to change the values of M, B, K and observe the response.

The main output of the simulation model in Fig. 4 is the displacement x. This variable will be used in the next section to animate the virtual reality environment. It is also displayed by the trend block to show the student the response of the system in real-time.

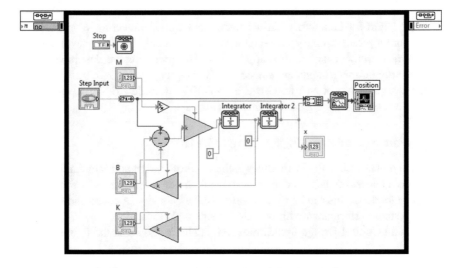

Fig. 4. Dynamic simulation engine of the virtual laboratory

3.2   3D Virtual Scene Creation and Animation Using LabVIEW

3D Virtual Scenes can be created using the 3D Picture Control Module. This module allows the user to design virtual worlds and draw/import 3D virtual objects.

The 3D picture control displays graphical representations of 3D objects. A 3D scene is a 3D object or a collection of 3D objects that can be viewed in the 3D picture control or in a separate scene window.

To design a 3D scene, multiple 3D objects can be generated. Their orientation, appearance, and relationship to other objects within the 3D scene can also be defined by the use. After a 3D object is created, the geometry of the object can be set using its drawable attributes, such as Create Box, or Create Sphere.

LabVIEW draws new objects with a default size and color and displays the objects in the center of the 3D scene. However, transformations, such as translation, rotation, and scaling can be used to change the size, orientation, or location of a 3D object in a 3D scene.

Already made VRML Files can also be loaded and added to a 3D scene. VRML files commonly appear with a.wrl file extension (Fig. 5).

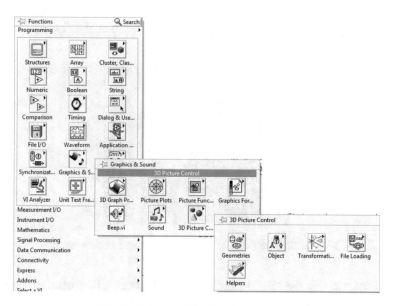

Fig. 5. Palettes of the 3D picture control

To build a 3D scene for the mass–spring–damper system, the code in Fig. 6 is used. The scene is shown in Fig. 7. The left part of the code is responsible for adding the walls, the spring, and the mass to the 3D scene. The right part of the code is responsible for continuously scaling the spring and translating the mass based on the actual displacement, x, that is generated by the simulation engine described in Sect. 3.1.

3.3 Virtual Laboratory Application Building Using LabVIEW

It is crucial to make the virtual labs easily accessible to students. The student must be able to run the VL on his local computer for free without requiring to install a large software, like LabVIEW development environment which is 5 + GB.

For that purpose, the LabVIEW software offers a significant advantage. It allows to build a stand-alone application from any LabVIEW project. This can be simply

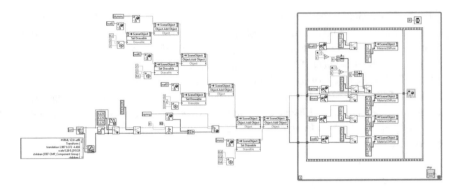

Fig. 6. Virtual reality animation of the mass–spring–damper system

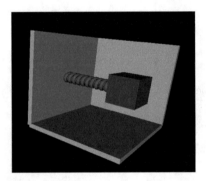

Fig. 7. 3D object created for a mass–spring–damper system

achieved by selecting the "Build Application (EXE) from VI..." from the tools drop down menu. This will generate a small executable application that can be run on any computer with LabVIEW Run-Time Engine (less than 650 MB) installed on it.

4   In-Class Use of the Mass–Spring–Damper System Virtual Experiment

The virtual experiment detailed in Sect. 3 was used in the class of PROCTECH 3CT3 in the fall of 2017. The students were notified to install the LabVIEW run-time engine and download the virtual experiment executable file to their laptop before the class. They were also notified beforehand to bring their laptops in the class.

The class started with an explanation of the theoretical concepts. Then, the in-class activity sheet in Fig. 8 was distributed to the students, and the students were requested to work in groups of two or three to solve the questions, and to perform the optional simulation using the virtual experiment on their local laptops.

The first question is an analysis one. They were given certain values for K, B, and M and they were requested to calculate some performance indicators that reflect the amount of oscillations in the system and how fast the system will settle. The parameters were chosen to yield a very lightly damped system. A screenshot of the user interface seen by the student for this particular simulation is shown in Fig. 9. The student is able to apply an external force and to observe the mass in 3D while significantly bouncing until it settles at the new position. He can also observe the time response on the right part of the interface. This user interface allows the student to correlate different theoretical concepts with what is actually happening on the real system.

The second and third questions are more elaborate. In the second question, the student is required to design the damper to obtain certain design specification. He can simulate his design and see the behavior as shown in Fig. 10. The third question allows the student to study the case where there is no damping at all and to watch a case of a sustained oscillation, as shown in Fig. 11.

5 Student Engagement

The use of the virtual experiment was a success. Almost all the students came to the class ready with their laptops and the necessary software installed. Although the simulation part of the activity was optional, most of the students were eager to try their designs and see it in 3D in the virtual experiments.

Statistical analysis of the impact of the virtual experiments was not performed yet and it is planned in the forthcoming semester. However, some initial indicators are worth mentioning. The course evaluations in the semester where the virtual experiments were used improved by 28% compared to the previous semester. Furthermore, even though the simulation part was optional, most of the students in the class were eager to complete the simulation part. In fact, I was able to observe that some groups even went further and they tested different values that were not required in the activity.

In-class Activity 7 – 3CT3
[Associated Virtual Experiment: MassSpringDamper Sim]

Student Name:	ID:
Student Name:	ID:
Student Name:	ID:

Case Study:
Consider the mass spring damper system below. Answer the following questions.

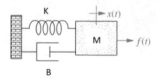

Questions	Optional Simulations
1. Let K=1 N/m, M=1 kg, and B=0.1 N.s/m. calculate the %OS, Ts, Tp	Simulate the system "MassSpringDamper.exe" with these parameters and confirm the results obtained
2. Let K=1 N/m, M=1 kg. Design B such that the system will have %OS = 10%.	Simulate the system "MassSpringDamper.exe" with these parameters and confirm the results obtained
3. Assume that you are allowed to pick the values of M, B, and K. What values will make the system undamped.	Simulate the system "MassSpringDamper.exe" with these parameters and confirm the results obtained

Fig. 8. In-class activity sheet for the mass–spring–damper system

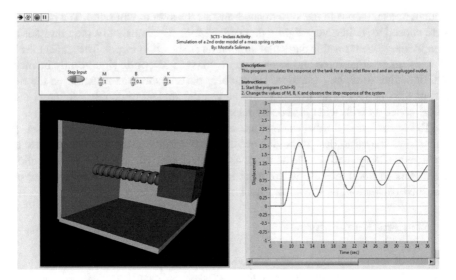

Fig. 9. Simulation scenario corresponding to question 1. The damping constant is 0.1

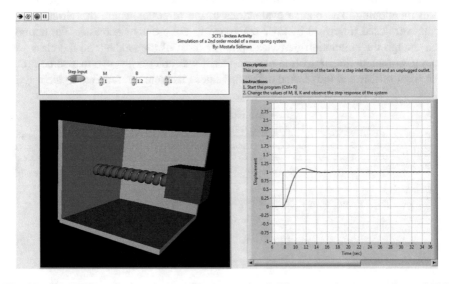

Fig. 10. Simulation scenario corresponding to question 2. The system has an overshoot of 10%

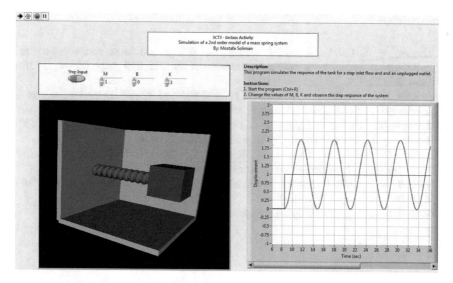

Fig. 11. Simulation scenario corresponding to question 3. The system is undamped

6 Conclusions and Recommendations

A methodology for developing virtual experiments that increase the student engagement and experiential learning is proposed in this paper. The method relies on using LabVIEW virtual reality tools and the Control Design and Simulation Module to develop 3D animations of different dynamical systems.

The virtual experiments are focused on strengthening specific learning outcomes where it is known that students often struggle. An example is illustrated in this paper, where the students can correlate some abstract mathematical quantities to the actual behavior of a 3D virtual mass–spring–damper system.

The proposed methodology offers many advantages compared to some existing technologies. The proposed method does not require the student to purchase any software license. Furthermore, the student does not require installing a large size software (Matlab or LabVIEW) on his computer to run the simulation. The virtual experiments are distributed to the students in the form of executables files that can be run on their own computers with a LabVIEW Run-Time Engine installed. The Run-Time Engine is free and it has much smaller size compared to LabVIEW. Finally, the proposed methodology requires a relatively small amount of programming expertise, and thus staff can quickly customize and produce such virtual experiments.

The ambition is to develop in the future an extensive library of virtual laboratories presenting advanced control algorithms and schemes from the field of automation, control, and robotics.

References

1. Rossiter, J. A. (2015). Enthusing students to engage and learn control with flexible multimedia assignments. *IFAC-PapersOnLine, 48*(29), 211–216. https://doi.org/10.1016/j.ifacol.2015.11.239.
2. Rossiter, J. A., Dormido, S., Vlacic, L., & Murray, R. M. (2014). Opportunities and good practice in control education : A survey. *Abstract : IFAC World Congress*, (i), 10568–10573.
3. Fabregas, E., Farias, G., Dormido-Canto, S., Dormido, S., & Esquembre, F. (2011). Developing a remote laboratory for engineering education. *Computers and Education.*
4. Hercog, D., Gergic, B., Uran, S., & Jezernik, K. (2007). A DSP-based remote control laboratory. *IEEE Transactions on Industrial Electronics.*
5. Uran, S., & Jezemik, K. (2008). Virtual laboratory for creative control design experiments. *IEEE Transactions on Education.*
6. Nise, N. S. (2015). *Control systems engineering.*

Mobile Health Care and Training

Mobile e-Training Tools for Augmented Reality Eye Fundus Examination

David Acosta[1], David Gu[2], Alvaro Uribe-Quevedo[3(✉)], Kamen Kanev[4], Michael Jenkin[2], Bill Kapralos[3], and Norman Jaimes[1]

[1] Universidad Militar Nueva Granada, Bogotá, Colombia
david.l.acostalaverde@ieee.org,norman.jaimes@unimilitar.edu.co
[2] York University, Toronto, Canada
david@davidgu.com,jenkin@eecs.yorku.ca
[3] University of Ontario Institute of Technology, Oshawa, Canada
{alvaro.quevedo,bill.kapralos}@uoit.ca,alvaro.j.uribe@ieee.org
[4] Shizuoka University, Shizuoka, Japan
kanev@inf.shuizuoka.ac.jp

Abstract. The direct fundoscopy examination procedure involves interpreting the intricate anatomy of the eye when viewed through the lens of an ophthalmoscope. Mastering this procedure is difficult, and it requires extensive training that still employs instructional materials including pictures, illustrations, videos, and more recently, interactive computergenerated models. With the goal of adding realism to eye fundus training and overcoming the limitations of traditional media, the simulators employing manikin heads can be used. Such simulations utilize interchangeable pictures and embedded displays that allow the presentation of various eye conditions. Modern simulators include immersive technologies such as virtual reality and augmented reality that are providing innovative training opportunities. Unfortunately, current high end virtual and augmented reality simulation is quite expensive and for more adequate experience, it engages only single trainee at a time. This paper addresses the question of whether lower end simulation systems could provide comparable training experiences at an affordable cost. With respect to this, we discuss the design and development of two augmented reality systems for eye fundus examination training employing low-cost mobile platforms. We conclude with reporting that some preliminary results of the experimental use of the systems include usability perception feedback and comparisons with the Eyesi simulator.

1 Introduction

Direct ophthalmoscopy (or fundoscopy) is a medical procedure that is difficult to master as it requires extensive practice to properly interpret the intricate anatomy of the eye, when viewed through the lens of an ophthalmoscope (also referred to as a fundoscope) [1]. The eye fundus examination (also known as

© Springer Nature Switzerland AG 2019
M. E. Auer and T. Tsiatsos (eds.), *Mobile Technologies and Applications
for the Internet of Things*, Advances in Intelligent Systems and Computing 909,
https://doi.org/10.1007/978-3-030-11434-3_13

fundoscopy and ophthalmoscopy) is a routine examination where the back part of the eye comprised of the retina, optic disc, choroid, and blood vessels, is observed to determine its health and diagnose any underlying conditions including head injuries, diabetes mellitus, increased intracranial pressure, and glaucoma [2].

As will be described in Section 2, traditional eye fundus examination methods, which include practicing on live patients and fellow trainees, viewing normal and diseased images and videos [2], or more recently, rendered interactive 3D models [3], of the fundus are problematic. The need to improve eye simulation training has led to the development of various simulators, including those projecting interchangeable images from printed or digital pictures through sockets simulating the eye, thereby allowing the trainee to diagnose the fundus of a manikin simulator head using an ophthalmoscope.

Prior work has examined the effectiveness of ophthalmic manikins when compared to traditional eye examination training methods. For example, Androwiki et al. [4], analysed the diagnosis skills of 45 medical students who trained with the Eye Retinopathy Trainer and compared this with 45 medical students who received a lecture instead. Upon completion of the training session, both groups diagnosed volunteer patients, and those who learned with the Eye Retinopathy Trainer, made a more accurate diagnosis. Moreover, Ricci and Ferraz [5], conducted a review of direct and indirect ophthalmic simulation. They observed that the majority of studies related to the topic focused on exploring the development of various training tools, and highlighted the lack of research oriented at studying the effectiveness and effects of eye fundus simulation tools.

Although medical simulation continues to improve and drop in price at providing better training practices, the costs associated with the commissioning, maintenance, inclusion, availability, and updating simulation technology can be prohibitive, thus limiting the intended impact [6]. However, there are low-cost alternatives. For example, employing a tube canister with eye pictures to simulate the fundoscope and a styrofoam head where images are inserted and projected into a socket for examination, Ricci and Ferraz (2014), developed an affordable eye fundus simulator [7]. As a complement to current medical training practices, virtual, augmented, and mixed reality (VR, AR, and MR respectively), approaches in conjunction with Makerspace trends (i.e., 3D printing, open electronics, and the internet of things), are providing consumer-level tools that can positively impact the virtual simulation domain. In this paper, we discuss the design and development of two mobile AR eye examination systems that can complement traditional laboratory eye fundus examination training. We have chosen AR as the underlying technology since it provides us with the ability to enhance traditional training materials through visual and tactile augmentation while employing lightweight commodity hardware. A preliminary usability study was conducted to explore the differences between the AR-based eye fundus examinations coupled with a physical fundoscope replica, and eye funduscopy training with the high end Eyesi direct ophthalmoscopy simulator.

2   Related Work

In the context of medical education, simulation is an 'educational technique that allows interactive and immersive activities by recreating all or part of clinical experience without exposing patients to any associated risks' [8]. Simulation is becoming the standard in medical education to assist trainees in developing transferable cognitive and psychomotor skills. Advances in electronics, mechanics, and computing are providing novel forms of safe, controlled, and simulated training. However, there are many challenges associated with effective simulation-based training including insufficient practice, lack of confidence, and the complexity of the procedure being simulated itself [5]. In the field of ophthalmology, and more specifically, direct ophthalmoscopy, extensive practice is required to master the skills associated with eye fundus examination while ensuring the comfort of the patient as exposure to the ophthalmoscope's light can be discomforting [9].

Direct ophthalmoscopy training requires the trainee to observe, navigate, and diagnose the eye fundus while operating the fundoscope [1]. Aside from the use of printed material discussed in the previous section, eye fundus training can also involve trainees learning on live patients or examining fellow trainees. Although the use of live patients increases realism, it also introduces challenges and limitations including those associated with potential issues with the patient's behaviour and a lack of variable eye conditions [9]. The patient's behaviour can also be affected by their level of comfort as longer examinations can lead to dry eyes, blinking, and light sensitivity [5].

Given the limitations associated with live patients, a common eye fundus examination training technique includes the use of static images whereby large galleries of eye anomalies initially found in expensive medical textbooks, but now more commonly accessed via the Internet are presented to the trainees [9]. While using these images, trainees can familiarise themselves with numerous eye conditions, without worrying about patient interactions, making this their preferred study method [10]. Current trends in digital media have provided multiple avenues for studying the eye fundus using large image databases and videos [11]. However, picture-based training fails to represent the challenges encountered while performing the real examination associated to the behaviour of the patient.

The popularity of simulation in ophthalmology, in general, is increasing as is the number of eye examination simulators [12]. Although simulation-based training is gaining momentum as part of medical education, the significant investments required at set-up and the increased running costs raise a prohibitive barrier of entry, particularly in the developing world. Eye fundus manikins are typically employed by one trainee at a time given the nature of the procedure, where one user examines one patient. Therefore, multiple simulators to address the needs of larger groups would be required, a situation leading to further costs. On the other hand, improper or insufficient training can lead to poor diagnoses in the future, a scenario that can ultimately put the health of a patient at risk. Notable ophthalmoscopy simulators include the Eye Exam simulator

Fig. 1. Eyesi manikin-based opthalmology training simulator. The user utilises the fundoscope mock-up to view the eye in a physical simulated head. The display embedded in the fundoscope simulates its real counterpart along with the operational inputs that allow adjusting focus and lighting. The external monitor presents the trainee's view and other information relevant to the training.

from Kagaku [13], and the Eyesi Direct Ophthalmoscopy simulator developed by VRmagic [14]. The Eye Exam simulator is a manikin designed for fundus examination training that employs an ophthalmoscope in conjunction with an adjustable pupil aperture that provides a realistic fundus view to the trainee using interchangeable photographs inside the head. Newer models employ displays embedded in the simulated manikin head and a fundoscope that provides higher photographic quality when compared to printed images [9].

Here we explore the usability performance of our AR mobile system relative to the Eyesi Direct Ophthalmoscopy simulator developed by VRmagic. EYEsi is a virtual reality-based training system for teaching direct ophthalmoscopy that allows high fidelity training consisting of: (i) the ophthalmoscope handpiece with a built-in screen simulating an ophthalmoscope, (ii) a model head representing the patient, (iii) an external display mirroring the view from the examination, and (iv) a computer running the simulation [14] (see Fig. 1).

3 Development

Through a thorough analysis of the eye fundus examination procedure, we identified the related cognitive and psychomotor tasks essential when making a diagnosis. Overall, the steps include explaining the process to the patient, providing instructions, dilating the pupil, and performing the physical assessment. Here, we focus only on the eye fundus observation portion and the steps associated with it as presented in Table 1.

During the examination, common ophthalmoscope operations include: (i) powering on and off the device, (ii) manipulating a focus dial (clockwise rotation zooms in and counterclockwise rotation zooms out), (iii) adjusting the light brightness dial at the base, and (iv) adjusting the light aperture dial to control shapes and sizes of the illumination with the application of green or blue filters. Figure 2 provides a graphical illustration of the fundoscope parts. A significant concern while performing the examination is the time required to complete the examination (the "extermination time"), as extended exposure to the light or an improper angle of approach can result in longer examination times, multiple attempts, and ultimately patient discomfort causing dry eye and involuntary blinking.

Table 1. Eye fundus examination sample procedure [15]

Step	Description
1	Locate the red reflex (the reflection of the ophthalmoscope light on the vessels) and maintain the retina within sight.
2	Approach the patient from arm's length up to 15 cm distance from the head and with the red reflex on sight, approach up to 2 cm until a clear view of the fundus is obtained.
3	Locate the optical nerve and if the anatomical landmark is lost, then restart the procedure.
4	Orientation of the ophthalmoscope is important to obtain a better view of the fundus. The patient can also be asked to look up, down, and at the light.

Fig. 2. Fundoscope parts.

Based on a careful study and consultations with two content experts from the Universidad Militar Nueva Granada Medical Simulation Laboratory, three eye conditions were identified for training: (i) a healthy eye, (ii) an eye exhibiting diabetic retinopathy, and (iii) an eye exhibiting papilledema. Eye funduses exhibiting these conditions were chosen as training scenarios and subsequently used in the preliminary usability studies. We employed the Unity3D game engine to develop the simulation as it supports the rapid prototyping of cross-platform mobile applications able to run on a variety of mobile devices with different operating systems. To explore the potential of consumer-level hardware to compete with high-performance simulators, two mobile applications were developed, one employing AR and a mobile phone as the ophthalmoscope, and the other using immersive mobile virtual-augmented reality in conjunction with a 3D printed ophthalmoscope wirelessly connected to a mobile phone. Each of these approaches is described in detail below.

(a) Augmented reality system in operation. The user holds a mobile phone in front of the physical target.

(b) The various displays available on the smartphone itself. Displays 01-06 show different user interface displays that the trainee uses. Once the scenario is chosen, the smartphone is pointed at the tracking image (screen 04), and the phone displays the view of the simulated patient (screen 05). As the trainee moves in towards the patient's eye, the back of the retina becomes visible (screen 06) and the trainee can use the fundus controls on the phone to adjust the view of the retina.

Fig. 3. Augmented reality simulation with the physical target approach.

3.1   Augmented Reality with Physical Target System

For this system, we employed a paper printed marker that serves as a reference target and used this target with a six degrees of freedom (DOF) tracking module accomplished using Vuforia [16]. The eye examination contents projected over the marker are visualised through the camera of the mobile device and displayed on the device's screen. In this application, the mobile device itself serves as the ophthalmoscope, whose functionality has been mapped to a touch-based graphical user interface (GUI). The GUI display provides the menus and tools required for the trainee to navigate the eye fundus through while rotating the mobile phone and changing the light intensity, filters, and focus. Once inside the examination menu, a list of eye fundus conditions and a set of specific tasks to achieve during the procedure are available. After the scenario has been set up, the simulation prompts the user to point the mobile device at the printed marker. A virtual head is then displayed on top of the marker, and the screen presents an adjustable lens whose light intensity, filter, and focus can be accommodated to the examination needs. The trainee is notified when the task is accomplished and provides feedback by showing the trainee where the condition was located within the eye. Figure 3 illustrates the stages associated with the eye fundus examination employing the physical target approach.

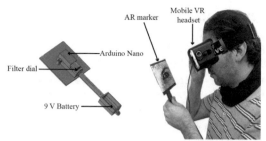

(a) The user views the world through a mobile VR headset powered by an embedded phone which provides both computation and display. Tangible interaction is provided by a 3D fundoscope simulator.

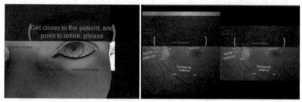

(b) The printed marker on the fundoscope replica displays within the VR headset the eye fundus examination scene.

Fig. 4. 3D printed ophthalmoscope operation.

3.2 Immersive Augmented Reality with 3D Printed Ophthalmoscope System

In this system, the trainee wears a mobile virtual reality (VR) headset powered by a mobile phone and operates a 3D printed ophthalmoscope to conduct the examination. The virtual environment presents a patient that must be examined, and it is projected on top of the printed marker situated in the upper area of the 3D printed user interface. The 3D printed stethoscope was designed to mimic the operation of a real fundoscope with reduced functionality. The replica connects wirelessly to the VR headset through Bluetooth, and it sends signals from the potentiometers that control both light intensity and focus adjustment. User interactions are captured, processed and sent to the simulation via an Arduino-Nano embedded within the printed user interface. Data transmission is achieved utilising an HC-05 Bluetooth module paired to the mobile phone and powered with a 9 V battery as presented in Fig. 4.

Through the operation of the user input device, the trainee can locate the eye fundus and approach it until fully visible for the examination. Once the back of the eye is in view, the simulation requires the trainee to find anatomical landmarks including the macula, the fovea, and the optic nerve. Selecting points of interest within the eye fundus is possible by tracking the orientation of the virtual camera. A ray cast allows detecting collisions with objects that are programmed to trigger events associated with the diagnosis.

4   Preliminary Usability Study

We conducted a preliminary usability study to determine the perceptional differences and potential preference bias towards either of our two developed systems and the Eyesi simulator. We employed the System Usability Scale (SUS) [17], a technique that measures the flow, ease of use, and complexity of a system. Participants are presented with a series of questions, and their responses are based on a 5-point Likert scale and a single score ranging from 0 to 100, representing the composite measure of the overall usability of the system being studied, is calculated. To calculate the total SUS score, one is subtracted from the score for each of the odd numbered questions in the SUS questionnaire, while for each even-numbered question, the resulting value is subtracted from 5. The updated values for each of the ten questions are added, and this total is multiplied by 2.5. Based on previous studies, a SUS score above a 68 is considered above average, and anything below 68 is considered below average [18]. A higher score indicates a higher degree of overall system usability.

We invited 17 random volunteers with Information Technology and Health Sciences backgrounds (13 male and four female, six of which were first-year medical students), aged between 18 and 22 years to participate in the preliminary study. The participants were divided into three groups to observe an eye fundus while using the AR with the physical target, the 3D printed ophthalmoscope, and the Eyesi simulator. Participants received a 15 minutes introduction to the eye anatomy and eye examination before they used the developed tools and the Eyesi simulator. Once the preliminary instruction was completed, the participants were asked to identify the macula, fovea, and optical nerve in the designated tool within a 10 minutes session.

5   Results

Results from the preliminary study indicate a trend whereby the participants perceived the physical target AR system to be more usable than the one with the fundoscope replica, but less usable when compared to the Eyesi simulator. Table 2 presents the individual SUS scores obtained from the participants and their mean values.

The small sample size prohibited the use of statistical techniques to more formally evaluate the differences between the various training approaches. Although our results are preliminary and greater work remains, the relatively close scores between the various approaches suggest that commodity hardware-based techniques can compete with the more expensive and sophisticated simulators.

6   Conclusion

Here, we have presented two mobile AR eye fundus examination training systems employing commodity hardware to the problem of using AR for eye fundus examination training. One method utilised a smartphone as the fundoscope while visualising a virtual patient projected on top of a physical paper printed marker.

Table 2. SUS results comparison.

	Physical target AR	3D printed fundoscope	Eyesi simulator
	87.5	67.5	75.0
	72.5	55.0	92.5
	60.0	65.0	82.5
	70.0	72.5	82.5
	77.5	77.5	47.5
	–	67.5	92.5
Mean	73.5	67.5	78.75

The second approach employed the mobile device inside a mobile VR headset while providing 3D stereoscopic AR visualisation of the virtual patient on top of a 3D printed fundoscope replica that allowed adjusting the light and focus settings during the examination. Both technologies are inexpensive, widely available and through advances in VR/AR software infrastructures can be readily employed to prototype and mimic eye examination interactions. Our instruments present an advantage when developing complementary training tools that can help overcome the limitations of existing expensive systems that require substantive training and investment in personnel. We studied the usability perception of the eye examination with using our two developed methods and compared the results to the high end and commercially available Eyesi eye examination simulator.

AR and VR provide the promise to revolutionise training in complex tasks such as direct fundoscopy. Can lower cost commodity hardware enhance the use of VR/AR in such practice? Although in our preliminary study no significant statistical usability differences were registered, the SUS did allow us to identify critical issues to address associated with the ease of use, required training, and assistance required to use these tools.

With regards to the 3D printed ophthalmoscope system, the primary challenge appeared to be the need to wear the smartphone with a mobile VR headset while at the same time operating the fundoscope replica. The experience was perceived as cumbersome and somewhat complicated as the participants familiarised themselves with the simulation and the augmented world. On the other hand, the participants perceived the AR mobile examination to be more usable as its operation was very familiar to what other applications offer in terms of touch-based interactions. Interestingly, most participants found the Eyesi simulator more usable, except for one who was not able to accomplish the simulation, with an overall perception of requiring further training to operate it properly.

Our small testing population makes it difficult to identify significant usability differences between the three tools; however, both commodity-based AR approaches were highly regarded by the participants, many expressing their excitement with this new experience and pointing out possible applications in other fields. We are thus planning to continue the software development, extend the content base, and conduct further more extensive experiments.

References

1. Yusuf, I., Salmon, J., & Patel, C. (2015). Direct ophthalmoscopy should be taught to undergraduate medical studentsyes. *Eye, 29*(8), 987.
2. Clark, V. L., & Kruse, J. A. (1990). Clinical methods: The history, physical, and laboratory examinations. *Journal of the American Medical Association, 264*(21), 2808–2809.
3. Barsom, E., Graafland, M., & Schijven, M. (2016). Systematic review on the effectiveness of augmented reality applications in medical training. *Surgical Endoscopy, 30*(10), 4174–4183.
4. Androwiki, J. E., Scravoni, I. A., Ricci, L. H., Fagundes, D. J., & Ferraz, C. A. (2015). Evaluation of a simulation tool in ophthalmology: Application in teaching funduscopy. *Arquivos Brasileiros de Oftalmologia, 78*(1), 36–39.
5. Ricci, L. H., & Ferraz, C. A. (2017). Ophthalmoscopy simulation: Advances in training and practice for medical students and young ophthalmologists. *Advances in Medical Education and Practice, 8*, 435.
6. Zendejas, B., Wang, A. T., Brydges, R., Hamstra, S. J., & Cook, D. A. (2013). Cost: The missing outcome in simulation-based medical education research: A systematic review. *Surgery, 153*(2), 160–176.
7. Ricci, L. H., & Ferraz, C. A. (2014). Simulation models applied to practical learning and skill enhancement in direct and indirect ophthalmoscopy: A review. *Arquivos Brasileiros de Oftalmologia, 77*(5), 334–338.
8. Perkins, G. D. (2007). Simulation in resuscitation training. *Resuscitation, 73*(2), 202–211.
9. Ting, D. S. W., Sim, S. S. K. P., Yau, C. W. L., Rosman, M., Aw, A. T., & San Yeo, I. Y. (2016). Ophthalmology simulation for undergraduate and postgraduate clinical education. *International Journal of Ophthalmology, 9*(6), 920
10. Mackay, D. D., & Garza, P. S. (2015). Ocular fundus photography as an educational tool. *Seminars in Neurology, 35*(5), 496–505.
11. Borgersen, N. J., Henriksen, M. J. V., Konge, L., Sørensen, T. L., Thomsen, A. S. S., & Subhi, Y. (2016). Direct ophthalmoscopy on youtube: Analysis of instructional youtube videos content and approach to visualization. *Clinical Ophthalmology, 10*, 1535.
12. Wallace, B. S., & Sabates, N. R. (2013). Simulation in ophthalmology. *Missouri Medicine, 110*(2), 152–153.
13. Kagaku, K. (2018). *M82 eye examination simulator.* Retrieved August 7, 2018 from http://www.kyotokagaku.com/products/detail01/m82.html.
14. VRmagic. (2018). *Eyesi by vrmagic: Direct ophthalmoscope simulator.* Retrieved August 7, 2018 from https://www.vrmagic.com/simulators/simulators/eyesir-direct-ophthalmoscope/.
15. Gagnon, L., Lalonde, M., Beaulieu, M., & Boucher, M.C. (2001). Procedure to detect anatomical structures in optical fundus images. In *Medical imaging 2001: Image processing* (Vol. 4322, pp. 1218–1226). International Society for Optics and Photonics.
16. PTC, V. (2018). *Vuforia.* Retrieved August 7, 2018 from https://vuforia.com/.
17. Brooke, J., et al. (1996). SUS-A quick and dirty usability scale. *Usability Evaluation in Industry, 189*(194), 4–7.
18. Sauro, J. (2011). *Practical guide to the system usability scale: Background, benchmarks & I.* CreateSpace Independent Publishing Platform

Use of Mobile Apps for Logging Patient Encounters and Facilitating and Tracking Direct Observation and Feedback of Medical Student Skills in the Clinical Setting

Anthony J. Levinson[1]([✉]), Jill Rudkowski[2], Natasja Menezes[3], Judy Baird[4], and Rob Whyte[5]

[1] Division of e-Learning Innovation, McMaster University, Hamilton, ON, Canada
levinsa@mcmaster.ca

[2] Faculty of Health Sciences, Departments of General Internal Medicine and Critical Care, McMaster University, Hamilton, Canada
rudkowj@mcmaster.ca

[3] Department of Psychiatry and Behavioural Neurosciences, McMaster University, Hamilton, Canada
menezes@mcmaster.ca

[4] Faculty of Health Sciences, Department of Family Medicine, McMaster University, Hamilton, Canada
bairdj3@mcmaster.ca

[5] Michael G. DeGroote School of Medicine, McMaster University, Hamilton, Canada
rwhyte@mcmaster.ca

Abstract. Undergraduate medical education consists of 15–24 months of clerkship, where medical students participate directly in patient care within various healthcare environments. It can be a challenge to document what patient encounters students have experienced, and that they have been directly observed in those settings. We set out to build mobile apps to improve the tracking of student–patient encounters, as well as their direct observation and feedback during the clerkship. Needs analysis was used to outline the current challenges and the requirements for each core rotation. We standardized the data collection across the different clinical rotations. We focused on tools for formative feedback only. Two apps were built. The first tracks Essential Clinical Experiences (ECE). For each rotation, students log required encounters that include clinical conditions or procedures. Students are required to log 267 ECEs throughout their clerkship. The second app records direct observation and feedback. To date, using the MacDOT app, over 11,000 direct observations have been logged, with over 2000 observers. These apps have improved the systematic approach to tracking medical student clinical encounters, as well as direct observation and feedback during the clinical clerkship. Ongoing enhancements are in progress.

Keywords: Medical education · Workplace assessment · Formative feedback · Blended learning

© Springer Nature Switzerland AG 2019
M. E. Auer and T. Tsiatsos (eds.), *Mobile Technologies and Applications for the Internet of Things*, Advances in Intelligent Systems and Computing 909,
https://doi.org/10.1007/978-3-030-11434-3_14

1   Context

1.1   The Challenge of Documenting Clinical Encounters

Undergraduate medical education typically consists of at least 15–24 months of clinical clerkship, where medical students participate directly in patient care within various clinical healthcare environments. It can be a challenge to document that students have been directly observed by supervisors in those settings, and been provided with formative feedback. To date, most solutions to try to capture this have been paper-based solutions: for example, clinical logbooks or "passports" that students carry with them, and supervisors fill out on paper when an observation has occurred. However, there have been a lot of limitations to paper-based methods, including the lack of centralized program tracking of clinical observations; the additional requirement to enter the observation data if it is to be further analyzed; the risk of students losing hard copy materials; the inconvenience of having to carry physical "encounter cards" or "passports" with them at all times during a clinical rotation; the lack of standardization of analog methods of recording observations across the various clerkship rotations; and others.

1.2   Clinical Clerkship

The Michael G. DeGroote School of Medicine at McMaster University's Undergraduate Medical Program admits 203 students to the program each year at each of three campuses in Hamilton, Niagara and Waterloo, Ontario, Canada. The 3-year program in Medicine uses a problem-based approach to learning that should apply throughout the physician's career. The components have been organized in sequential blocks with early exposure to patients and case management. The academic program operates on an 11 months-a-year basis and students qualify for the Medical Doctor (MD) degree at the end of the third academic year. The 63-week clinical program (or clinical clerkship) consists of rotations in medicine, general surgery, orthopedic surgery, family medicine, anesthesia, psychiatry, pediatrics, obstetrics and gynecology, and emergency medicine. There is also elective time, one half of which must be spent in clinical activity. The compulsory components of the clinical program are carried out in teaching practices and in all the teaching hospitals in the regions of each campus: Waterloo, Niagara, and Hamilton. As well, many clinical placements occur in the communities associated with the Rural Ontario Medical Program.

2   Purpose

We set out to design, build, test, implement, and evaluate mobile apps to improve the systematic approach to tracking medical student–patient encounters, as well as direct observation and feedback during the clinical clerkship. Our goals were to ensure that students were experiencing essential clinical encounters in each of the clerkship rotations, throughout all teaching locations and campuses; as well as to ensure that students were receiving direct observation and feedback from supervisors, across a range of different skills.

3 Approach

3.1 Design

Extensive needs analyses were undertaken to outline the current challenges, and the technical and nontechnical requirements for each of the clerkships.

Nontechnical design approaches included outlining and refining the needs and requirements for each of the clerkships, in particular, harmonizing and standardizing an approach that would be consistent across the different clinical rotations. Needs analysis also took into consideration potential barriers and facilitators, such as challenges in having busy clinicians use a new software platform, and the need for a mobile solution. A decision was made to have the Essential Clinical Encounters app be entirely student managed, without requiring faculty authentication of encounters. For the Direct Observation and Feedback app, the student would initiate an observation; and the app would be used for formative feedback only, rather than summative assessment of the student's clinical skills.

Technical design methods included using an iterative, agile approach given the very tight timelines for the projects. We used iterative reviews of design mock-ups and functional prototypes to refine the app design. Challenges such as the lack of a definitive database of observers and the need for students to be able to drive the process were incorporated into the design.

3.2 Build

We used an open-source mobile framework using JavaScript (NativeScript) [1]. Student authentication was integrated with our existing "medportal" intranet using Google's Firebase [2]. The medportal website has a service layer built with ServiceStack [3] and it accepts JWT web tokens to validate a user's identity. A database to store the records of the (1) essential encounters and (2) observations and feedback was built using Microsoft SQL. A web app version was also developed as an additional alternative to the mobile app. Native apps were provisioned to the Google Play and iTunes app stores for students to download. A pragmatic approach to testing the apps was adopted using internal quality assurance testing, followed by release of beta versions of the apps, and pilot testing with a subset of learners, observers, and clinical rotation sites, prior to launch. Iterative refinements occurred based on feedback.

3.3   Implement and Evaluate

Nontechnical aspects of the implementation included extensive communication with the clerkship academic leads and students prior to launch. Mixed quantitative and qualitative methods were used to evaluate the projects including the number of app downloads from the app stores and number of encounters and observations logged. Qualitative methods included student ratings on the benefits of the observation, and both student and observer ratings and comments on the experience of using the app. Additional methods of feedback were solicited through face-to-face meetings and conversations with curriculum leads and students. Administrative dashboards and reporting screens were developed to monitor adherence with curriculum goals and accreditation standards.

4   Outcomes

4.1   Essential Clinical Experiences App

Two separate apps were designed, developed, implemented, and evaluated. The first app tracks Essential Clinical Experiences (ECE Tracker) [4]. For each clerkship rotation, students log required encounters that include core clinical presentations or conditions, procedures, and professional competencies. Students also note the patient type/location and the level/depth of the clinical experience. Students are required to log 267 ECEs throughout their clerkship, with the expectation that specific encounters are logged during specific rotations. While initially developed as a web-based application, it was later developed as native apps for both Android and iOS. To date, there have been 53 installs of the Android app and 64 installs of the iOS app. Detailed reports can view metrics on students, as well as reports for each of the clerkships by encounter type or campus location, to ensure that students are getting the required breadth of clinical exposures regardless of clerkship rotation site. Students can also view their progress on dashboards within the medportal website, as well as within the app. Clerkship directors and administrators typically review the status of encounter logging both at mid-rotation and end-of-rotation to ensure mandatory completion of the breadth of encounters per rotation. Some rotations have enhanced the formative pedagogical value using the mid-rotation review as an opportunity to discuss particular encounters, or by including "reflective questions" within the app related to particular encounters. While the app is used by students for logging encounters, faculty can also use the app to view the lists of required encounters for their relevant clerkship rotation. See Fig. 1 for example screenshot.

4.2   Direct Observation and Feedback App

The second app records direct observation and feedback of medical students by faculty, residents, or allied health professions staff observers. The MacDOT app [5] is a formative educational tool for direct observation of clinical clerks (senior medical students) in clinical settings with actual patients. Clerks receive immediate feedback from faculty and other observers in the domains of history taking, physical examination, procedures, and other professional competencies such as communication and critical appraisal skills. See Fig. 2 for screenshot.

The MacDOT has been created as a tool to facilitate honest and open "in the moment" feedback from observers to students. No grade is given to the student, only narrative feedback for a selection of clinical encounters students need to do, and need to be observed doing. Each core clerkship rotation has a specified number of MacDOTs that must be done in order for that rotation to be considered complete.

Each observation event is initiated by the learner, who first authenticates to the app using their existing McMaster medportal account credentials. The learner then selects the clerkship rotation and type of clinical encounter that will be observed, before choosing the faculty or other observer from a list. At that point, the observer will use the learner's smartphone or tablet to provide formative feedback, before handing the device back to the learner. The observer may choose to write comments, or choose from a standardized "pick-list" of typical feedback. The learner receives verbal feedback from the observer while they are completing the MacDOT. Prior to submitting the observed encounter, the learner discusses a potential educational learning plan with the observer, makes a note of that action plan, and submits the encounter.

Students also rate whether the feedback they received from the observer was helpful for their learning on a 5-point Likert scale with an optional comment. A confirmation e-mail following submission is sent to both the student and the observer, and both participants are also invited to provide a rating of their experience using the app using a 5-point Likert scale with an optional comment.

Students are able to track their progress with respect to completion of the required set of observations within the app, as well as on the medportal website. Clerkship directors and program administrators are also able to view or export status reports to monitor student completion of the required observations. Students are also able to track and easily recall their previous feedback to incorporate into their goals for subsequent encounters.

The MacDOT app was officially launched in November 2017 with the entire cohort of medical students entering their clerkship. To date, there have been 73 Android and 84 iOS installs of the app. As of August 22, 2018, over 11,300 direct observations have been logged, with over 2000 observers. Each student has logged about 56 MacDOTs to date, with an average of 5.7 observations per observer; and about 40.7 logged per day. For 92% of the logged observations, students found the feedback they received helpful to their learning. Comparable uptake, utilization, and ratings of observations have occurred across all three campuses.

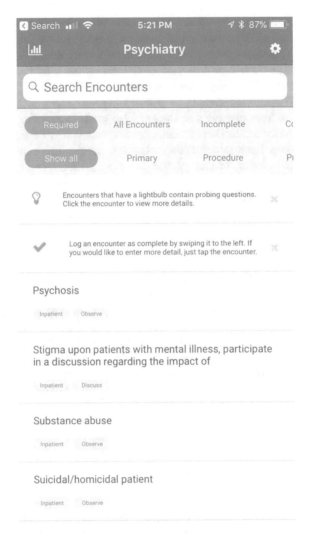

Fig. 1. ECE Tracker app screen for listing encounters related to a particular clerkship. The learner can search or filter encounters, quickly log an encounter by swiping, or enter more details about the encounter by selecting it from the list.

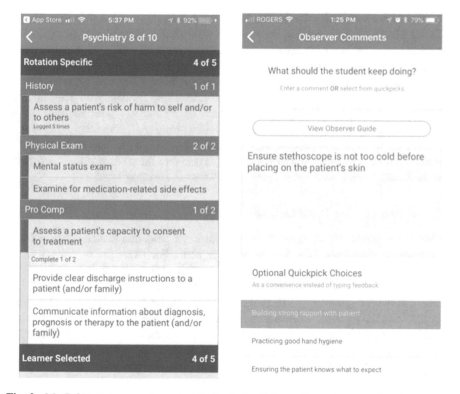

Fig. 2. MacDOT app example screenshots. Left, listing of rotation-specific observations including student status of completions. Right, example of observer feedback screen showing both free text entry and observer-selected "quickpick" feedback.

5   Discussion

We designed, built, tested, and implemented apps to improve the systematic approach to tracking medical student clinical encounters, as well as direct observation and feedback during the clinical clerkship. The program is better able to document that students are getting essential clinical experiences, and that they are being observed across a range of key clinical tasks throughout their clerkship rotations. Students find the observation and feedback helpful to their learning.

5.1   Pedagogical Considerations

Throughout the design, development, and implementation of these apps, several pedagogical issues and options have emerged. First, the decision should be made whether to use these types of apps for either formative or summative purposes. Initially, the MacDOT was conceptualized as more of a summative assessment of student skills; but this was later changed to an entirely formative tool. Second, whether to use student self-report only (as with the ECE Tracker app), or some form of confirmation/authentication by faculty (as in the e-mail "receipt" sent to the observer with the MacDOT app). Adding in an e-mail validation to the observer brings with it the requirement of having a faculty/observer database with validated e-mails, which does not always exist in a format that is easy to integrate with app function. Third, in our experience, having these apps—in particular, the MacDOT—can act as a driver to facilitate and increase direct observation and feedback by faculty. We have also considered other pedagogical enhancements, such as adding in "positive triggers" or notifications related to specific encounters—for example, additional educational resources or references related to an encounter type. It is important to consider the "signal-to-noise" of these types of triggers to ensure that the student is not overloaded with notifications that may overshadow their value as learning tools.

5.2 Implementation Considerations

In the design phases, it is crucial to work with key curriculum stakeholders to try to harmonize requirements prior to app development. For example, in our case, we worked closely with clerkship directors to try to standardize the list of encounters and observations, as well as any rules or requirements with respect to "completion" per rotation. It is very important to be pragmatic with respect to the efficiency of logging encounters on the fly in the clinical setting: there may be additional data that curriculum planners might wish students to log, but it is vital that students can log encounters or observations as quickly as possible. For the ECE Tracker app, we were able to include a "quick add" function, which would rapidly log an encounter with "default" parameters.

It is also very important to develop and implement policies around expectations for "completions" and monitoring of student status, as well as clear communication with both students and curriculum planners around these policies. To facilitate dissemination and uptake, we also made both apps "student driven," so that the student could quickly log an encounter or initiate the observation, without adding additional burden on the faculty observer to download the apps or learn a new tool. We did not have a preexisting observer database to tap into for our MacDOT observers, so elected to build our own observer database using some preexisting sources of faculty and resident data. We opted to include two unique identifiers where possible (e-mail address and physician license number). Students can add new observers within the app on the fly, if the observer cannot be found in the database already. Where resources allow, it would be ideal to have ongoing review of the database of observers to validate new entries and synchronize duplicate entries.

5.3 Technical Considerations

The native app versions integrate with our existing student authentication architecture, which speeds up logging in the clinical setting versus having to authenticate to a website prior to logging an encounter. Local caching/storage within the app is also useful in clinical areas where there is poor Wi-Fi or cell signal. (The app will automatically update the database when connectivity is restored.) Web-based versions of the apps have also proved useful, however. In some hospital environments, desktop computers are ubiquitous, so the student might just as easily log an encounter that way. Having a web-based version also provides a backup to students, in case of device failure or unavailability. In addition, meeting accessibility requirements is sometimes easier using web standards versus native apps. We do not collect any patient health information or identifiers related to logging encounters or observations; this would add an additional overlay of privacy and security requirements in order to meet health information protection standards. For the MacDOT app, we chose to standardize "pick-lists" across all observation types, but many observers have requested more tailored, "observation-specific" feedback pick-list options. This highlights some of the pragmatic technical decisions that must be made given constraints of timelines or app complexity, which sometimes limits the flexibility of an app. Hard-coding components of the app to accelerate development may also limit the repurposing of the app for other contexts. For example, because many of the components of these apps are specific to the undergraduate medicine clerkship, it is not straightforward to repurpose the apps for tracking or logging of postgraduate medicine clinical encounter requirements. It is also important to factor in ongoing maintenance for apps, as new smartphone operating systems and versions appear. We opted to create native apps for the two most popular smartphone platforms in use by our medical students. The use of NativeScript did make this process more efficient.

5.4 Comparisons with Previous Work

Ferenchick and Solomon described using a cloud-based mobile technology for the assessment of competencies among medical students, and also found that it was feasible in capturing discrete clinical performance data with a high degree of user satisfaction [6]. Their platform was built as a cloud-based responsive website, rather than a native app, and was used for summative assessment rather than formative feedback purposes.

Chamney et al. developed a Resident Practice Profile application to facilitate residents in family medicine logging their experiences and generating reports, similar to the ECE Tracker app [7]. Their forms-based application was also built for web rather than a native app, and focused on resident clinical experience tracking rather than medical student. Similar to the ECE Tracker, it was designed for rapid data entry, so the learner could log an encounter in less than 30 s. Our data indicate that many students often use the "quick add" feature of the ECE Tracker and batch enter encounters for efficiency. Those that log them in real time "on the fly" typically take less than 20 s.

Robertson and Fowler reported on medical student perceptions of learner-initiated feedback using a mobile web application [8]. As with our MacDot app, students perceived using their application to initiate feedback to be valuable, but in their pilot study, it did not guarantee timely feedback. While our app is designed to have the observer provide feedback in real time on the learner's mobile device, their application triggered an e-mail to the evaluator. This time delay likely leads to attrition in faculty feedback. It is also likely that our program's policy to mandate the logging of direct observations and feedback plays an important role with respect to student and faculty compliance.

6   Conclusions

Tracking encounters as well as recording medical student observations and feedback are feasible using native apps, and benefits from a program-wide design and implementation strategy. Native apps do have some advantages over web-based tools for these purposes in the clinical setting, including rapid user authentication and their ability to log experiences in areas with poor Wi-Fi or cell signal.

Ongoing enhancements to these apps are in progress, including the eventual consolidation of these and other program functions into a single clerkship app. Further research on the impact of these apps on teaching and learning within the medical program is planned.

Acknowledgements. Dr. Levinson would like to acknowledge support from the John R. Evans Chair in Health Sciences Educational Research and Instructional Development. We would also like to thank the Division of e-Learning Innovation for their work on the design and development of the apps. In addition, we are grateful to feedback from clerkship administrators and directors, as well as all of the medical students who have contributed to app testing and feedback.

References

1. NativeScript. https://www.nativescript.org/.
2. Firebase. https://firebase.google.com/.
3. ServiceStack. https://servicestack.net/.
4. ECE Tracker app for iOS, https://itunes.apple.com/ca/app/ece-tracker/id1244494664?mt=8. ECE Tracker app for Android, https://play.google.com/store/apps/details?id=ca.medportal. ecetracker&hl=en.
5. MacDot app for iOS. https://itunes.apple.com/ca/app/macdot/id1249758503?mt=8. MacDot app for Android, https://play.google.com/store/apps/details?id=ca.medportal.macdot&hl=en_ CA.
6. Ferenchick, G., & Solomon, D. (2013). Using cloud-based mobile technology for assessment of competencies among medical students. *PeerJ, 1,* e164. https://doi.org/10.7717/peerj.164.
7. Chamney, A., Mata, P., Viner, G., Archibald, D., & Peyton, L. (2014). Development of a resident practice profile in a business intelligence application framework. *Procedia Computer Science, 37,* 266–273.
8. Robertson, A. C., & Fowler, L. C. (2017). Medical student perceptions of learner-initiated feedback using a mobile web application. *Journal of Medical Education and Curricular Development, 4,* 1–7.

MyVitalWallet—I Bring My (Own) Health

Filipe Neves[4], Micaela Esteves[1,4(✉)], Angela Pereira[2,4],
Olga Craveiro[3,4], Bruno Rodrigues[4], and Hugo Gonçalves[4]

[1] CIIC - Computer Science and Communication Research - ESTG, Leiria,
Portugal
micaela.dinis@ipleiria.pt
[2] CiTUR - Tourism Applied Research Centre - ESTM, Leiria, Portugal
angela.pereira@ipleiria.pt
[3] CISUC-University of Coimbra, Portugal and Algoritmi-University of Minho,
Braga, Portugal
olga.craveiro@ipleiria.pt
[4] School of Technology and Management, Polytechnic Institute of Leiria, Leiria,
Portugal
{filipe.neves, micaela.dinis, angela.pereira, olga.
craveiro}@ipleiria.pt, {2120667, 2162204}@my.ipleira.
pt

Abstract. Nowadays, the widespread use of mobile technologies boosts the development of a great number of mobile applications related to health care. Although, the majority of them are not adapted to the end-users needs. This way, many of them are not being used since the users do not consider these applications useful to their daily life. In this context, this paper describes a mobile application that intends to stimulate users to correct and maintain their healthy conditions by keeping vital and analytic parameters under surveillance. The researchers' main purpose was to develop an application adapted to the end-users needs, so a user-centred design approach was used. The application enables users to register, check, analyse and keep track. Also, the application gives feedback about values related to biometric and biomedical parameters, like glycaemic index, blood pressure level or weight. All these results have further shared them with healthcare professionals. The end-user tests show an effective, efficient and user-friendly application.

Keywords: Mobile health care and training · E-health technologies ·
User-centred design · Mobile application

1 Introduction

Health is considered by the overwhelming majority of Humanity a greater good. However, some modern lifestyle constraints push people to acquire bad health habits. Among a vast number of vital and analytics values, high blood pressure, overweight, high blood sugar values, chronic inflammation or low white blood cells may announce a coming disease, that may lead to catastrophic results such as, cardiovascular diseases, diabetes or anaemia [1].

© Springer Nature Switzerland AG 2019
M. E. Auer and T. Tsiatsos (eds.), *Mobile Technologies and Applications
for the Internet of Things*, Advances in Intelligent Systems and Computing 909,
https://doi.org/10.1007/978-3-030-11434-3_15

Strategic management as a mean to health care philosophy is nowadays universally accepted [2]. With this in mind, governments implement their national health services policies. People buy health insurances, gym programs, try to keep themselves informed about good care practices and check their vital parameters with healthcare professionals, among others. Awareness of their own role in ensuring good health condition is becoming more and more a reality [3]. In fact, the costs to prevent diseases are less than the costs of treating them [4].

In recent years, the huge development of the Information and Communication Technologies through the use of versatile and smart pocket devices, such as smartphones and electronic notepads are being extensively explored and applied in vast areas, in which the healthcare has a prominent position. An example of this, is the use of micro sensors to measure and transmit vital values in real time. Additionally, the use of mobile applications has revealed to have great acceptance whilst bringing good benefits [5].

Mobile applications do not only have great potential to improve self-management in health status, but also instigate positive health behaviours. Furthermore, one cannot neglect how their use can change the users' attitudes [6, 7]. In this context, people desire to take control of their health and easily track their vital parameters through a mobile application. There are a huge number of health care applications, but none of them is tailored and dedicated enough to fit users' needs.

In this context, the paper describes "MyVitalWallet", a mobile application that was developed to incentive and alert users to be aware and take care of their health. The novelty is that the user followed up the application development by presenting his daily needs and giving suggestions whilst User-Centered Design and Extreme Programming approaches were used in order to get an app that meet the users' requirements.

By using this app, users can be attentive to their clinical vital parameters' history and trends. In a pedagogic attitude, this app is intended to lead users to face the tracking of their vital parameters as a natural behaviour and so, feel themselves more conscious and vigilant. Maintaining users responsible for such parameters, correcting bad habits and reinforce good ones in order to maintain respective values inside healthy ranges in the daily life is a pedagogical aim of this application. The application also allows to share such vital values with healthcare professionals while bringing their own vital parameters in their pocket.

"MyVitalWallet" is not intended to be used specially by patients of any special disease. Its purpose is to provide the user with vital and analytical parameters in a preventive maintenance attitude. However, it can also be used by people to self-manage chronic conditions or diseases by controlling the parameters that are related with it.

This paper is organized as follows: in Sect. 2, our motivation and related work is outlined; Sect. 3 describes the methodology applied to the project; Sect. 4 presents the development of the mobile application; Sect. 5 discusses the results obtained; and, in the last section, we present the conclusions.

2 Related Work

The rapid growth of mobile technologies and their regular use have boosted the development of mobile applications in all areas. One area with a great number of developed applications is health, which, even in the early 2000s estimated the existence of more than 40,000 mobile health applications [4].

In general, mobile health applications have the ability to engage patients in their own health care. Since mobile devices have become part of people's daily moments, they can be used as a way to control the health.

By doing a simple search on Google Play or on Apple Store using the 'Health' keyword, it is possible to find out many mobile health applications. The applications that are found are mainly generic and designed to control diabetes, weight, blood pressure, among others [8–10]. For this research, the authors only considered and analysed the applications that have similarities to the proposed application and the most popular software tools according to the user reviews and suitability. The analysed applications were Blood Glucose Tracker [11], 'Diário de peso-IMC' (Weight Diary) [12] and Samsung Health [13].

The Blood Glucose Tracker allows the user to register, edit, visualize and analyse only the glycaemic index values. Despite the simple interface of this application, it can only be used to control the blood glucose. The 'Diário de peso-IMC' has similar functions, but only regards weight control. Samsung Health provides basic features that help the users to improve his health since it allows them register and analyse the diary activities and routines as, for example, automatically record the user's daily activities, monitor the eating habits or the sleeping patterns. Nevertheless, these applications are not specific to control the biometric parameters as MyVitalWallet app is.

Despite the strengths of these applications, the majority of them are more concerned on delivering information than targeting the specific needs of the users [8]. According to Schnall et al. [5], the creation of applications to support healthful behaviours in consumers' daily lives demands the use of design processes centred in the end-users. These processes result in apps more robust, usable and adapted to people's needs. Thus, it is crucial to know the characteristics of the end-user [14] to develop a user-friendly interface.

The UCD methodology is an iterative design process in which the user characteristics, the usability goals, the environment, tasks and workflow are crucial to develop the interface design [15].

In [16], Norman argues that users should be in the centre of the design process and this should match the users' needs. With this process, the system is designed to be adapted to the users' needs rather than users adapt to the interface system [15].

3 Methodology

On developing MyVitalWallet, the researcher's main concern was to develop an application that does not only include powerful functionalities but that also has a user-friendly interface. With this concern in mind, a UCD was used to develop the interface. UCD focuses on the end-users needs and in the context of the application. Indeed, user satisfaction is its main goal. The end-users are in the centre of the whole development process and are involved in all steps process, which are: (i) conception, (ii) prototype development, (iii) tests and (iv) result analyses, as illustrated in Fig. 1. These steps are repeated until users become fully satisfied with the functionalities and interface provided by the system under development.

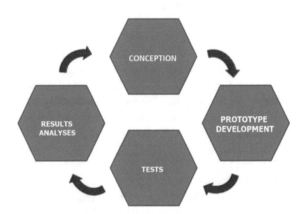

Fig. 1. User-centred design methodology

In the current project, several prototypes were created and evaluated by the users during the design and development process. The prototypes were improved following the evaluation carried out by the end-users. The design process finished only when a good satisfaction level was achieved.

After this phase, the EXtreme Programming (XP) methodology was used to develop the app 'MyVitalWallet'. This methodology is designed to produce high-quality software, adapting to the changing needs of the client which is part of the team. Indeed, he is a central member of the team. In XP, it is the client who must approve or disapprove the progress of the project [17] and so was in the current app development.

In this project, the client was also one of the end-users. He has followed the development of the app, testing, and suggestions to enhance the app, as well as add new features. He participated in all the decisions taken along with the development, namely, 'What the system should do', 'When the system must be ready', 'When the programmers must work in a certain task' and 'what is the next task'. All the team, including the client, met weekly. In each meeting, the work done by programmers during the week was presented, the decisions taken about possible changes were made and tasks to carry out next week were addressed.

To go deeper in the agile methodology, the team applied pair programming, one of the XP main practices [8]. The code was written by a pair of programmers working together using only one device. The code becomes more organized and readable, and so easier to understand for other person and the problems are also solved faster.

The combination of UCD with XP allowed the end-users to have a thorough knowledge of the whole process, and more important, the application was tailored to their needs, without any surprise. For programmers, it was also very important as far as they always knew if they were following the correct path since they frequently received feedback from the users. This combination was a good choice since the main objective was achieved in the specified deadline and with good results.

4   MyVitalWallet Application

'MyVitalWallet' is a mobile application that responds to the people's need of keeping informed about their vital and analytic parameters, in order to reinforce or correct the user's daily habits with the purpose of maintaining a good health condition. This app responds, specifically, to the needs expressed by people to register and continuously consult this kind of information. Designed under user specifications, the app is characterized by the simplicity and the sobriety, characteristics not usually present in the market of apps [5].

With this application, the user can record, check and analyse the values for biometric and biomedical parameters like glycaemic or cholesterol indices, blood pressure levels or the weight. The application also allows the user to visualize the recorded data under the form of tables, graphics and trends, while having a critical view relative to their own registered values. The main novelty of this application is its customization ability by which it is possible to create innumerable new parameters with different thresholds for healthy and unhealthy values. The healthy range values of created parameters can be taken by the users as a target to achieve and maintain, intended to promote a healthy lifestyle by either taking correcting measures or by maintaining the correct ones. Apart from stimulating the user good fitness, the application also allows to send the biometric and biomedical parameters data to the healthcare professionals that are accompanying the user. By their turn, they can give some feedback on interpreting of the tracked values and how to take appropriated actions.

According to a surrounding environment study [18], the Android operative system represents 85% of the smartphones market. Based on this fact, the app was developed for Android platforms.

4.1   Features and Functionalities

The 'MyVitalWallet' smartphone application was developed with simplicity and sobriety in mind. It permits to define (and further edit) the user profile by specifying exclusively the parameters considered influent in the health results. A certain minimalism has been adopted to be simple. Starting with the logo, three elements were crucial in order to represent vitality: (i) the heart, which is, undoubtedly, the ex-libris representation of vitality; (ii) the shape, which suggests the healthy people radiating dynamism; and (iii) the red hue as a vital sign of life.

Relative to the offered functionalities, the main view, as shown in Fig. 2, acts as a menu whose buttons permit to initiate the management of values, parameters, statistics, backups, sharing and settings. Since the most frequent action is to introduce values, a dedicated button has been added where the thumb tends to press, i.e. on the bottom of the window.

Fig. 2. 'MyVitalWallet' main view

The 'Value' button opens a new window with three value tasks available, where the user can create, select, edit and delete parameters. For each registered value, a feedback message is shown stating a healthy appreciation of that value.

The 'Parameter' button permits to select which parameters to monitor through the button 'Select parameter' (Fig. 3). Through the button 'Create parameter' (Fig. 4), it is possible to create a new parameter. Depending on the parameter type, it is possible to specify one threshold value or two threshold values. For example, in the parameter HDL cholesterol, the normal value for men must be greater than 40 mg/dl (Fig. 5).

Fig. 3. Parameter management view

The 'Statistics' button in the main window (Fig. 2) permits to visualize the evolution of the selected parameter over the time as a graphic. Furthermore, weekly and monthly zooming views are available to choose details (Fig. 6). It is also allowed to visualize data as a list of values sorted by date and time and appreciation (low, normal or high) is also displayed. A button to share the information is also available (Figs. 6 and 7).

The 'Backup' button of the main view (Fig. 2) permits to upload and download a backup file with all the introduced values. The 'Share' button permits to share introduced data with third parties, as for example the healthcare personal coach. Data is shared as Comma Separated Value (CSV) format. The 'Settings' button permits the user profile to be changed and provides a tutorial of the app.

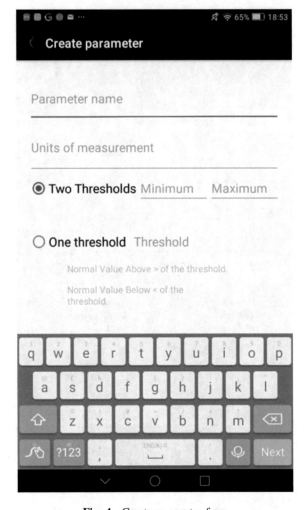

Fig. 4. Create parameter form

4.2   Usability Tests

In order to evaluate the design, effectiveness, efficiency and satisfaction, usability tests were carried out in three steps. In the first step, the instructions about the conduction of the tests were given to users, which included the tasks that users had to accomplish using MyVitalWallet app, such as creating a new parameter and entering respective values. Defined tasks are shown in Table 1.

In the second step, usability tests were conducted with the aim to evaluate the mobile application and its concept by accomplishing the predefined tasks. Once the tasks were accomplished, the third step has been initiated by giving users an inquiry.

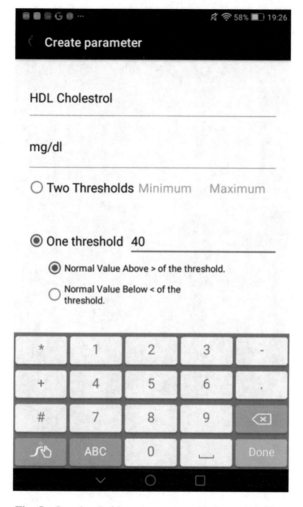

Fig. 5. One threshold parameter normal above definition

Concerning storage capacity, the application is very frugal since data is saved in CSV format. For example, registering 4096 values occupies ~1 MB, which is a very affordable value for the contemporaneous devices.

Inquiry

The inquiry was organized in five parts. The first part consisted of characterizing the participants in terms of age, literacy and some habits on using mobile devices. The second part aimed to evaluate how user-friendly and well designed the application is. For this, a Likert opinion scale with five levels in which 1 means 'not user-friendly' and 5 means 'very user-friendly' was used.

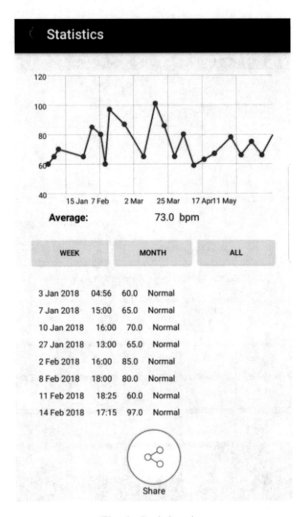

Fig. 6. Statistics view

The third part of the inquiry aimed to evaluate how easy the accomplishment of the tasks was felt. The same Likert opinion scale with five levels, from 1 (very hard) to 5 (very easy), was also provided to users in order to evaluate how easy the app was.

In order to evaluate the efficiency, the time users took to complete each task was also measured. For timing reference, the same tasks were completed by the developers/programmers since they are the most acquainted with the app.

The fourth part of the inquiry just consisted on evaluating the satisfaction and how useful the application would be in daily life. The fifth part consisted on users' suggestions aiming to enhance the application. In any part of the inquiry, sensitive information, as health condition or identity, has been asked.

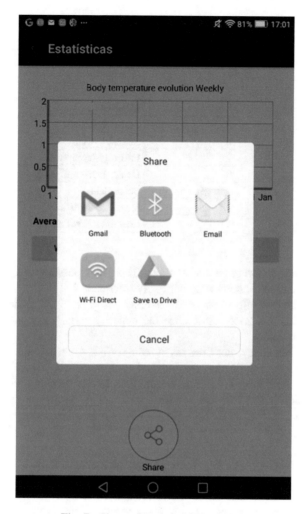

Fig. 7. Share of statistical information

5   Results and Discussion

In usability tests, 36 people participated, between the ages of 21–40. Regarding to qualification, 61% of participants have higher education degree and 39% have high education. All of them have a smartphone and use mobile applications daily. Only 16% of the inquired additionally use their tablet too. Related to smartphone operating system, 22% use exclusively iOS and 78% prefer Android. This result is in line with the initial study.

Concerning the interface design, the average opinion is 4.41 and the standard deviation (SD) is 0.59 (Table 2), which is considered by researchers as a good result. This low standard deviation value represents a very good uniformity of opinions, which

Table 1. Tasks given to users

Task #	Task
1	Register profile
2	Edit profile
3	Select parameter
4	Create new parameter
5	Edit parameter
6	Register value
7	Edit registered value
8	Delete registered value
9	Share data
10	Visualize statistics
11	Load/download data to/from cloud

confers confidence to the gathered opinions. Thus, it is possible to conclude that the developed application presents a user-friendly interface.

Regarding the task effectiveness, the average value is 4.25 and the standard deviation is 0.62. This is a good result since the majority participants considered easy to accomplish the tasks.

To evaluate efficiency, the time taken to carry out each task was measured. The participants used the application for the first time and from task #3 until task #11 (Table 1), they took from 6 to 25 s to complete it. The average value is 17 s and standard deviation is 6 s. The average time achieved by developers was 14 s and the standard deviation is 5 s. The ratio between values is less than 20%, which represents good values, considering that participants can improve timings as they become acquainted with the application.

Regarding tasks #1 and #2 (Table 1), the participants took 1 min and 20 s to complete them. Despite this time is greater than previously presented ones, it is not considered critical since these tasks are rarely carried out.

Concerning application usefulness, the average of the opinion is 4.25 and its standard deviation is 0.60. This result confirms that participants considered the application useful. Table 2 summarizes the results obtained from the usability tests.

Concerning satisfaction, participants' opinions were registered. The majority of them considered the application interesting and very useful and intuitive, as reported in the following comments:

Table 2. Average and standard deviation scores

	Average	SD
Interface design	4.41	0.59
Effectiveness	4.25	0.62
Efficiency (all tasks)	17″	6″
Usefulness	4.25	0.60

"Appreciated the design very much", "Very easy to use and intuitive", "Useful, especially for people with diseases that require continuous or frequent monitoring", "Useful for elderly people".

Some participants suggest more functionalities including to export results in PDF format instead of just CSV, allow login with Google or Facebook accounts or develop the app for iOS platform.

6 Conclusions and Future Work

Mobile technologies have been a crucial main drive for development of useful healthcare applications which have a great potential to support patient's self-management. However, some applications do not serve the users' needs since they lack personalized feedback and usability issues, particularly for elderly people.

This article points out the contribution a UCD approach had on developing a mobile health care. The main motivation was to create an easy interface and useful application. This should match with the tasks that end-users consider important to track their health.

For this purpose, usability tests were done, which results show that the users highlight the user-friendliness interface as well as its usefulness. Users, also, mention some additional functionalities that should be integrated in 'MyVitalWallet' application.

As future work, the researchers intend to endow the application with some statistical utilities that include the possibility of analysing and correlating results from different parameters, endow the application with health advices triggered from analysing the history and trending values according to medical criteria. Endow the app with the ability to import data automatically and develop the application for iOS.

References

1. Miller, B. (2016). Silent inflammation. *The single phenomenon behind many of the most feared diseases of middle & old age.* Oak Publication Sdn Bhd.
2. Ginter, P. M., Duncan, W. J., & Swayne, L. E. (2018). *The strategic management of health care organizations.* Wiley.
3. Stoddart, G. L., & Evans, R. G. (2017). Producing health, consuming health care. In *Why are some people healthy and others not?* (pp. 27–64). Routledge.
4. Hale, I. (2011). The greatest good. *Canadian Family Physician, 57*(8), 868–869.
5. Schnall, R., Rojas, M., Bakken, S., Brown III, W., Carballo-Dieguez, A., Carry, M., et al. (2016). A user-centered model for designing consumer mobile health (mHealth) applications (apps). *Journal of Biomedical Informatics, 60*, 243–251.
6. Sarkar, U., Gourley, G. I., Lyles, C. R., Tieu, L., Clarity, C., Newmark, L., et al. (2016). Usability of commercially available mobile applications for diverse patients. *Journal of general internal medicine, 31*(12), 1417–1426.
7. West, J. H., Belvedere, L. M., Andreasen, R., Frandsen, C., Hall, P. C., & Crookston, B. T. (2017). Controlling your "App" etite: how diet and nutrition-related mobile apps lead to behavior change. *JMIR mHealth and uHealth, 5*(7).

8. Pereira, A., Esteves, M., Weber, A. M., & Francisco, M. (2017, November). Controlling diabetes with a mobile application: Diabetes friend. In *Interactive Mobile Communication, Technologies and Learning* (pp. 681–690). Cham: Springer.
9. Stoyanov, S. R., Hides, L., Kavanagh, D. J., Zelenko, O., Tjondronegoro, D., & Mani, M. (2015). Mobile app rating scale: A new tool for assessing the quality of health mobile apps. *JMIR mHealth and uHealth, 3*(1).
10. Qiang, C. Z., Yamamichi, M., Hausman, V., Altman, D., & Unit, I. S. (2011). *Mobile applications for the health sector* (p. 2). Washington: World Bank.
11. Blood Glucose Tracker. (2018). http://www.littlebytes.mobi/. [Consult in 13 May 2018].
12. Diário de peso, IMC. (2018). https://www.easycreation.net/. [Consult in 14 May 2018].
13. Samsung Health. (2018). https://www.samsung.com/global/galaxy/apps/samsung-health/ [Consult in 13 May 2018].
14. Nielsen, J. (2003). Usability 101: Introduction to usability.
15. Abras, C., Maloney-Krichmar, D., & Preece, J. (2004). User-centered design. In Bainbridge, W (Ed.), *Encyclopedia of Human-Computer Interaction* (Vol. 37(4), pp. 445–456). Thousand Oaks: Sage Publications.
16. Norman, D. (2014). Things that make us smart: Defending human attributes in the age of the machine. *Diversion Books.*
17. Lindstrom, L., & Jeffries, R. (2004). Extreme programming and agile software development methodologies. *Information Systems Management, 21*(3), 41–52.
18. IDC. (2018). "IDC-Analyse the future". https://www.idc.com/promo/smartphone-market-share/os. [Consult in Feb 2018].

PTGuide—A Platform for Personal Trainers and Customers

Rosa Matias[1,3]([✉]), Micaela Esteves[1,3], Angela Pereira[2,3],
Beatriz Piedade[1,3], João Marques[3], Tiago Sousa[3], Ruben Gonçalves[3],
and Ruben Carreira[3]

[1] CIIC - Computer Science and Communication Research - ESTG, Leiria,
Portugal
{rosa.matias,micaela.dinis,beatriz.piedade}
@ipleiria.pt
[2] CiTUR - Tourism Applied Research Centre - ESTM, Leiria, Portugal
angela.pereira@ipleiria.pt
[3] Polytechnic Institute of Leiria, Leiria, Portugal
{2150643,2150682,2150650,2151575}@my.ipleiria.pt

Abstract. Sports are a vital factor for the well-being, contributing for a good health. As it can be observed by amount of mobile applications related to sports available in the market. Moreover, sportspeople have distinct training objectives and purposes, consequently customization and adaptation to different individuals' profiles are gaining attention. Certified personal trainers are the correct professionals to create customized fitness training plans. Meanwhile, to maximize results they need to know their customers' habits and monitor their progress. In this work, it is presented the Personal Trainer Guide (PTGuide) platform formed with a mobile and web application for personal trainers and their customers, respectively. The aim of PTGuide is to identify digital mechanisms where the services provided by certified personal trainers can be enhanced. A study was conducted on the analysis of similar mobile applications and a survey questionnaire. The study showed that the platform should include a web application for personal trainers and a mobile application for their customers. Among others, the web application intends to manage certified personal trainers, customer management, individual trainings and nutrition plans configuration. The mobile application permits customers to be guided through the individual fitness training and nutrition plans. The tests showed that users, namely, personal trainers are very comfortable and satisfied with the system. It was considered useful and interesting.

Keywords: Mobile applications · Personal trainer · Sports · Mobile health care and training · E-health technologies

1 Introduction

Mobile health (mHealth) applications have a tremendous success, specifically, for biometric data monitoring management, nutrition applications or applications to control diseases such as diabetes or high blood pressure [1]. Also, a spreading idea is that

© Springer Nature Switzerland AG 2019
M. E. Auer and T. Tsiatsos (eds.), *Mobile Technologies and Applications
for the Internet of Things*, Advances in Intelligent Systems and Computing 909,
https://doi.org/10.1007/978-3-030-11434-3_16

physical inactivity should be fought with strength. Moreover, sports have many benefits such as combat stress, increase energy or combat obesity. Fortunately, in recent years, Portugal, like most Western countries, has been experiencing exponential growth in physical activity. It is undeniable that the number of fitness gyms and healthcare clubs has increased significantly. Meanwhile, people have different human body structures and different goals and so the tendency is to customize fitness plans and adapt them to each person. In this way, the services provided by personal trainers are been widely requested. Moreover, due to its ubiquity mobile devices opened the doors to a more effective sport individualization where users have access to specific contents and may monitor their evolution and improvement even during training. Also, personal trainers want to monitor their clients' e

volution and motivate them. With the exponential growth of personal training, the problem is how to find a personal trainer with the necessary characteristics and qualifications for an individual. Nowadays, such search is based on social networks like Instagram, Facebook, YouTube, where sometimes the transmitted information is not the most suitable. Some personal trainers' social media networks convey the idea that all the clients have excellent results, in incredible times. In the case of customers who need/want a personal trainer for personal assistance, mobile applications can turn out to be more effective and efficient. In this way, a group of researchers proposed to some computer science students a project to develop a platform to enhance the search of personal trainers and improve the online communication between personal trainers and their customers. The system was named Personal Trainer Guide (PTGuide). PTGuide's main objective is to provide personal trainers and their customers a system that demystifies their interaction.

This article is organized as follows: In the next section, the related work is described; in Sect. 3, the methodology has been described; Sect. 4 describes the PTGuide development process; Sect. 5 is dedicated to tests; and finally, the conclusion and future work.

2   Related Work

Mobile health (mHealth) descends from electronic health (eHealth) [2] a broad term to designate electronic systems related to health care and supports the achievement of health objectives using mobile and wireless technologies [3, 4]. There is substantial enthusiasm around the concept of mHealth used to describe the mobile telecommunication technologies for healthcare delivery and support of wellness [5, 6]. The applications related to sports may also be considered mHealth applications, since the practice of physical exercise contributes to a good health. Meanwhile, humans have different body structures, this way sports individualization and personalization is gaining attention [7]. The personal trainer can adapt sports to each person and provide professional directions in a one-on-one exercise program [8]. Personal trainers may also be defined as certified professional of gym fitness clubs [9] and they offer professional instructions in fitness clubs and customized training courses. They are also considered motivational leaders and may deliver services outside gyms and fitness clubs. As a high-level trainer, a personal trainer works to establish and promote professional customized sports programs [10]. Both, medicine and sports science can be considered as two linked fields of research since they require a close study of the body functions [5]. As opposed to medicine where graduated professionals are certified for practice, there are many non-certified personal trainers. This reinforces the need of an online platform with mediation for these certified professionals where customers have access to personal trainers' profiles. This system should assure the customer that the health fitness instructor has a professional training curriculum [11].

Through a research on Google Play and Apple Store, it was possible to find out many mobile applications related to health and fitness. In this research, the authors only considered three applications that have similarities to the purposed application and the most popular software tools according to the user reviews and suitability. Three successful training applications were identified, and their features were analysed (Table 1). The applications are *Freeletics* [12], *Workout Trainer* [13] and *Trainerfu* [14].

Freeletics supports nutrition and exercise training plans. However, the plans are not defined by a certified personal trainer but, instead, by machines using data mining over the web. Consequently, the customer hardly gets a personalized training and motivational feedback. Also, *Freeletics* does not integrate a single platform nutrition and exercise plan, being necessary to instal different applications to do so.

Workout Trainer application, in contrast, has many workout plans predefined and allows the user to search a personal trainer. Through the PRO+ version, it is possible to create workout plans. The workout trainer has a similar system to social network having followers, news feed, chat and a leader board, in which users evaluate their personal trainers and this can be shared with friends. The application has a forum with contents about exercise, nutrition and frequently asked questions.

Both *Freeletics* and *Workout Trainer* applications have video demonstrations. But, when it comes to training feedback, these applications are somewhat poor, because whenever a user change to a new exercise, the applications assume that the user has completed the exercise, which is not always true. In addition, at the end of a training session, the user is asked to enter the overall evaluation using a proprietary scale.

The *Trainerfu* application is much more efficient, regarding training feedback. Users can introduce measures in training. This metric is wonderful for personal trainers, as they have a deeper knowledge about customers' performance, and can tailor the training plan and maximize results. Finally, it is important to reinforce the idea that *Trainerfu* is the only application that clients can access only after the personal trainer creates the client's credentials. However, the client cannot search for a personal trainer and does not have individual nutrition plans.

All applications mentioned above have individual plan training configuration and feedback capacities. *Freeletics* has nutrition plans, but it is necessary to instal a different application. *Workout Trainer*, in opposite, has the possibility to search for personal trainers, even though none of them support multiple trainers for one customer. Both *Trainerfu* and *Workout Trainer* have chat support. Table 1 presents a comparison between the three applications.

Table 1. Comparing training applications

Feature/Application	Freeletics	Trainerfu	Workout trainer
Individual plan training	✓	✓	✓
Train feedback	✓	✓	✓
Individual nutrition plans	✓	✗	✗
Search for personal trainers	✗	✗	✓
Support for multiple personal trainers	✗	✗	✗
Chat	✗	✓	✓

3 Methodology

In this context, the researchers began to study the answers for the following questions: (1) What are the most appreciated applications functionalities for both personal trainers and fitness customers? (2) What mobile operating systems should be taken into consideration?

The first question is important since the application should have the end-user's acceptance. The second question is relevant since it was necessary to define the development platform and establish the system architecture.

To answer the above-mentioned questions, a study was conducted. It consisted of analysing the mobile applications market and a survey.

Later, the software life cycle methodology was identified for software development. At a first stage, the software development life cycle adopted was the waterfall model [15], as the team was small and the project predictable. The waterfall model is the classical model of software engineering and its stages are requirements gathering, prototyping, implementations, test and validation [16]. The development stages occur sequentially. However, the pure waterfall methodology presents a problem, because the phases must be totally completed one after the other, an extension of the waterfall model was adopted, namely, an incremental/iterative model. Based on the requirements, multiple development cycles took place where cycles are divided into smaller ones, so it is easier to manage iterations [16, 17]. It is commonly named as multi-waterfall model.

Also, for the design phase it was considered a user-centred approach. The process began with the design of low fidelity prototypes. Then tests were made with end-users which approved them. The next step was the development of high-fidelity prototypes. End-users had an important collaboration in this phase.

The first weeks consisted mainly of defining the methodology, project requirements, prototypes design, user stories narrations and architecture design and afterwards, the implementation and test phases. The personal trainers accomplish the project as part of the development team.

4 PTGuide Development

The PTGuide platform consisted of two related application, one web responsive for personal trainers and other mobile application for their customers.

4.1 Requirements Identification

For the requirements identification, it was necessary to analyse similar systems and identify their strengths and weaknesses. The analysed applications were *Freeletics* [12], *Workout Trainer* [13] and *Trainerfu* [14]. Table 1 summarizes these applications functionalities. Interviews were also conducted with personal trainers, which were crucial to help researchers to systematize important features that the platform should implement. Resulting in the following features: individual plan training configuration, train feedback, individual nutrition plan configuration, search for the right personal trainers, support multiple personal trainers and chat.

Another element that contributes to the requirements identification was a survey applied to personal trainers and fitness gym clubs' customers. The survey started with the individuals' characterization and then, regarding each individual profile, questions were made. The population survey was mainly from Leiria, a small town near Lisbon, in Portugal. Leiria has a population of around 127 thousand inhabitants and it has around 20 fitness gym clubs. From the 419 participants, 11.7% were personal trainers and 88.3% were fitness clubs' customers, who answered the survey. Refer to gender, 60.8% were female and 39.6% male with an average age of 31 years old. Regarding to the qualifications, 42.6% have higher education.

It was found that only 5.4% did not practice exercise with regularity, which is normal considering the audience profile. The main reasons why individuals did not practice physical exercise with regularity were due to not having time, the lack of company and the lack of motivation.

The main device used by personal trainers and customers is Android device followed by personal computer (Fig. 1). So, it was considered important to develop the application for mobile devices and a responsive website for pc's.

Concerning the use of fitness applications, 230 fitness gyms clubs' customers refer that did not use any (Fig. 2).

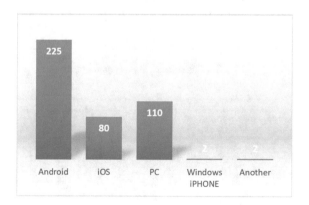

Fig. 1. The main devices for Internet access

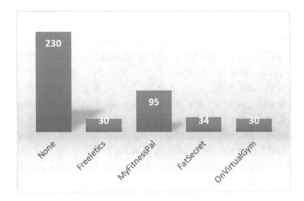

Fig. 2. Gym clients' applications

The participants had to classify in a Likert scale from 1 to 5, 1—not important; 5—very important, the importance of a set of fitness application functionalities.

The participants highlight the videos with exercises and the nutrition accompaniment, as the most important features to be presented in a fitness application (Fig. 3).

Trainers considered the nutrition accompaniment and payment management as very important functionalities (Fig. 4).

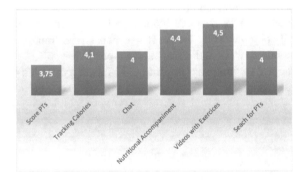

Fig. 3. Customers' important features

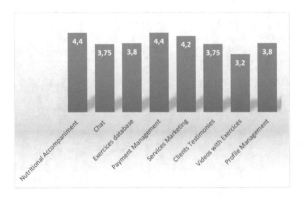

Fig. 4. Personal trainers' important features

It was established that PTGuide platform should allow data to exchange between the two applications, web and mobile. The responsive website is for personal trainers where they can configure, manage the training and nutrition plans for each client.

The mobile application is for personal trainer customers, which allows mobility.

Based on the analysis of similar systems and on the answers to the survey, the researchers defined a list of requirements for PTGuide platform presented in Tables 2 and 3.

Table 2. Requirements for PTGuide website (personal trainer)

PTGuide website—Requirements
Configure personal trainer profile
Create individual training plans
Create individual nutrition plans
Customer management
Training schedule management
Nutrition schedule management
Chat
Video database explaining exercises
Payment management
Marketing

To guarantee that the system is restricted to certify personal trainers it was introduced the mediator role. It is a superuser or an entity responsible for verifying the personal trainer curriculum and certificates. Personal trainers only have access to the platform after the superuser validation.

Table 3. Requirements for PTGuide mobile application customers

PTGuide mobile application—Requirements
Visualize training plans with multimedia support (video and images)
Visualization of the main objectives and already achieved objectives
Visualize nutrition plans
Visualize alerts sent by personal trainer
Visualize personnel trainers rating system
Searching mechanism to find the ideal personal trainer
Chat

4.2   System Architecture

The PTGuide platform is formed by two distinctive interrelated applications. Therefore, rising a question on how to create a communication channel between them, the main components defined for the overall platform architecture were the mobile application, the application server and the website. The application server addresses issues such as databases, authentication, messaging and web application hosting (Fig. 5). Since there was a requirement stating that "(...) *The application should notify the customer when it has new messages* (...)", servers of Expo were included as well as a notification service. A common data storage layer was created and implemented using a MySQL database server. An API was developed and was responsible for services consumed by both the mobile application and the website. Users use both Android and iOS mobile devices (Fig. 1).

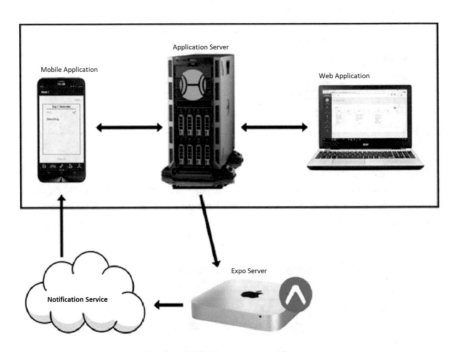

Fig. 5. PTGuide system architecture

The researchers considered not to split the development for specific and native devices and not to develop separate applications. Therefore, React Native [18], a cross-platform development framework was chosen. React Native enables the development of a solution that will work in both Android and iOS mobile operating systems. The website was developed using Laravel which is a framework for the development of web pages, based on the PHP language. Laravel is built on one of the best known standards for web development: *Model–View–Controller* (MVC). The above-mentioned API was developed using Laravel providing Create, Read, Update and Delete (CRUD) methods.

4.3 PTGuide Platform

Since the mobile application development contains several screens with dependencies, an *n* hierarchy structure of the screens (Fig. 6) was made. In Fig. 6, it can be observed an authentication, the workout plan and nutrition modules. There are also modules related to the chat and instant communication between parties (personal trainers and customers).

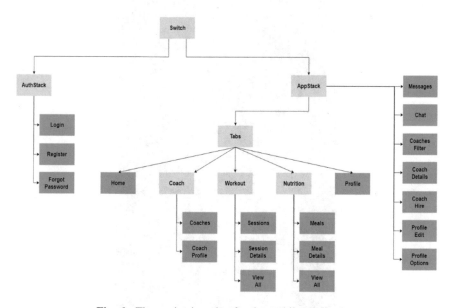

Fig. 6. The navigation plan for the mobile application

In Fig. 6, each blue rectangle represents a screen, while each grey rectangle represents a navigation component. The switch is the tree root component and each time the app is open, if the user is not authenticated, the screens from AuthStack are presented. Otherwise, the screens of AppStack are presented.

In the mobile application development, several *Design Patterns* were used. The main purpose of this choice was to facilitate the communication between programmers, by improving the software quality and increasing the productivity, since it was not necessary to create new solutions that have already been developed [19]. The *Design Patterns* used were given as follows:

- Lazy Load,
- Optimistic UI,

- Offline first,
- Singleton,
- Observer.

Figures 7 and 8 present two screen examples of mobile application.

Fig. 7. Authentication screen

In the website, Design Patterns were also used, such as Lazy Load, Model–View–Controller and Optimistic UI.

Figures 9 and 10 present two screens of the website. In Fig. 10, the main working area of the personal trainer is presented where he can configure the workout and the nutrition plans.

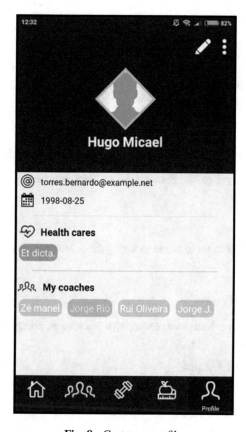

Fig. 8. Customer profile

Fig. 9. Personal trainer profile

Fig. 10. Personal trainer main work area

5 Tests

In this phase, several tests were made to verify the software compatibility and end-user acceptability.

5.1 Compatibility Testing

To guarantee the best performance, responsive layout and compatibility were tested in the following devices:

- Xiaomi Redmi note 4/3 pro,
- Apple iPhone 6,
- Weko Lenny2,
- White brand Android tablet.

Due to the different scenarios tested (recent Android smartphones, old Android smartphones, iOS smartphone and Android tablet), many adjustments were made in the application. By this way, it can be guaranteed that PTGuide platform responds equally well in all devices. Namely, there was a need to create custom components for different operating systems (iOS and Android) and screen resolutions.

5.2 Manual Tests

The manual tests were performed by all parties (developers, researchers and a personal trainer) as functionalities were being implemented. It consisted in simulating real scenarios and confirming if results were the expected. These tests were done by trying to simulate a limited set of situations that could cause any errors. This was the only type of testing performed by the application developers. Thus, in this phase, several problems were found, which were solved, and afterwards contributed very positively to lower the percentage of problems faced in the following test types.

5.3   Acceptance Tests

Black-box tests were performed at the end of each iteration phase not only by personal trainers, who were integrated in the development team, but also by gym fitness customers. The tests consisted of allowing users to explore the applications and asking them to perform certain test cases, derived from the requirements list. The tests were always followed up by one element team.

One of the most valuable results obtained with these tests was at the usability level. This way, the interface was redefined considering some of Nielsen's main heuristics [20, 21] and matches the user's expectations.

6   Conclusion and Future Work

The developed PTGuide platform has the intention of being a system for certified personal trainers as well as their customers. So, the platform has a mediator, an entity responsible for the validation of certified personal trainers. The potential personal trainer customers can search in the platform for a suitable personal trainer by observing their profiles, certificates and assigned scores.

The identification of the most important platform features was done through end-users meetings and interviews, the analysis of similar mobile applications and by a survey. As a result, many important features were identified like the configuration of personalized training and nutrition plans. Personal trainers can schedule and configure the exercises and meals for customers after the communication of the clients' objectives.

The tests show that the end-users appreciated the PTGuide platform and highlighted its usability.

As a future work, the researchers intend to upgrade the applications and make it available in the market.

References

1. Vandelanotte, C., Müller, A. M., Short, C. E., Hingle, M., Nathan, N., Williams, S. L., et al. (2016). Past, present, and future of eHealth and mHealth research to improve physical activity and dietary behaviors. *Journal of Nutrition Education and Behavior, 48*(3), 219–228.
2. Armstrong, S. (2007). Wireless connectivity for health and sports monitoring: A review. *British Journal of Sports Medicine, 41*(5), 285–289.
3. Steinhubl, S. R., Muse, E. D., & Topol, E. J. (2013). Can mobile health technologies transform health care? *JAMA, 310*(22), 2395–2396.
4. Pereira, A., Esteves, M., Weber, A. M., & Francisco, M. (2018). Controlling diabetes with a mobile application: Diabetes friend. In M. Auer & T. Tsiatsos (Eds.), *Interactive mobile communication, technologies and learning*. IMCL 2017. Advances in intelligent systems and computing, vol 725. Cham: Springer.
5. Mata, F., Torres-Ruiz, M., Zagal, R., Guzman, G., Moreno-Ibarra, M., & Quintero, R. (2018). A cross-domain framework for designing healthcare mobile applications mining.

6. Istepanian, R. S., Jovanov, E., & Zhang, Y. T. (2004). Guest editorial introduction to the special section on m-health: Beyond seamless mobility and global wireless health-care connectivity. *IEEE Transactions on Information Technology in Biomedicine, 8*(4), 405–414.

7. Mata, F., Torres-Ruiz, M., Zagal, R., Guzman, G., Moreno-Ibarra, M., & Quintero, R. (2018). A cross-domain framework for designing healthcare mobile applications mining social networks to generate recommendations of training and nutrition planning. *Telematics and Informatics, 35*(4), 837–853.

8. Jiang, H. L. (2005). Present operation and future management of manpower in the fitness industry. *National Physical Education Quarterly, 145,* 76–82.

9. Chiu, W. Y., Lee, Y. D., & Lin, T. Y. (2011). Innovative services in fitness clubs: Personal trainer competency needs analysis. *International Journal of Organizational Innovation (Online), 3*(3), 317.

10. Chen, Z. L. (2005). Fitness club athletic instructor competency. *National Physical Education Quarterly, 145,* 71–75.

11. Malek, M. H., Nalbone, D. P., Berger, D. E., & Coburn, J. W. (2002). Importance of health science education for personal fitness trainers. *The Journal of Strength & Conditioning Research, 16*(1), 19–24.

12. FREELETICS—Functional High Intensity Bodyweight Training. (2018). Retrieved from https://www.freeletics.com/en.

13. Workout Trainer by Skimble—Top Free Fitness App Coached by Certified Personal Trainers. Online Personal Training Software. Download on Apple Watch, iPhone, iPad, Android Devices, Smartwatches (2018). Retrieved from https://www.skimble.com/.

14. Best App for Personal Trainers, & Clients. TrainerFu. (2018). Retrieved from http://www.trainerfu.com/.

15. Royce, W. W. (1987, March). Managing the development of large software systems: Concepts and techniques. In *Proceedings of the 9th International Conference on Software Engineering* (pp. 328–338). IEEE Computer Society Press.

16. Larman, C., & Basili, V. R. (2003). Iterative and incremental developments. A brief history. *Computer, 36*(6), 47–56.

17. Munassar, N. M. A., & Govardhan, A. (2010). A comparison between five models of software engineering. *International Journal of Computer Science Issues (IJCSI), 7*(5), 94.

18. Build Native iOS, Android and Windows Apps with JavaScript. (2018, June). *Use react native.* Retrieved August 3, 2018, from http://www.reactnative.com/.

19. Fowler, M. (2002). *Patterns of enterprise application architecture.* Addison-Wesley Longman Publishing Co., Inc.

20. Nielsen, J., & Molich, R. (1990, March). Heuristic evaluation of user interfaces. In *Proceedings of the SIGCHI Conference on Human Factors in Computing Systems* (pp. 249–256). ACM.

21. Esteves, M., & Pereira, A. (2015, November). YSYD-you stay you demand: User-centered design approach for mobile hospitality application. In *2015 International Conference on Interactive Mobile Communication Technologies and Learning (IMCL)* (pp. 318–322). IEEE.

Development of a VR Simulator Prototype for Myocardial Infarction Treatment Training

Juan Sebastian Salgado[1], Byron Perez-Gutierrez[1(✉)], Alvaro Uribe-Quevedo[2], Norman Jaimes[1], Lizeth Vega-Medina[3], and Osmar Perez[4]

[1] Universidad Militar Nueva Granada, Bogotá, Colombia
jsebastiansalgado@hotmail.com, byron.perez@ieee.org,
norman.jaimes@unimilitar.edu.co
[2] University of Ontario Institute of Technology, Oshawa, ON, Canada
alvaro.quevedo@uoit.ca
[3] Universidad El Bosque, Bogotá, Colombia
lizvega@ieee.org
[4] Fundación Clínica Shaio, Bogotá, Colombia
oaperezs2@hotmail.com

Abstract. A medical virtual reality simulator provides a learning and practicing tool where trainees are exposed to diverse medical conditions including, the treatment of high-risk pathologies like the myocardial infarction, to develop cognitive and psychomotor skills. Additionally, simulators provide controllable patient conditions that can be configured to improve decision-making in time-sensitive medical procedures. The study of myocardial infarction treatment cardiology relies on analyzing medical cases and patient examination under medical supervision. However, the lack of practice can result in a human error that can be minimized with the simulation. In this paper, a virtual reality simulator prototype that allows students to practice the myocardial infarction procedure based on the American Heart Association guidelines is presented. The proposed prototype uses immersive virtual reality technology to create an engaging situation where the students will be able to recreate the treatment on a virtual patient. The prototype presents an immersive and interactive approach that can benefit the adherence to current myocardial infarction guidelines.

Keywords: Virtual reality · Myocardial infarction

1 Introduction

The heart is the muscle responsible for pumping blood to the body. However, there can be abnormalities where the blood flow does not meet the desired conditions, thus resulting in a lack of oxygen, nutrition and waste disposal, that can negatively affect bodily functions leading to heart infarction. There are two

© Springer Nature Switzerland AG 2019
M. E. Auer and T. Tsiatsos (eds.), *Mobile Technologies and Applications for the Internet of Things*, Advances in Intelligent Systems and Computing 909,
https://doi.org/10.1007/978-3-030-11434-3_17

types of myocardial infarction, the STEMI (ST—Elevation Myocardial Infarction), which is when there is a full obstruction of the artery and it manifest on the elevation of the ST segment on the electrocardiogram, and the NSTEMI (non-STEMI), which is when the artery has a partial block and it manifests in a depression of the ST segment or a T inversion.

The American Heart Association (AHA) has developed several recommendations to treat myocardial infarction [1,4], and academic institutions have adopted these for the training of their residents. The guidelines indicate the steps, the pharmacology treatment, and the contradictions to consider. According to World Health Organization, cardiovascular diseases are the number one cause of death globally [8], and several initiatives have been adopted to face this problem including the improvement of medical education [2,5].

Myocardial infarction treatment is taught through medical case studies followed by patient examination under medical supervision. However, this approach introduces limitations associated with few cases and/or patients available leading to scarce conditions and scenarios to practice with, hindering the development of expertise [3]. To improve cognitive and psychomotor skills, simulation is currently employed as a standard learning method, where controlled and safe situations present enriched and multiple cases to increase the trainee's response readiness [7]. In this paper, the development of a virtual reality (VR) simulator prototype for myocardial infarction treatment training that provides the opportunity for the students to follow the AHA guidelines on a virtual patient is presented. The goal of the research is to present an initial VR prototype to help users to improve the adherence to cardiology guidelines.

2   VR Prototype Development

To develop the VR simulator prototype, the STEMI and NSTEMI pathologies were characterized according to the AHA guidelines [1,4]. The guidelines provided information employed to determine the system's inputs and outputs that determine the parameters for the intended VR practice. A virtual patient was designed and developed within an examination room to provide an adequate and realistic virtual training scenario, where the user can monitor the vital signs and apply the pharmacotherapy required to treat myocardial infarction.

2.1   Guidelines for STEMI and NSTEMI

The STEMI and NSTEMI are pathologies where a full or partial block in a coronary artery is present. In both cases, the symptomatology consists of pain chest, hard breathing, and dizziness, for this reason, the default state of the virtual patient and trainee's first pharmacotherapy response are the same in either pathology giving us the opportunity to merge both guidelines [1,4] as Fig. 1 depicts. The next step is to identify the type of infarction (STEMI or NSTEMI) by the computation of GRACE score [6] as an estimator of prognostic factor. After applying the recommended treatment and considering the time lapse, the patient is transferred or not to Percutaneous Coronary Intervention (PCI) in order to perform an angioplasty to remove a obstruction in a coronary artery.

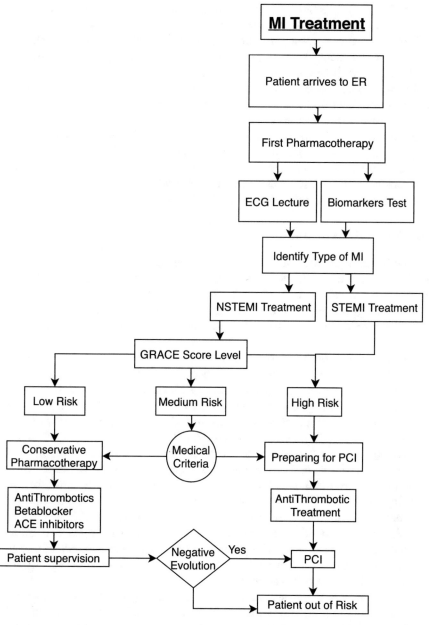

Fig. 1. Myocardial infarction treatment using AHA's STEMI and NSTEMI joint guidelines

2.2   System Architecture

The proposed prototype is based on the user interaction through sight, touch, and hearing senses with a VR system following the AHA guidelines as its main feature in order to generate a life-like emergency room environment. The information flow is managed by a scene graph layer comprising the logic of guidelines and the state of the system, an application layer for handling the VR user interface and a rendering layer responsible of handling the graphics, haptics, and auditory aids with the external equipment.

In the application layer, the user interacts with the virtual patient, available medical equipment and medication, and medical orders of specialized laboratory exams. The scene graph layer verifies all the steps performed by the user and keeps updated the state of the system, is responsible for generating the time stamp of all actions of the user for the medical report. For this prototype, all the computation and rendering are proposed to be made in a single computer with consumer level VR equipment to reach mainstream audience as residents of cardiology, cardiologists, and medical schools.

2.3   Virtual Environment

For the initial prototype, a 75 year old virtual patient was developed from real anonymized data. The medical condition for this patient was a 30-min evolution of oppressive chest pain radiating from the neck, diaphoresis and nausea. An electrocardiogram (ECG) and vital signs information (heart rate, oxygen saturation level and blood pressure indicators) was included in the virtual medical record.

Having into account the two stages of the treatment (pharmacotherapy and PCI procedure), two scenes were developed. The first one includes all the main props and assets of an emergency room and includes a user interface with the time lapse visual cue, as a decision-making tool for the treatment, medical orders selection for medication, tests and exams ordering, vital signs monitor and ECG. The user can transfer the virtual patient to PCI for the angioplasty when founds it pertinent. Figure 2 presents the user interface including a Heads-Up Display (HUD) with medical orders and vital signs information, the user can move around the entire room an interact with objects using a virtual hand, i.e., select medicines, calculate GRACE score, order lab exams for biomarkers, apply oxygen, measure blood pressure, view ECG, view the patient medical record and view the final report. To reinforce the immersion in the emergency room, audio cues with the sound of medical equipments and background sounds of a hospital facility were included in addition to touch feedback as a response when the user interacts with elements of the interface.

The second scene, presented in Fig. 3, is the hemodynamic room for PCI procedure including all the necessary instruments and medication, and a HUD with medical orders and vital signs. The user has to insert a balloon catheter into the radial artery to be passed over a guide-wire into the narrowed vessel in the heart and then inflated to a fixed size to permit the normal blood flow. For the

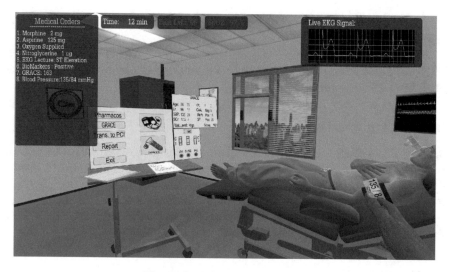

Fig. 2. Initial treatment scene

initial prototype, the interaction with the catheter and guide wires are handled by means of a panel where the user selects the desired action. As visual cues, the user has X-ray monitors to track the movement of the guide wire within the vessels and sound of medical equipment.

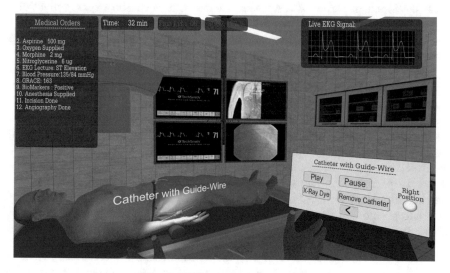

Fig. 3. PCI—angioplasty scene

Finally, the simulation provides a report based on the performance and decisions taken during the VR training session. This includes all the medical orders (medications administered to the patient and blood tests) with its corresponding time stamp.

All the virtual environment was programmed in Unity 3D based on its capabilities of integration of consumer level VR equipment and multiplatform compatibility for future prototype versions.

2.4   VR Equipment

Because of the single computer architecture for this prototype, a VR Ready computer was selected. An Oculus Rift CV1 [(consumer level head mounted display (HMD)] was selected for the VR immersion with its Oculus touch controllers, and the decision was taken based on the capability to simulate both hands, finger movement, and touch feedback based on the vibrotactile actuators. This HMD also has integrated stereo headphones for the audio cues.

3   Results

A preliminary validation of the simulator prototype was carried out by a cardiologist and a medical doctor who is an expert in clinical simulation, who expressed interest in the potential of the prototype as a complementary tool in cardiology simulation practices.

For the evaluation of the simulator prototype, it was proposed to analyze the user's experience with it. In order to analyze the user's perception with the tool developed, a series of tests were carried out with the experts mentioned above, which allowed verifying whether or not it contributes to the learning of the myocardial infarction treatment. In Fig. 4, a cardiologist using the prototype is shown.

To carry out the verification, the evaluation was focused on the following points: accuracy and precision of the information, usability, user experience, immersion, and environment.

At the time of the evaluation, the experts were informed about the use of the software, giving a brief introduction to it and its functionalities. Then, the experimenter instructed him to sit down and put on the HMD. Throughout the test, the application indicates to the user the necessary instructions that it has to perform. At the end of the test, the experts were thanked for their participation and were asked to give feedback on aspects of the evaluation to get their opinion on the application and their satisfaction. The evaluation asked about the following aspects obtaining these results:

- **Accuracy and precision of the information**: The information presented in the system is truthful. It was consulted from specialized literature and guides and is precise in what the simulator should transmit. Rating: 4.5/5.0.
- **Usability**: It is intuitive to use the application in terms of common actions in a emergency room. The use of equipment requires an introduction or explanation from trained personnel so that the person who will use them knows how

to do it because not all the potential users in medical training have access to this type of devices. After receiving this information, it is easy to use and entertaining. Rating: 4.5/5.0.

- **User experience**: It is very satisfying to use the simulator. The user has the opportunity to interact with the patient in an unconventional way but it helps to see the concepts and guidelines from another perspective. The final report requires an expert evaluation at the end of the session. Rating: 4.0/5.0.
- **Immersion**: It is a highly immersive application due to the use of virtual reality devices contrary to what happens with smartphones or tablets in a 2D viewing. The models that are shown quite close to the real structures and the possibility of listening to sounds and navigating with the hands make it an interesting and novel tool for medical learning and gain adherence to AHA guidelines. Rating: 4.8/5.0.
- **Environment**: The environment helps the user to improve the experience and immersion. The visual aspects of the environments like the atmosphere, the realism of the virtual patient, the animations and the quality of the renders, the information and indicators displayed on-screen during the treatment, and how close those environments are similar to a real live hospital facility were highly appreciated. Rating: 4.7/5.0.

The qualifications of the specialists were obtained at the qualitative level and improvements or perspectives of the simulator and an average quantitative grade of 4.5/5.0.

The experts contributed with suggestions to improve the way that the user interacts with the virtual environment, and also suggested changes about medical

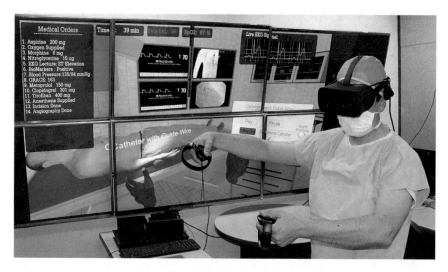

Fig. 4. Sim under test by a cardiologist physician

information that will be applied before the test with the students, for example, some available drugs in the simulator that are not for common use in some countries and the applicable dose range of some others. The most relevant aspect was the ease of being able to follow the treatment suggested by the AHA for the myocardial infarction.

4   Conclusions and Future Work

The VR simulator prototype for myocardial infarction treatment proposed offers an opportunity to asses a cardiologist resident technical expertise following the AHA guidelines. By analyzing the report, timeline is possible to figure out if the pharmacotherapy was performed correctly and on time.

The future development is the integration of rubrics in the simulator, where the users will be able to see the metrics to identify the mistakes and hits of the treatment followed by them. These results will be used to generate a report where the user will get a final score by the hand of tips to improve its performance. Besides, the virtual environment will be improved by enhancing the object's render and the implementation of a full body avatar to increase the feeling of ownership in the virtual world.

Acknowledgements. This research was funded by Universidad Militar Nueva Granada's Research Vice-rectory, under project INV-ING-2633/2018.

References

1. Amsterdam, E. A., Wenger, N. K., Brindis, R. G., Casey, D. E., Ganiats, T. G., Holmes, D. R., et al. (2014). 2014 AHA/ACC guideline for the management of patients with non-ST-elevation acute coronary syndromes: A report of the American College of Cardiology/American Heart Association Task Force on Practice Guidelines. *Journal of the American College of Cardiology, 64*(24), e139–e228.
2. Lüscher, T. F. (2017). Acute coronary syndromes: Early diagnosis, management, and risk prediction. *European Heart Journal, 38*(41), 3037–3040.
3. Mylopoulos, M., & Regehr, G. (2007). Cognitive metaphors of expertise and knowledge: Prospects and limitations for medical education. *Medical Education, 41*(12), 1159–1165.
4. O'gara, P. T., Kushner, F. G., Ascheim, D. D., Casey, D. E., Chung, M. K., De Lemos, J. A., et al. (2013). ACCF/AHA guideline for the management of ST-elevation myocardial infarction: Executive summary: A report of the American College of Cardiology Foundation/American Heart Association Task Force on Practice Guidelines. *Journal of the American College of Cardiology, 61*(4), 485–510.
5. Senior, J., Lugo, L., Acosta, N., Acosta, J., Díaz, J., Osío, O., et al. (2013). Guía de práctica clínica para pacientes con diagnóstico de síndrome coronario agudo: atención inicial y revascularización. *Revista Colombiana Cardiología, 20*, 45–85.
6. Tang, E. W., Wong, C. K., & Herbison, P. (2007). Global registry of acute coronary events (grace) hospital discharge risk score accurately predicts long-term mortality post acute coronary syndrome. *American Heart Journal, 153*(1), 29–35.

7. Westerdahl, D. E., & Messenger, J. C. (2016). The necessity of high-fidelity simulation in cardiology training programs. *Journal of the American College of Cardiology, 67*(11), 1375–1378. http://www.onlinejacc.org/content/67/11/1375.
8. World Health Organization, W., & Organization, W. H., et al. (2016). *The top 10 causes of death.* http://www.who.int/news-room/fact-sheets/detail/the-top-10-causes-of-death.

Game Based Learning

LeARning—An AR Approach

Mihai Albu, Scott Helm, Thanh-Dat Cao, George Miltenburg$^{(\boxtimes)}$,
and Angelo Cosco$^{(\boxtimes)}$

Mohawk College, 135 Fennell Ave W. Hamilton, Hamilton ON L9C 0E5,
Canada
{mihai.albu, scott.helm, thanh-dat.cao, george.
miltenburg, angelo.cosco}@mohawkcollege.ca

Abstract. A remote, instruction-based approach allows the students and teachers to interact in an AR environment that does not require their physical presence in the classroom, nor the physical presence of an object of study. In many laboratory environments, students are present in the same physical location and are instructed to interact with real devices. The aim of this research is to create a general development software tool that allows importing 3D models of laboratory devices; set up specific script actions that students must perform; and allow AR 3D interaction and communication with the instructor. The system allows one-to-one or one-to-many AR communication that will eliminate the barriers of audio calls or manual video-instructed actions. The application has a natural extension in any industrial environment that requires remote assistance with visual interpretation of problems and solutions.

Keywords: Augmented reality · Distance-based learning

1 Introduction

With a general agreement that online learning is an advantageous method over in-class teaching [1], experiential learning needs additional resources. Typical apprentice training will also require in-class presence and exposure to possible failures [2]. Moreover, many experimental lab activities cannot be practiced at home, not even talking about having virtual assistance at all. Many debates propose to make it easier for apprentices to access their in-class learning, with a standard wait time for training, online training options, and support for apprentices in rural and remote communities. Studies showed that augmented reality (AR) can benefit from mental models, spatial cognition, and social constructivist learning theories to improve learning outcome [3]. AR is described as "the fusion of any digital information with physical world settings, being able to augment one's immediate surroundings" [4].

Hands-on augmented reality (AR) classroom experiences are rare. Current technological advances in AR provide massive opportunities for distance-based teaching with hands-on experimental learning. Various AR tools allow loading augmented reality objects into the view scene and interactively manipulating them. Referring to AR learning, [5] described pointed that "there is evidence that specific skills can be improved, that learners were motivated and challenged through the interactive

© Springer Nature Switzerland AG 2019
M. E. Auer and T. Tsiatsos (eds.), *Mobile Technologies and Applications
for the Internet of Things*, Advances in Intelligent Systems and Computing 909,
https://doi.org/10.1007/978-3-030-11434-3_18

problem-solving activities and that the technology offered many opportunities for collaboration". While AR viewers range from basic mobile phones that can load AR applications, to smart glasses systems, to advanced AR devices that allow user interaction, the applicability in-class activities are still reduced. The AR implementation of a class lesson benefits from the lack of physical presence but still includes student–teacher interaction, with the "gaming" manipulation of objects belonging to the lesson without physical danger, plus the correct estimation of a student's performance. Taking advantage of the Unity3D scripting environment, projects can be implemented to allow manipulation of augmented objects that are part of a laboratory/lesson, check that all interactions were performed correctly, and finally display and report the results of the work. On the same approach, audio instructions can be recorded and offered as responses to user interactions. One significant advantage of AR learning is the possibility of "learning by mistakes" (Fig. 1).

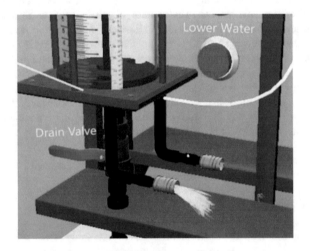

Fig. 1. Simulated failures in a lab experiment.

A student can try as many times to perform the tasks, without the fear of failing, which will increase their confidence and effectiveness. Moreover, several laboratories or lessons may result in dangerous or life-threatening situations. An AR simulation will avoid those, and may even offer examples of such cases to showcase what could happen in negative circumstances.

2   Approach

Electrical Engineers and Instrumentation Technicians study Instrumentation and Control in several courses and in many labs. In a specific laboratory experiment, students are required to follow step-by-step instructions to complete a laboratory experiment, where they may be assisted by an instructor. This project uses Microsoft's HoloLens AR development kit and implements the interactivity with Unity3D, to replace the traditional lab with an AR version. The system requires the import of a 3D model, together with a list of actions to perform in the form of a text-based script. This allows for a multifunctional, interchangeable, and universal approach that can be applied to any 3D model and teaching experiment (Fig. 2).

Fig. 2. Typical experimental AR view

This approach allows not only script-based teaching, but also for remote advising while students perform the laboratory assignments, thereby allowing for distance education. The student can be away from the school, and/or the professor can be away from the class. Users set up the application for students to perform. They may request AI or remote assistance as instructions for the completion of the project. Student interaction allows click/press, click/drag/connect, click/record measurements, and visualize results. Typical mistakes are flagged by the system automatically. Reports are generated upon completion of the experiments, allowing for easier examination of the results. Remote assistance allows for "physical" pointing to parts of the laboratory device. In addition, the students receive audio advice on specific actions to perform.

3   Methodology

The basic approach follows the below pattern of actions:

1. Loading model

The model is loaded as the scene is entered (Fig. 3), but how things are displayed deviates at this point. If working in the sandbox, the model appears right as the scene has finished loading. Given that this was our test environment, where the user's height is not being accounted for, we approximated a default height for rendering the model.

The second path involves spatial perception from within Microsoft's own Mixed Reality Tool Kit; their scripts are intended to help developers make use of the Holo-Lens. However, we found their tutorials cumbersome and often quite outdated, leading to the creation of custom object placement scripts. Using their default settings for spatial perception/understanding, a custom placement script was written to ensure that game objects were rendered at a specified height relative to the floor, so that it would be consistent every time (in retrospect, it may have been better to render it as a percentage distance from the floor to the user's eye line rather than static numbers). Once the user's environment has been scanned by the device, one of two things occurs. We either match the sandbox's default height (if the environment was deemed unsuitable, or was not scanned enough), or every object shifts into its new suitable position and becomes visible to the user at this point.

(a)   **(b)**

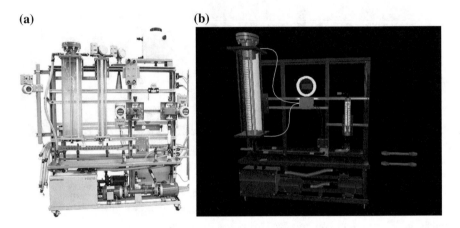

Fig. 3. **a** LabVolt 353x series. **b** 3D model

2. Interactions

We created code that would control which part of the machine could advance through a set of instructions once the current active goal was met. Housed in an array, each instruction was its own object. When the step was advanced, information was pulled from the target object at the current point in the array: text for the user with an objective, and a target object to enable if required for that stage of the experiment. These instructions were generated from a text file (Table 1) that was read through a script in the editor that created a specific "asset" resource on demand. This asset would then be loaded depending on the required experiment to be run (Fig. 4).

Table 1. Script-based elements

Action	Element	Command
Connect pipes	Pipe 1	Drag and drop pipe 1 to low left hose
Connect pipes	Pipe 2	Drag and drop pipe 1 to up right hose
Turn on valve	Valve	Click and rotate left valve 1

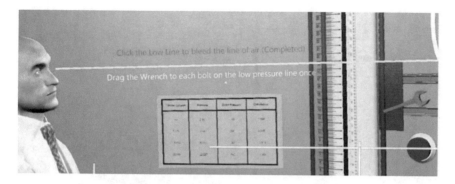

Fig. 4. Instruction-based activities from a script

In the future, this would allow for flexibility where a series of experiments could be generated for the same machine while being housed within one scene. Data is recorded, analyzed, and reported back to the user (Fig. 5), highlighting errors or successful completion of the laboratory task.

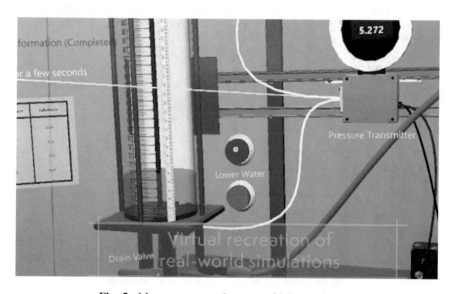

Fig. 5. Measurements and reports of lab experiments

Each object that the user could interact with for the experiment had (at minimum) two scripts on them.

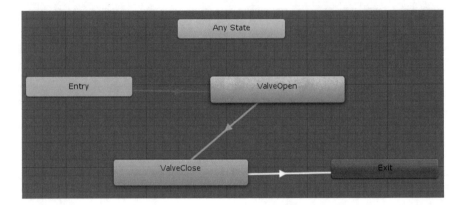

Fig. 6. Animation pane for valve opening

One user altered the color of the part in the question and played a sound when the correct object was interacted with. The other user was responsible for feeding information back to the manager who handled which instruction the user was currently working on, updating it with a "(Correct)" beside it while also advancing one step further in the list, or an "(Incorrect)" mark that kept the user in their current stage (Fig. 6, Sample code 1).

```
/// Activate light source and change the col-
our of the material when the pointer enters
/// the trigger.
public void Enter()
{
    targetMat.color = newMatColour;
    targetMat.EnableKeyword("_EMISSION");
    HoverSounds(); // play the audio clip once
    StartCoroutine(HideText());
}
```

Sample code 1. Allow color interaction to object

We wanted failure states to play an important part of the lab process, both because it made the experience feel more "game-like" but also because we believe making mistakes is an important part of learning—and our choice of technology allows the user to fail in a way that results in no serious consequences.

The scene is designed to be adjustable/scalable for easier user manipulation and to accommodate various interaction styles (Sample code 2).

```
public void AdjustInteractiveSize(float size)
    {
        Vector3 sizeVector = new Vector3(size, size, 1);
        foreach (GameObject thing in exampleManipulable)
        {
            thing.transform.localScale = sizeVector;
        }
        canChange = true;
    }
```

Sample code 2. Resize objects

3. Networking

One of the primary goals of this project is allowing students and instructors to connect to the same lab. This will aid the learning process, where students can work on their unique lab instance, and instructors can switch between labs, providing assistance to the students as needed. As the result, the Water Lab application supports networking capability, which enables connectivity between users.

The networking backend follows a client–server model. The server infrastructure is comprised of a Windows server running Sharing Service from Microsoft's Mixed Reality Toolkit. A server can host many sessions, each of which may contain one or more rooms. The lobby functionality allows clients to select and hop around sessions and rooms, in addition to creating their own room. Once connected to the server, each running instance of Water Lab is registered as a client. In term of scalability, real-life testing shows that a server can support up to five active clients at a time.

In the context of this project, each room can be considered an instance of the lab. When a client creates a new room, a new lab instance is initialized. Each lab instance contains data about the state of that lab, such as water level and valve position. Those data are distributed to all clients in that room to ensure that they all interact with a synchronized lab (Sample code 3).

```
public virtual XString GetRemoteAddress() {
    global::System.IntPtr cPtr = SharingClient-
    PINVOKE.NetworkConnection_GetRemoteAddress(swigCPtr);
    XString ret = (cPtr == global::System.IntPtr.Zero) ? null : new
    XString(cPtr, true);
    return ret;
}
```

Sample code 3. Retrieving remote address procedure

To reduce the load on the server and prevent conflict errors among users, changes made by the user must be submitted manually, via a designated button in the lab. The lifecycle of a room ends when the last client (not necessarily the host) exits the room.

To simulate a real-life collaboration experience, Water Lab assigns every remote client a 3D model as their avatar. Visually, the 3D model shows the client's name, their position and their line of sight (in the form of a white line starting at their head). Based on that, each client in the room has the following data: position of the client within the room, head rotation and cursor position (i.e., where the client is looking).

These data are created when a client joins or creates a room, along with the avatar. While the application is running, data are sent to the server, which processes, and subsequently distributes them to other users. Once it receives the data, Water Lab will update the 3D model of the respective client to reflect the changes. Upon leaving the room, the avatar and all related data to the user are discarded. Since the process happens in real time and continuously, everything is seamless from the user's perspective.

4   Conclusions and Future Work

The implementation of the system allows for remote experimental teaching of laboratory material. The project clearly eliminates any laboratory experiment-based risks while providing a "close to reality" experimental approach to learning. In addition, it allows for "gamification of learning," which is learning by making mistakes, but doing so in a safe, controlled manner.

The final goal of this research is to create a system that allows any laboratory experiment to be implemented in augmented reality in a user-friendly environment.

Acknowledgements. The research was supported by Mohawk College School of Engineering Technology and the Marshall School of Skilled Trades and Apprenticeship.

References

1. Reuter, R. (2009). Online versus in the classroom: Student success in a hands-on lab class. *American Journal of Distance Education, 23*(3), 151–162.
2. Cosco, A. (2018). Mohawk college pilots innovative instructional technology. The Canadian Apprenticeship.
3. Cheng, K.-H., & Tsai, C.-C. (2013). Affordances of augmented reality in science learning: Suggestions for future research. *Journal of Science Education and Technology, 22*(4), 449–462.
4. Fitzgerald, E., Ferguson, R., Adams, A., Gaved, M., Mor, Y., & Thomas, R. (2013). Augmented reality and mobile learning: The state of the art. *International Journal of Mobile and Blended Learning, 5*(4).
5. Luckin, R. (2011). Limitless or pointless? An evaluation of augmented reality technology in the school and home. *International Journal of Technology Enhanced Learning, 3*(5).

Gamifying History Using Beacons and Mobile Technologies

Alexiei Dingli[✉] and Daniel Mizzi

University of Malta, Msida 2080, Malta
{alexiei.dingli,daniel.mizzi.14}@um.edu.mt

Abstract. This project focuses around incorporating multiple elements of gamification (Kapp Few Ideas 13, 2014 [1]) within the capital city of Malta. Often, tourists, or even natives, find it tough to become knowledgeable about a city's history. However, through the implementation of games, anything could be learned much more easily! By using transmitters (Gomez et al. Sensors 12(9):11734–11753, 2012 [2]) scattered around the city, every corner is transformed into an intractable entity. Through an application developed solely for smartphones, players join together to embark themselves on an exciting adventure which takes them around the city. The application consists of a single role-playing game, which introduces interactions with various parts of Valletta, along with some lesser known history facts. Unknowingly, players would be digesting information about the city while both having fun and competing with others. The game is developed specifically in a way to encourage knowledge gathering and understanding. As the player evolves within the game, more knowledge is absorbed through quests. In this virtual world, the player is able to live life as it in was in the past, replicating any skill capable of being done in the real world. Players get stronger as they complete quests and roam around the capital city, slaying monsters that have broken free, and training by themselves. Experience and in-game value are given to players as they level up through the game. More knowledge is gathered, as players unravel the history of the city of Valletta. The main objective, therefore, will be to encourage and lessen the struggle required to learn massive chunks of information. By including a whole virtual world, players synchronize their knowledge gain with their playing experience (Childress and Braswell Dist Educ 27(2), 187–196, [3]). This will require a rich system with all the features possible to make player engaging even easier.

Keywords: Gamification · Beacons · Role playing · Valletta · Learning · Virtual · Knowledge

© Springer Nature Switzerland AG 2019
M. E. Auer and T. Tsiatsos (eds.), *Mobile Technologies and Applications for the Internet of Things*, Advances in Intelligent Systems and Computing 909, https://doi.org/10.1007/978-3-030-11434-3_19

1 Introduction

Beacon technology is becoming extremely important in today's world due to its ever-growing list of benefits. Acting as a stationary transmitter, devices which support Bluetooth Low Energy technology will immediately realize once they are within a certain range. This enables for real-time interaction as a device nears the beacon's radius of action.

Pairing up this technology with a city results in an effective way of exposing tourists or other unaware individuals. Learning historical facts is not always the most fun experience, and hence this would be ideal to spice up the whole process. Not only would this concept increases history knowledge, but it would also increase general knowledge regarding the map, making navigating around the city much more easy and fluent.

Gamification [4] is the concept of introducing game concepts into nongame related contexts. This idea is often introduced to aid in some learning curve and make the whole process easier. For this project, gamification will be adapted in the sense of a virtual world, where the player is able to level up and achieve more as the story evolves.

Hence, this project aims at facilitating the learning curve of the city's historical facts and locations of interest, all the while ensuring a fun and productive experience. At the same time, the project also encourages teamwork and proper coordination. Working as a team is of high importance, especially when facing a boss that one character cannot face alone. The greater the risk, the greater the reward.

To achieve what was mentioned above, a mobile application will be developed through which any individual can register to and login with their personal details. Upon logging in, a user will be able to customize his in-game character as he wishes. Within the game, the player will be able to evolve his character and complete quests, which are based on the history of the city.

The beacons spread throughout the city will act as a medium for the users to be able to interact within the game. Non-playable characters, as well as objects, will all find their place in the game, in both static and dynamic forms. When it comes to static non-playable characters (NPCs) and objects, beacons will be used as a means for the player to communicate. For instance, should the player wish to access his bank, he will simply have to make his way to the nearest banker, and once within the banker's radius, a conversation will initiate. Boasting low cost from both a developer's and player's view, beacons are a perfect fit.

The player is assumed to be using an Android mobile device supporting Bluetooth Low Energy technology. An Internet connection will also be required, in order to provide the full online experience. Last but not least, the device's location system will also be used as a fallback for more accurate position tracking of the player.

Piecing all the information mentioned together, the system will consist of a fun and online experience for all kind of players. The activities within the game are full of action, and hence providing the users with some slight form of exercise at the same time. Players will have the chance to compete with one another, with the main aim of placing first within the ranking ladder.

A fully interactive system which incorporates knowledge of the city through itself will, therefore, be developed. Within this virtual world, players are able to role-play their own character, while uncovering information about the walls of Valletta.

2 Background Research and Literature Review

2.1 Bluetooth Low Energy

Bluetooth Low Energy (BLE) is an emerging wireless technology developed by the Bluetooth Special Interest Group (SIG) to promote short-range communication [2]. When it comes to controlling and monitoring applications, BLE requires an extremely low amount of power. BLE is ahead of its time by making use of a single-hop solution, unlike other low-power wireless solutions, such as ZigeeBee, 6LoWPAN, or Z-Wave [5,6].

BLE Beacons A beacon is a small device (roughly $3\,cm \times 5\,cm \times 2\,cm$ varies depending on the company) that constantly transmits radio signals to nearby smartphones and tablets, containing only a very small amount of information. Both the signal strength and the time taken between each signal transmission can be adjusted to give the desired coverage. Mobile applications are able to listen to the signals being broadcast by the beacons, and once a relevant signal is heard, can trigger an action on the phone. Although phones have this capability, it is not mutual; beacons are not able to read data off of phones. Beacons make use of the previously mentioned Bluetooth Low Energy (BLE). Due to being specifically designed to use very little power and send less data, battery life can range from 1 month to 2–3 years. What is more is that all modern phones can support BLE once the Bluetooth option is turned on. Also, in the same way that BLE uses a small amount of the power of full Bluetooth in the beacon, it uses much less power on a phone, resulting in very efficient usage [7].

Beacon protocols are standards of radio communication, each describing the structure of a data packet beacons transmit [8]. The two top contenders are Eddystone and iBeacon developed by Google and Apple, respectively. Whereas iBeacon is officially supported by iOS devices only, Eddystone has official support for both iOS and Android. Eddystone is also an open protocol, meaning that its specification is available to everyone for public use and viewing. Apple's iBeacon was the first official BLE beacon technology to come out, so most beacons take inspiration from the iBeacon protocol format. iBeacons are enabled in a multitude of Apple SDKs (Software Development Kits) and can be read and broadcast from any BLE-enabled iDevice. Unlike Eddystone, the iBeacon is a proprietary, closed standard [9].

Over the years, beacons have massively increased in popularity, and are gradually being implemented in more places of interest. This is mainly because of all the possibilities one can do with beacons, which are constantly ever growing as people imagine new uses. Currently, beacons are most commonly used for the following:

- Positioning and navigation indoors,
- Tracking of people or possessions,
- Location-based advertisements or notifications,
- Security and automatic locking/unlocking of a computer system,
- Trigger requests such as payments.

As time passes by, more implementations are being advertised, with the US leading the way in beacon integration. Marketing and loyalty are among the most common purposes at the moment, with some examples being:

- Major League Baseball is using beacons within their ballparks to offer fans the ability to check in and possibly get additional content on their own phones,
- San Francisco airport is using beacons to aid blind people in navigation,
- CES 2014 used beacons to simulate a virtual treasure hunt,
- BeHere application allows students to register attendance automatically with their phones in their pockets.

Without a doubt, Bluetooth Low Energy beacons have opened up a new path when it comes to interaction in our daily lives, and it is only getting better day by day [7].

2.2   Gamification

Gamification is the process of game adaptation to a context which does not include any sort of games whatsoever. Typically, gamification applies to nongame applications and processes, in order to encourage people to adopt them. The technique in question encourages people to perform chores that they usually consider to be boring, such as filling in surveys, shopping. By involving some simple game mechanics, a task immediately becomes more interesting. In its simplest form, it means that you get some kind of reward as a result of completing a task. The reward may be something physical such as a trophy or even as vague as just having fun while completing the task [4]. Cechanowicz et al. [10] have concluded from their own conducted research that the motivational benefits of games increase with the level of gamification, and that these benefits apply regardless of the age, gender, and game experience.

The idea of including games into nongame contexts has been around for quite a while now. In fact, when it comes to marketing, many business organizations are discovering and embracing the use of the concepts of gamification for their products. Some examples of gamification in the business world include the following:

- The Samsung Nation [11],
- Jillian Michaels Fitness Program [12],
- Keas [13],
- Progress Wars [14],
- Verizon Insider [15].

All the businesses mentioned above have adopted gamification concepts to highly motivate their users to get more interactive and productive at the same time

[16]. Outside of marketing, gamification can also be used in other areas such as tourism and promoting learning. In the case of e-learning, the basic idea is to uncover content progressively, which has a decent focus on exercises and offers points for correctly solving them. The main goal here is to motivate the students to learn the available material as best as possible in order for them to perform satisfactorily during the evaluation stage and advance successfully through the course [17]. On the other hand, gamification can be used by tourism for marketing, sales and customer engagement, training, productivity enhancement, and crowdsourcing. Compared with other fields, however, the use of gamification in tourism is still in its early stages.

2.3   Localization

Detecting the actual position of an individual on a world map may sometimes prove to be quite difficult. Often, the signal we rely on loses focus, and results in inaccurate data. Such a case could occur especially in the capital city of Malta. For this reason, localization of players will not always depend on a data signal. Instead, beacons can also be used as a fallback in the case of a lost signal. Localization using beacons is a method that has been on the rise, mainly due to beacons offering features such as low cost, and hence being deployable in large numbers.

Chawathe [18] speaks about the various methods of localization available using Bluetooth beacons. Unlike other efforts, the method proposed does not rely on the Received Signal Strength Indication (RSSI) or other indicators of signal strength. Instead, it uses only the visibility factor of the beacon itself. The main method used is called cell-based localization. This method consists of having a beacon set, calculating the intersections based on which beacon is able to see the player.

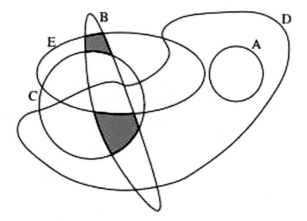

Fig. 1. Cell-based localization: a mobile device that is in the range of beacons B, C, and D but not in the range of A and E must be located in the lower shaded region.

As Fig. 1 shows, by knowing which beacons are able to contact the mobile device, we can conclude an estimate (region) of where the user currently lies. The only drawback, also pointed out in [19], is the time taken in order to finalize the set which is able to contact the mobile device in question. This high latency would hinder the experience, as the location estimate would take too long to update. Hence, multiple solutions are provided, involving the following:

- Localization by Beacon Visibility,
- Localization by Beacon Probes.

Should the abovementioned methods prove to be an issue when implementing, a vast amount of alternatives has also been provided by [20]. Dahlgren and Mahmood [20] aim at using Bluetooth technology in order to provide positioning based on signal strength parameters that can easily be obtained. The positioning techniques mentioned here involve the following:

- Trilateration using RSSI,
- Filter technology,
  - Particle filter for positioning,
- Fingerprinting,
- Cell-based positioning,
- Triangulation,
- Time of flight.

Two tests were conducted, one within a smaller confined space, and the other being in a larger area. Since this project resides in the city of Valletta, we will look into the latter test. Four methods were chosen; iterative trilateration, particle filtering, trilateration, and fingerprinting. According to the results, the fingerprinting approach provides both the best accuracy and precision. The worst approach, on the other hand, was the particle filter approach, giving an average error in estimation close to 7m. These results will, therefore, be taken into account for this project.

2.4 Aim

For this project, the idea of gamification will be adopted to the concept of learning. This will be achieved by making use of the technology discussed previously, that is, the Bluetooth Low Energy beacons. By including the gamification element, users will be more motivated to learn the necessary knowledge aimed within the project.

3 Proposed Solution

The final implementation will include a mobile application through which users are able to login and register using their mobile device. The mobile application will be split up into two main interfaces; the puppet master and the player. The game itself will be a role-playing game, partaking during Valletta's prominent times.

3.1 The Puppet Master

The puppet master is provided with a simple nontechnical and easy-to-use interface through which the game can be modified to full extent. This means that the game can be customized in whatever way the puppet master wishes, and hence being very adaptable should the game require deployment in a different environment. The game master would simply have to edit the game files and adjust the information accordingly. Once the game information has been edited as required, all the game master would have to do is deploy the game. As a result, the puppet master is said to be able to alternate the virtual world within the game, as the story deeply depends on his design choices [21].

3.2 The Player

The player will play himself as a character wandering around the various parts of the city. In this virtual world, the player is able to collect resources, complete quests, and make himself stronger as time goes on. A map is given to the player, outlining the various areas along with the location of static objects and characters. A tutorial will also be given to fresh players, dictating how the game works, along with any other useful game mechanics.

Monsters wander around, constantly posing a threat to the individual in question. It is up to the player to do his best and survive within this virtual world. The possibilities are endless, simulating what one is able to do in real life. Resources spawn at random locations, some harder to obtain than others; the higher the risk, the better the reward.

3.3 The Interface

Figure 1 displays the master panel on the left. Through such an interface, all of the game information can be edited in a simplistic manner; no code involved. This allows any type of individual to customize the game as required, with full control. Briefly, this interface holds the key for the puppet master to alternate the story.

On the right, however, is the player panel. This panel provides game information for the player. Everything involving the game will be achievable throughout this interface, be it talking to characters, slaying monsters, or even purchasing items (Fig. 2).

Fig. 2. The master and player interfaces on the left and right, respectively.

4   Evaluation Plan

The main aim of the system is to help users engage themselves with learning about the history of the city, without boring themselves to death. By implementing the gamification element, users will be able to increase their knowledge while still having fun.

The system will be a successful one if players continue playing the game actively, and at the same time learn the expected knowledge. Hence, when the time comes for an evaluation test to be conducted, its focal point should be that of testing the players' knowledge, in order to understand whether by introducing the gamification element, players actually managed to take in information. Another test to be taken into consideration, therefore, would be to take notice of how long a player stays active for. If players stop playing the game within a few days of registering, only to never play again, this means that the game was not interesting enough, and knowledge is essentially lost.

For the mentioned tests, it would be best to look at the following protocols:

- Question-asking protocol—users are prompted direct questions about the product. Their inability to answer questions gives an indication of where potential problems lie.
- Thinking-aloud protocol—serving as an extension to the above protocol, users are asked to perform tasks and explain their thought process while doing so.

4.1   Testing the Knowledge

In order to evaluate whether knowledge is being absorbed through gamification elements, the question-asking protocol should be utilized on users, identifying to what extent the knowledge was learned. If users receive a relatively low score, this means that the main aim of the game was not achieved as required. The knowledge, which the storyline was based on, was either too vague, or not given enough emphasis.

4.2   Durability of the Game

If the game is not fun enough, and the players feel it is not engaging and interactive, or it is simply plain out boring, the path laid out in order to learn knowledge will immediately fall. Having a good game concept which is enjoyable is the pinnacle of the system. When saying this, multiple factors have to be taken into account as given below:

- Interface design—a good and simplistic yet flavourful design attracts popularity and does not bore players.
- Gameplay—if the main gameplay is stale, players are left with no choice but to stop playing the game.
- Storyline—a good storyline often attracts more players; in this case, an ineffective storyline would result the main aim of this project to fail.

In this case, the thinking-aloud protocol would be used. This will be used to evaluate various parts of the system that users might give more attention to rather than developers. Game players will immediately notice if the storyline is off, if the game interface is buggy and does not always work as intended, or even if the game simply does not feature enough content. Through this protocol, all these factors could be improved upon, adjusting the path through which players should be learning more knowledge about the city.

4.3   Similar Systems

Although systems which make use of the exact same concept are not present, we can still look systems which are somewhat relatable, including their own version of gamification.

Koeffel et al. [22] discusses the concept of evaluating a user experience in games. The paper states currently, few approaches link results of heuristic evaluation methods to user experience. Heuristic evaluation is conducted on several games and the resulting data is compared to user-experience-based game reviews. The paper then assesses its own method and improvements and future perspectives are offered.

Moving on, [23] mentions that there exist three methodological categories for experiences that surround digital games as given below:

- The quality of the product,
- The quality of human–product interaction,
- The quality of this interaction in a given social, temporal, spatial, or other context.

The paper goes on to mention various evaluation methods for assessing individual player experience as given below:

- Qualitative interviews and questionnaires—user feedback is gathered in game development. Surveys can be deployed regarding the actual gameplay.
- RITE Testing—Rapid Iterative Testing and Evaluation specifies data analysis at the end of a testing day once changes to the interfaces or the game design is made.
- Game Metrics Behavior Assessment—the system logs any action the player takes while playing, such as input commands, time taken to travel from one location to another, or even interaction with in-game entities.

Apart from individual player experience, the player context experience should also be looked at. The same paper provides various solutions as follows:

- Cultural debugging—testing conducted to assess how and if cultural conventions are understood properly across different contexts.
- Playability Heuristics—can be implemented quickly and cheaply directly into the game development process.
- Qualitative interviews and questionnaires—assess context and social impact on individual player experience.
- Multiplayer game metrics—the interaction of several players is studied.

4.4 The Plan

The following aspects of the system will be evaluated:

- Players will be asked to test the interface and outlining their thoughts. By combining the question-asking protocol together with the thinking-aloud protocol, any unappealing features will be updated accordingly.
- Knowledge of fresh players will be surveyed before playing the game. As players advance, they will be given a small questionnaire asking them knowledge questions which they should be able to answer given the progress they have covered within the game.
- Interviews will be conducted during gameplay (by using the previously mentioned protocols) to ensure that monotony is not present.
- Game metrics will also be taken into account, providing a clear picture as to where the system might falter.

5 Conclusions

This project aims at making it easier for individuals to learn knowledge about the city of Valletta through the concept of gamification.

5.1   Expectations

The game will be expected to be interactive enough to engage its players properly, providing them a rich experience full of knowledge. By integrating gamification elements, knowledge learning becomes less of a bore to some. This element of gamification also to a certain point adds a decent amount of replay value. A good storyline, a simplistic yet interactive interface and a functional game will all be part of the plan to make the system work as required.

5.2   Challenges

Replay Value Creating a game from scratch is certainly no easy task, especially keeping in mind that the base storyline has to not only attract, but must also boast of replay value. If players are already bored of the game by day one, only a portion of the knowledge intended is learned, if any. Obviously, besides the storyline in itself, game functionality and features must also be taken into account, as all these factors make up a part of a player's replay value.

Evaluation Due to the limitations of such systems presently available, certain aspects of evaluation have to be decided without taking into account other systems. Despite this, factors of similar systems will still be taken into account, and instead adapted to this system.

References

1. Kapp, K. (2014). What is gamification. *Few Ideas, 13.*
2. Gomez, C., Oller, J., & Paradells, J. (2012). Overview and evaluation of bluetooth low energy: An emerging low-power wireless technology. *Sensors, 12*(9), 11734–11753.
3. Childress, M. D., & Braswell, R. (2006). Using massively multiplayer online role-playing games for online learning. *Distance Education, 27*(2), 187–196.
4. Marczewski, A. (2013). *Gamification: A simple introduction.* Andrzej Marczewski.
5. Gomez, C., & Paradells, J. (2010). Wireless home automation networks: A survey of architectures and technologies. *IEEE Communications Magazine, 48*(6), 92–101.
6. Ludovici, A., Calveras, A., & Casademont, J. (2011). Forwarding techniques for ip fragmented packets in a real 6lowpan network. *Sensors, 11*(1), 992–1008.
7. Beacons: Everything you need to know. Pointr blog. Retrieved on December 13, 2016, from http://www.pointrlabs.com/blog/beacons-everything-you-need-to-know/.
8. What is a beacon protocol? can beacons broadcast multiple packets simultaneously? estimote community portal. Retrieved on December 13, 2016, from https://community.estimote.com/hc/en-us/articles/208546097-What-is-a-beacon-protocol-Can-beacons-broadcast-multiple-packets-simultaneously.
9. Understanding the different types of ble beacons—mbed. Retrieved on December 13, 2016, from https://developer.mbed.org/blog/entry/BLE-Beacons-URIBeacon-AltBeacons-iBeacon/.

10. Cechanowicz, J., Gutwin, C., Brownell, B., Goodfellow, L. (2013). Effects of gamification on participation and data quality in a real-world market research domain. In *Proceedings of the First International Conference on Gameful Design, Research, and Applications* (pp. 58–65). ACM.
11. Samsung nation—gamification world map. Retrieved on December 13, 2016, from http://www.gamificationworldmap.com/project/samsung-nation/.
12. The challenges. Retrieved on December 13, 2016, from http://www.jillianmichaels.com/fit/the-community/the-challenges.
13. Product: Keas population health management. Retrieved on December 13, 2016, from http://keas.com/product/.
14. Progress wars. Retrieved on December 13, 2016, from http://www.progresswars.com/.
15. Verizon fios & custom tv—internet, cable & phone. Retrieved on December 13, 2016, from http://www.verizon.com/home/verizonglobalhome/ghp_landing.aspx.
16. 10 amazingly successful examples of gamification—create your customer community with crezeo. Retrieved on December 13, 2016, from http://crezeo.com/blog/10-amazingly-successful-examples-of-gamification/.
17. Xu, F., Weber, J., & Buhalis, D. (2013). Gamification in tourism. In *Information and communication technologies in tourism 2014* (pp. 525–537). Springer.
18. Chawathe, S. S. (2009). Low-latency indoor localization using bluetooth beacons. In *2009 12th International IEEE Conference on Intelligent Transportation Systems* (pp. 1–7). IEEE.
19. Chawathe, S. S. (2008). Beacon placement for indoor localization using bluetooth. In *2008 11th International IEEE Conference on Intelligent Transportation Systems* (pp. 980–985). IEEE.
20. Dahlgren, E., & Mahmood, H. (2014). Evaluation of indoor positioning based on bluetooth smart technology. *Master of Science Thesis in the Programme Computer Systems and Networks.*
21. unfiction.com blog archive undefining arg. Retrieved on December 13, 2016, from http://www.unfiction.com/compendium/2006/11/10/undefining-arg/.
22. Koeffel, C., Hochleitner, W., Leitner, J., Haller, M., Geven, A., & Tscheligi, M. (2010). Using heuristics to evaluate the overall user experience of video games and advanced interaction games. In *Evaluating user experience in games* (pp. 233–256). Springer.
23. Nacke, L., Drachen, A., & Göbel, S. (2010). Methods for evaluating gameplay experience in a serious gaming context. *International Journal of Computer Science in Sport, 9*(2), 1–12.

A Framework for Preadolescent Programmers to Create Cooperative Multiplayer Reading Games

Christopher Kumar Anand$^{(\boxtimes)}$, Curtis d'Alves, Yumna Irfan, Biya Kazmi,
Stephanie Koehl, Stephanie Lin, Christopher William Schankula,
Chinmay Jay Sheth, Pedram Yazdinia, and John Zhang

McMaster University, Hamilton, ON, Canada
anandc@mcmaster.ca

Abstract. Many schools are transitioning to 1:1 iPad use, starting in the primary grades. These devices have many capabilities which create the potential for new types of social learning. Our goal in this project is to demonstrate that, with the right framework and instruction, middle-school children can create an interactive and collaborative mobile learning game, learn related physics and mathematics, reason using fundamental computer science concepts and structure team-based problems for younger children. Through making a multiplayer game, students will build relationship skills, social awareness, critical thinking skills and display their creativity. By building an educational game, they will see that software is an effective tool for solving social/cultural problems. Instead of taking a test, our students create a game to demonstrate their mastery of computer science. **Topics** Dynamic learning experiences, Interactive and collaborative mobile learning environments, Game-based learning.

Keywords: Reading game · Mobile game · Multiplayer game · Cooperative game · Social learning · Newtonian physics

1 Introduction

Our Computer Science Outreach program at McMaster University has been hosting workshops and visiting local schools for over a decade, and we have developed several iPad apps and low-tech interventions to explain the Information Revolution to K-8 children. A few years ago, we started an experiment teaching Functional Programming with Elm, focussed on graphics. This approach was much better received by children and educators than previous approaches to teaching programming. Our explanation for this success is that it builds Algebraic Thinking [6], on top of a natural childhood activity, namely drawing.

This year, we decided to try leveraging other successful ideas from the field of education, and we will now describe our adoption of Jigsaw, an approach to

© Springer Nature Switzerland AG 2019
M. E. Auer and T. Tsiatsos (eds.), *Mobile Technologies and Applications for the Internet of Things*, Advances in Intelligent Systems and Computing 909,
https://doi.org/10.1007/978-3-030-11434-3_20

building social cohesion and teamwork in schools. Rather than digitizing the Jigsaw approach—an effort worth pursuing—we have adapted our web-based IDE to allow teams to cooperatively build levels of a multiplayer, cooperative reading game. Different team members have different design responsibilities, but must work together to make a playable game. Since game objects collide elastically, we also developed a series of very simple programming exercises to develop an understanding of velocity and acceleration in the plane.

But before we describe our framework, and the evaluation of its first implementation in a summer camp, we describe the language Elm, our graphics library and the Jigsaw method for readers who are not familiar with them.

Elm Programming Language: Elm (http://elm-lang.org) is a language designed for the development of web application front ends [3]. The language syntax is deliberately simple and is essentially a subset of the Haskell syntax, excluding, for example user-defined type classes. While being strictly typed just like Haskell, the Elm compiler also forces programmers to follow best practices while developing code, by not accepting incomplete case coverage in case expressions. It also uses a model-view-update paradigm that keeps pure code separate from code with side effects without the need for monads.

Elm code compiles down to JavaScript which provides many practical advantages, such as the ability to easily create web tools for the language and the ability to run on any device with a web browser, avoiding the need to navigate school board policy about the installation of software on school computers.

While at first, it may seem that this type of language is more suitable for expert users, many of the features useful to experts (strict types, pure functions) are equally useful to beginners. In addition, Elm's simple syntax and purity (i.e. lack of side effects) matches what students have learned or will learn about algebra. The Bootstrap program makes use of the latter fact [10].

GraphicSVG Library: Our graphics library is structured as a domain-specific language (DSL) with two main types, meant to match students' intuition from visual arts. A `Stencil` functions as a physical stencil. We provide familiar and novel constructors:

```
circle, square, triangle: Float -> Stencil
rect, oval, wedge: Float -> Float -> Stencil
ngon: Int -> Float -> Stencil
roundedRect: Float -> Float -> Float -> Stencil
line: (Float, Float) -> (Float, Float) -> Stencil
polygon: List (Float, Float) -> Stencil
text: String -> Stencil
```

Unlike many graphics library types, `Stencils` do not have an associated fill or outline colour, and thus cannot be drawn to the screen, in the same way that a plastic stencil does not change the student's paper without the use of a pen or pencil. In order to make a visible shape described by a `Stencil` on to the screen, a `Stencil` must be converted into a `Shape msg`, where msg is a type parameter specifying a user-defined message type required for interaction. Just as in the physical world, there are two ways of doing this:

```
filled: Color -> Stencil -> Shape msg
outlined: LineType -> Stencil -> Shape msg
```

Once a Shape is created, students can apply familiar transformations:

```
move: (Float, Float) -> Shape msg -> Shape msg
scale, rotate: Float -> Shape msg -> Shape msg
group: List (Shape msg) -> Shape msg
```

Elm's left-to-right pipe function (|>), analogous to a Unix pipe, allows students to write these transformations in a modular fashion with a low syntax barrier:

```
circle 10
   |> filled red
   |> move (10,10)
   |> scale 1.3
```

This is equivalent to scale 1.3 (move (10,10) (filled red (circle 10)), but is much more readable. This pipe operator has been explored by other researchers, including [9], who developed a visual representation of the operator for students learning functional programming.

This is the first example of how Elm's type system enabled our library to leverage students' pre-existing knowledge about drawing to learn how to use the DSL. It is worth noting that they do not need to have explicit knowledge of algebraic data types or Elm's type system to get started or even go far with the library. Instead, the words Shape and Stencil themselves are used as a teaching tool to help them understand which functions to apply when. It also has the advantage of producing legible error messages when, for example, a student forgets to fill in a Stencil.

The GraphicSVG package is described in more detail in [4].

Online Mentoring and Compilation System: We have a web interface that allows students to compile Elm graphics in a web browser. Students are assigned logins and can save their work in different slots, chosen from the login screen, see Fig. 1. Students can also chat with mentors online (partly visible in the bottom right corner of Fig. 1), who can also view their code and point the students in the right direction.

Related Work: Jigsaw Teaching Method: The jigsaw teaching method was first introduced in 1971 by psychologist Dr. Elliot Aronson [2]. Prior to 1954, public schools were separated by race; however, after a ruling by the Supreme Court of the United States, it became a legal requirement for public schools to be integrated. This created many difficulties (such as discrimination, fighting, hate crimes, etc.) which caused students to feel unsafe, which in turn caused problems for teachers and harmed the students' abilities to learn. Dr. Aronson was brought into a school district in Austin, Texas in hopes of weakening the racial cliques. He wanted to create a learning environment where students had to pay attention to each member, thus making everyone in the group important. His method also teaches students to rely on each other rather than compete with each other, reducing the competitive nature between students.

The jigsaw teaching technique can be broken down into the following general steps:

1. Divide the subject material into five or six segments and divide students into groups of the same size (i.e. if the subject material is split into five parts, then there would be five students per group).
2. Assign each student to learn one segment, ensuring that they have direct access only to their own segment.
3. Once students have had enough to go over their segment and become familiar with it, form temporary 'expert groups'. These groups are formed by grouping students together based on the segment they have been assigned (i.e. each member of the new group has been assigned the same topic).
3. Have the students discuss the main points of the segment and rehearse the presentations that they will give to their original (jigsaw) groups.
4. Bring students back to their jigsaw groups and have each student present their segment to the group (group members are able to ask questions for clarification if necessary).
5. Finally, test the students on the material they have just learned, ensuring the test material is distributed relatively evenly between the segments.

Multiple studies have shown that classrooms using the jigsaw teaching technique showed a decrease in prejudice and stereotyping and a reduction in the number of absent students. They also showed that students had higher self-esteem, performed better on standardized exams, and liked school more than those in classrooms using traditional techniques [1].

2   Design

Camp: To implement the multiplayer game, we advertised a summer camp to students who had completed an 8-week in-class workshop using ElmJr, our projectional editor for iPad. As a result, they had experience writing Elm graphics code, but no experience of syntax and type errors. Fifteen children attended the camp, which ran for 5 days, from 9 a.m to 4 p.m.

Multiplayer Game Architecture: To allow multiple children to contribute to a single game, we created predefined tasks and distributed them among five team members. After team members type in their username, the login screen enables buttons for their assigned tasks, see Fig. 1 (left). For our experiment, we gave campers additional login credentials assigned to one of the team roles. In the future, we will allow teachers to assign roles.

Fig. 1. *Left* login: having entered the name for member four of our test team, the slots for the player, map quadrant, key, barrier, wormhole, recycle bin and switch are enabled, in addition to the first row of slots: picture, game, Wordathon, orbs game template, maze game template and physics (available to everyone). *Right* editing a slot with a map quadrant, showing the code on the left, the compiled code, which in this case displays the map holder interface, with the barriers created by this player visible, the footer pane (empty) and the help pane. Being the fourth player, this player edits the south-east quadrant.

The design was influenced by our existing 'slot' system, in which programming activities were divided into types of growing complexity, with ten slots of each type, mimicking the way that many game systems allow players to save a fixed number of games and assign them to slots. On the server side, each slot type as code fragments for an invisible header, an initial version of the editable code, and a visible footer. Every time the child clicks the compile button, the edited code is sent to the server, where it is concatenated with the header and footer, saved to the file system, and compiled in a directory with the required manifest, and a pre-populated directory of library packages. If the compilation is successful, the resulting JavaScript program is displayed in the second pane, otherwise the compiler errors are displayed, see Fig. 1 (right). Third and fourth panes contain the uneditable footer code and a chat window where children can ask for help, and receive answers from mentors who see the same code, in the same interface as the children, and who can send backlinks to differences between the child's code and their modified version of the code.

Slot System Design: Fig. 2 shows the dependency diagram for slots. By leveraging slots, we enable staged programming. Each slot starts with functional code, and every time a child's code compiles and passes basic tests, that code is saved for inclusion in downstream modules. All of the slots other than the map quadrants are included in the map quadrants, and all slots are included in the final game. Currently, the map quadrants do not include each other, but we will modify the build system so that the latest passing versions of MapNW, MapNE, MapSW are included in MapSE, and vice versa. Children had trouble figuring out how their map quadrant would fit in with the other quadrants, especially if they did not have a good overall paper design to work from, or were deviating

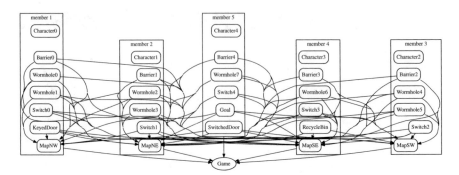

Fig. 2. Each child has a set of slots they can edit. Four of the children on each team can edit a map quadrant, and see the results of the other slots. All of the slots get compiled together into the game.

from that plan. When the map quadrant code compiles, we display the map holder interface, in which the full map is shown, as well as a movable inset is drawn at half the scale of the normal player view. The inset can be used to inspect the map, since even at half-size, it represents a fourfold magnification. This map holder view differs from the game version by (1) not including the word animation, (2) not including players and (3) including letters 'L' and 'P' for letter and player spawners. Spawners are where these objects enter the game board.

As Fig. 2 shows, each team member has specific responsibilities, which have been distributed in order to require cooperation. For example, teams are likely to want to use a common design language for the barriers, wormholes, etc. In addition, wormhole and switch functioning make coordination even more essential. Each quadrant has two optional wormholes, and each wormhole can target any other wormhole. If member 1 places Wormhole1 targeting Wormhole2, they are relying on the fact that member 2 will place Wormhole2 (which is optional) in an appropriate position. To achieve an interesting game, the team needs to work together to make the wormhole network work.

Client: A single Elm client app presents a login screen, from which children can log in as the map holder or a left- or right-handed player. Only one child can log in as the map holder, and once they are accepted, they see a list of players who have logged on, so they know when to start the game. During gameplay, the map holder sees a level map, with which to coordinate the other team players who each have their own player-centered view together with controls for acceleration and braking, see Fig. 3. To simplify the build system, and allow new levels to be created without modifying the server, the level design is compiled into the client, and is transmitted to the server from the client used by the map holder, at the time of login. The child-created drawing functions are compiled into the client, but so are the boolean functions which determine whether doors are open or closed. The map holder client determines which switches have players on top of them, evaluates the door functions with this information, and transmits the

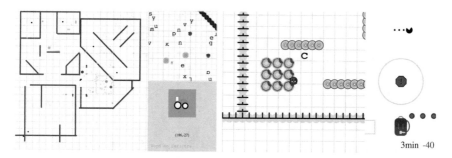

Fig. 3. *Left* one player coordinates the team. Their interface includes the full map, and a magnified view with the same extent as a player view, to help them understand what their team is seeing. *Right* all other players have a player-centric view. On the right there are controls; the stop sign is to brake, and closer proximity of the player's click to the outer circle increases acceleration.

result to the server which uses this information in calculating collisions, and which retransmits it to the other clients.

Server: The server is written in Haskell using the Warp application server framework [11]. Warp takes advantage of low-overhead green threads, and we use TChan channels from the STM package [8] to handle communication between threads. Currently, we only handle one game at a time, and there is a single thread to hold the game state and do collision detection and handle other game mechanics. In the future, we can easily support one thread per concurrent game. When a client first contacts the server, we create a thread to decode messages from the server and a thread to encode messages to the server. In our case, these threads do no other work, because all work involving the game state is handled in the main thread. Communication is over WebSockets using text messages, because this is well supported by both Warp and an Elm package.

Physics Challenges: To incorporate physics learning into the camp, we created physics challenges which allowed them to progressively discover position, velocity, acceleration, projectiles, gravity on the earth and the moon and basic collision as shown in Fig. 4.

3   Results

Fifteen campers created four levels of a multiplayer reading game. On a technical level, this was a success, although we noted areas for improvement. The build system allowed campers to develop modules independently, and linked them together seamlessly. On a social level, we identified a number of ways we need to scaffold teamwork in the future, so that the campers take better advantage of the opportunities for teamwork.

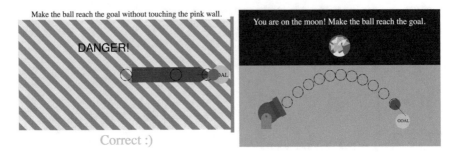

Fig. 4. Physics slots allowed campers to discover Newtonian mechanics in two dimensions. The challenge on the left required them to use a negative acceleration so the ball would stop and turn just in time to reach the goal but not hit the wall. The challenge to the right required them to use a negative acceleration (downwards), mimicking gravity on the moon.

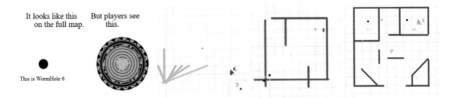

Fig. 5. *Left* an example Wormhole, showing a good use of colour. Most of the wormholes had rotating elements, which helped this wormhole achieve a score of 4/5. *Right* three different map quadrants, scored as 1/5, 3/5 and 5/5, respectively. Note that, the final map quadrant includes several goals arranged in an inverted 'U'. This shows that the camper thought about the difficulty of getting letters into the goals, and used the map design to make it easier.

Critical Thinking/Problem Solving/Discovery Thinking: Learning to program and learning to create graphics within the limits of a graphics library are both challenging activities, but two activities required children to think another level ahead. The first was in the design of their map quadrants to take into account the difficulty in manoeuvring letters on the field and the second was in using a stepping stone to higher order functions:

```
barrierOneTwoEtc : BarrierNumber      -- barrier to be replicated
                -> Float              -- bounciness of barriers
                -> (Float,Float)      -- position of first barrier
                -> (Float,Float)      -- position of second barrier
                -> Int                -- number of barriers
                -> List (BarrierNumber,(Float,Float),Float)
```

Campers who made complex map quadrants did take playability into account, as in the clustering of goals in the magnified section of Fig. 3. However, no campers figured out how to use barrierOneTwoEtc without one-on-one explanations. We had expected that they would figure it out from the name and provided examples.

Creativity: In Fig. 5, we see a typical design panel on the left, showing what the camper would see when a wormhole slot compiles. The text appears to explain to the camper that each object has a summary view for the full map, and a detailed view for the players and inset views. On the right of Fig. 5, we see map quadrants scored as 1, 3 and 5, respectively. The level of creativity did not seem to be affected by the complexity of the starter code. The starter player (the first screenshot in Fig. 6) was complex, but the remaining screenshots demonstrate the same range in creativity as in other slots.

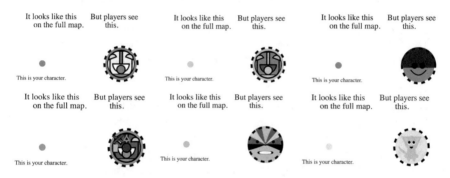

Fig. 6. These are the characters that some of the students created. They were rated based on creativity and aesthetics. The top left character was scored as 0/5 points as it was the default character. Top middle: 1/5, top right: 2/5 , bottom left: 3/5, bottom middle: 4/5, bottom right: 5/5.

Engagement: In Fig. 7, we see that engagement was consistent across activities, from the physics slots to the creative slots, but where completion was clear (physics) or the task was most open-ended (maps), the most-engaged campers excelled.

Computer Science Concepts: Thirteen of the students had previously attended 7 or 8 weeks of workshops using the iPad app ElmJr. ElmJr is a projectional editor, rather than a textual editor—context-sensitive menus allow the user to transform one working Elm program into another. As a result, the campers had some familiarity with Elm programs, but no experience with type and syntax errors. They seemed to quickly adapt to this new environment and were able to fix errors as they arose. It seemed to the experienced instructors that they required less help with syntax and type errors than expected. A more

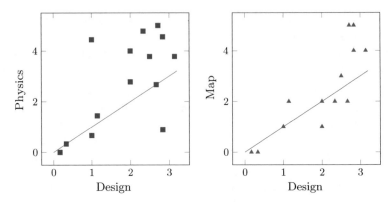

Fig. 7. To measure engagement, we scored all of the slots on a 0–5 scale. In the plot on the left, we see that the average score for the physics slots is correlated with the average score on design slots, but there is a large group of students who did much better on the physics slots. In the plot on the right, we see that the average map score also correlates well with the average design score, but there were a smaller number of campers who really put a lot of effort into their map quadrants.

subtle difference between these children, and older children, is that they are used to typing on glass screens and not on keyboards. Their typing was slow, but they did not express frustration with this. If anything, they were excited to be working with 'real' computers in a novel university lab environment.

Other concepts which they used fluently: types, functions, the use of time as an input variable in creating animations, chaining functions together with pipes and grouping of shapes to make compound shapes.

They did not use boolean logic although every door took a function

```
isOpen : (Bool,Bool,Bool,Bool,Bool) -> Bool
```

as an input. In previous workshops, children frequently ask about such functions, so we were surprised that none of the campers used them, but this may be because effectively using them required collaboration, since, by design, the switches were distributed among different map quadrants.

Physics Learning: Programming challenges took much less time and required less explanation than we expected.

Teamwork: All of the teams set out to work together on their levels. Although the build system could have allowed it, we did not show teammates' progress in the map quadrant design, see the right half of Fig. 1. Campers found this frustrating, and you can see that as a result, the different quadrants on the left side of Fig. 3 do not line up well. But the larger issues were social. Some campers did not have strong team-working skills and were sensitive to criticism. In the future, we will build in 'brainstorming' sessions where every camper is expected to contribute. This will help all of the campers take ownership for the level, and make compromise easier. We will also give campers a checklist, with things like

'Do your wormholes connect where they need to? Does the destination wormhole exist on the map?'

Social Awareness: Although we did get positive comments about making friends at camp, we do not have data on social awareness. Earlier in the year, we did create a visibly positive experience when 12-year-old children presented single-player reading games to 7-year-old children. The older children were among classes from 8 schools collaborating on the games, and our instructors and observing teachers were impressed by the excitement and positive attitudes of both groups. We did not have a younger group of children to whom to present the multiplayer game, but we are confident that in a future iteration, we will be able to arrange this, and we will use surveys to quantify the change in social awareness in the older children, and the ambitions of the younger children.

4   Discussion and Future Work

We have demonstrated the feasibility of having teams of 10-year-old children work together to create games, and learned a lot by observing the joys and frustrations of the children.

Other researchers have demonstrated the value of online exercises in **physics education** [5,7], but we were still surprised by the acceptance of these tasks by the students. We will build on this success by developing more challenges, including additional physics concepts and combining concepts in interesting ways. We will also work with physics teachers to develop paper tests to determine whether the children are really learning physics, or learning to game the system, but our feeling is that they are really learning physics by building on the familiarity with Cartesian coordinates gained by programming graphics.

Although the structure of the game design and gameplay both required **teamwork**, the campers did not take full advantage of this, and in the future, we will add offline team-building activities, have them play a pre-constructed game and discuss the team strategies in this context, and have them develop and present to the camp a project plan which explains how their individual contributions will be woven into a whole.

Some campers were dismayed by noticeable network delays. Given the high quality of console and mobile games, we need to create **fluid gameplay** by reducing network overhead and compensating for variable delays in message delivery. The lab computers were underpowered virtual clients, and many school environments have older equipment and network issues, so we need to work in these situations. Our first step will be to add timestamps to messages so that clients can compensate for variable transmission delays. Our second step will be to use more efficient text-based encoding of numerical data transmitted via websockets. If these steps are insufficient, we will switch to binary data using an analogue of User Datagram Protocol (UDP).

References

1. Aronson, E. (2002). Building empathy, compassion, and achievement in the jigsaw classroom. In *Improving academic achievement: Impact of psychological factors on education* (pp. 209–225).
2. Aronson, E., & Bridgeman, D. (1979). Jigsaw groups and the desegregated classroom: In pursuit of common goals. *Personality and Social Psychology Bulletin, 5*(4), 438–446.
3. Czaplicki, E. (2012). *Elm: Concurrent FRP for functional GUIs*. Senior thesis, Harvard University.
4. d'Alves, C., Bouman, T., Schankula, C., Hogg, J., Noronha, L., Horsman, E., et al. (2018). Using Elm to introduce algebraic thinking to k-8 students. In S. Thompson (Ed.), *Proceedings Sixth Workshop on Trends in Functional Programming in Education*, Canterbury, Kent UK, June 22, 2017. *Electronic Proceedings in Theoretical Computer Science* (Vol. 270, pp. 18–36). Open Publishing Association.
5. Graf, S., & List, B. (2005). An evaluation of open source e-learning platforms stressing adaptation issues. In *Fifth IEEE International Conference on Advanced Learning Technologies, 2005, ICALT 2005* (pp. 163–165). IEEE.
6. Kieran, C. (2017). *Teaching and learning algebraic thinking with 5-to 12-year-olds: The global evolution of an emerging field of research and practice*. Springer.
7. Kortemeyer, G., Kashy, E., Benenson, W., & Bauer, W. (2008). Experiences using the open-source learning content management and assessment system lon-capa in introductory physics courses. *American Journal of Physics, 76*(4), 438–444.
8. Marlow, S. (2013). *Parallel and concurrent programming in Haskell: Techniques for multicore and multithreaded programming*. O'Reilly Media, Inc.
9. Mongkhonvanit, K., Zau, C. J. Y., Proctor, C., & Blikstein, P. (2018). Testudinata: A tangible interface for exploring functional programming. In *Proceedings of the 17th ACM Conference on Interaction Design and Children* (pp. 493–496). ACM.
10. Schanzer, E., Fisler, K., Krishnamurthi, S., & Felleisen, M. (2015). Transferring skills at solving word problems from computing to algebra through bootstrap. In *Proceedings of the 46th ACM Technical Symposium on Computer Science Education* (pp. 616–621). SIGCSE '15, ACM, New York, NY, USA (2015). https://doi.org/10.1145/2676723.2677238.
11. Snoyman, M. (2018). *Warp: A fast, light-weight web server for WAI applications*. https://hackage.haskell.org/package/warp-3.2.23.

Using an Augmented Reality Geolocalized Quiz Game as an Incentive to Overcome Academic Procrastination

María Blanca Ibáñez[(⊠)], Jorge Peláez, and Carlos Delgado Kloos

Universidad Carlos III de Madrid, Madrid, Spain
{mbibanez, cdk}@it.uc3m.es, jorpelae@pa.uc3m.es

Abstract. Academic procrastination is associated with low academic achievement, thus it is considered an important challenge in education. Lack of self-regulation and motivation are among the main problems associated with procrastination. The aim of this study was to determine the effectiveness of a quiz game in Java programming to avoid procrastination. The sample was 45 students, who were randomly assigned to 2 groups. The control group used a Web-based tool and the experimental group used an augmented-based tool. The finding of this study suggests that the game activity embedded in the augmented reality tool helped the students to avoid procrastination.

Keywords: Academic procrastination · Augmented reality · Self-regulation

1 Introduction

The failure rates in programming courses at the university level are evidence to the fact that learning to program is a difficult task. Students that fail are not likely to register for a follow-on programming course. Besides, these students are not likely to try harder in other courses to get a good cumulative gross product average which tends to affect their capability in their field of concern [1]. In spite of research on factors that influence the enrollment and success of students in programming, it is still not fully understood what makes computer programming difficult and frustrating for some students.

Researchers emphasize the importance of self-regulation to improve academic achievement [2]. Self-regulation of learning behavior among students is associated with higher student grades and long-term retention [2]. On the other hand, procrastination—behavioral tendency to postpone tasks or decision making—is a central facet of conscientiousness and indicative of self-regulatory limitations associated with low academic achievement [3]. Therefore, academic procrastination is considered an important challenge in education [4].

Researchers agree on the significant effect of motivation in self-regulation and procrastination. Baumeister et al. [5] stated that the four components of self-regulation theory are (1) standards of desirable behavior, (2) motivation to meet standards; (3) monitoring of situations and thoughts that precede breaking standards, and (4) willpower, or the internal strength to control urges. Besides, high procrastination is associated with a low degree of self-determined motivation [6].

© Springer Nature Switzerland AG 2019
M. E. Auer and T. Tsiatsos (eds.), *Mobile Technologies and Applications for the Internet of Things*, Advances in Intelligent Systems and Computing 909, https://doi.org/10.1007/978-3-030-11434-3_21

In the educational arena, learning technologies are tools for mediating the practice of learning [7]. On one side, online learning makes education accessible to all and might support traditional learning [8]. On the other side, the introduction of technologies with a disruptive potential such as augmented reality (AR) has brought changes to common education practice [9]. Among the most promising affordances of AR technology is its possibility to motivate and engage students in learning activities [10, 11].

In response to the aforementioned issues, this study attempts to assess the effectiveness of an AR-based platform element in comparison to a Web-based platform to promote an early engagement in learning activities, and consequently overcome academic procrastination. The activities were carried out by first-year engineering students in a computer-programming course and they were designed according to the curricular objectives of the course.

2 Background

2.1 Procrastination, a Challenge in Education

Procrastination is the lack or absence of self-regulated performance and the tendency to put off or completely avoid an activity under one's control [12].

Sendcal et al. [13] have suggested that academic procrastination is a motivational problem that involves more than poor time management skills or trait laziness. Procrastinators are difficult to motivate and, therefore, are likely to put off doing school assignments and studying for exams until the last possible moment [14]. They may have difficulty acquiring new knowledge if steps are not taken to enhance their motivation [6]. Consequently, it might be useful to provide students with activities that push them to study in a motivating way.

2.2 Location-Based Augmented Reality in Education

Augmented reality is a 3D technology that enhances the user's sensory perception of the real world with a contextual layer of information [15]. AR has become a popular topic in educational research in the last decade [16, 17], mainly due to the availability of low-cost handheld devices with innovative features that allow the deployment of AR-based applications.

Applications based on this technology rest on three pillars: tools to track information about real-world objects of interest; hardware and software to process information; and devices to show the user the digital information integrated into the real environment [18, 19]. AR technology is often described with reference to its two predominant modes of tracking information from the physical world. The first mode is location-based AR, which makes use of a device's GPS to identify locations at which computer-generated information should be superimposed. The second mode is image— or maker-based AR, which requires recognition of a marker or specific object to bring up digital information.

There have been few but successful experiments demonstrating location-based AR usefulness in the education arena. Most of location-based AR interventions deploy scientific discovery learning strategies where learners should construct knowledge by themselves [20, 21]. Pokémon Go, the interactive mobile game launched in July 2016 has inspired some AR-based learning environments that combine augmented reality with gamification techniques [22, 23]. On the other hand, image-based AR has been used more extensively in education than location-based AR and result suggest that promotes motivation [10, 24], engagement [11], and better learning outcomes [25]. While studies suggest that augmented reality may be useful for the purposes of our research, to the best of our knowledge, these studies are one-time interventions. Therefore, they do not explore the AR possibilities to motivate students to persist in learning activities.

3 Method

3.1 Participants

This study was conducted on two sections of a programming undergraduate level course in spring 2018. The sample comprised 45 students (aged 18–25, $M = 19.75$, $SD = 1.82$). Among the study participants, 13 out of 45 were female and 32 out of 45 were male. Students had basic computer skills but they never had used an augmented reality application before.

3.2 Procedure

Students received traditional instruction from their teachers and completed a knowledge test. Before starting the intervention, students were assigned do one of the two groups: control (CG) or experimental (EG). The intervention lasted five days. Students' daily interactions with the tools were logged for the purposes of this study. After the completion of the interventions, the students completed a knowledge posttest questionnaire (see Fig. 1).

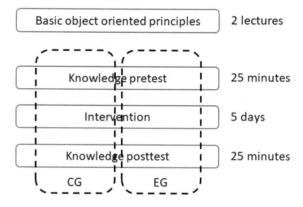

Fig. 1. Procedure of the experiment

3.3   Materials

In the experimental group, students used Multiple quEstions AcTivity Augmented Reality (MEgsTAR), an AR-based learning environment where students had to catch three digital animated characters placed in three points of interest (POIs) distributed throughout the campus of the University. In order to catch an animated character, students had to answer correctly four multiple-choice questions in a POI (see Fig. 2). Augmented reality technology was used to augment the campus with the characters and the questions students had to answer.

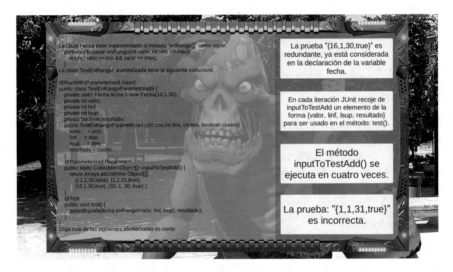

Fig. 2. Campus of the University augmented with a digital character to catch by answering correctly a Java question

MEgAsTAR was implemented using Vuforia within the Unity 3D SDK. The architecture has an interaction engine coupled with a workflow engine and an assessment engine (see Fig. 3). The interaction engine triggers an event when the student arrived at a point of interest and recover a character from the Media Assets repository and a question from the Questions database, then both informations are displayed into the student's smartphone. When the student chooses her answer, the workflow engine sends to the assessment engine the answer that then is validated with the information stored into the Questions database. Every event is stored by the interaction engine into the Logs database for the purposes of this study.

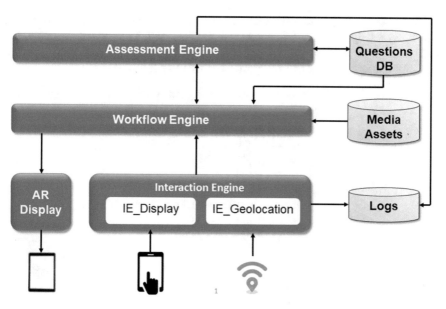

Fig. 3. MegAsTARs architecture

The control group used a Web version of MEgAsTAR which works on iOS tables or mobile phones. The Web version of MEgAsTAR included all the MEgAsTAR functionality but does not have its AR features.

3.4 Measurement Instruments

To ensure that students of both groups were representative sample, a pretest was conducted and analyzed. The pretest was comprised of 10 multiple-choice questions, each worth one point. Questions were addressed to assess their knowledge of basic principles of programming. In that follows, we present an example pretest question.

"Q. What is the output of the following program?.

```
class C {
    public static void main(String args[]) {
        int array_var[][] = {{ 1, 2, 3}, { 4 , 5, 6}, { 7, 8, 9}};  int s = 0;
        for (int i = 0; i < 3; ++i) {for (int j = 0; j <  3 ; ++j) {s = s + array_var[i][j];}
        }
        System.out.print(s / 5);
    }
}
```
Possible answers:
(a) 8
(b) 9
(c) 10
(d) 11"

To assess the difference on effectiveness of learning for control and experimental groups, knowledge posttest was conducted and analyzed. The posttest was comprised of 10 multiple-choice questions about object-oriented basic concepts, each worth one point. An example questions from posttest is listed as follows:

"Q. What is the value of "s" and "t" in "***"?

```
public class A{
    public static int s = 0;
    public int t = 0;
    public A() {t=s; s++;}
}
public class Main {
    public static void main(String[] args) {
        A a1 = new A();
        A a2 = new A();
        // ***

    }
}
```

Possible answers:
(a) a1.s=2, a1.t = 0, a2.s = 2, a2.t = 1
(b) a1.s=1, a1.t = 2, a2.s = 2, a2.t = 2
(c) a1.s=2, a1.t = 2, a2.s = 2, a2.t = 1
(d) a1.s=1, a1.t = 0, a2.s = 2, a2.t = 2 "

The interactions of students with the platforms were logged weekly for statistical analysis. For each student, the logged events were related to academic effort understood as the number of questions answered per day.

4   Results and Discussion

4.1   Sample Equivalence

To ensure that students of both groups were representative sample in terms of pretest scores, a Wilcoxon–Mann–Whitney test was conducted. The results show that learners from both groups had no statistically significant difference in their pretest scores ($Z = 1.42$, $\rho = 0.16$), indicating that both samples come from the same population.

4.2   Academic Effort

The Adjusted Rank Transform (ART) test [26] was applied to the 2×5 mixed factorial design. Platform type included two levels (AR and Web) and days consisted of five levels (day 1, day 2, day 3, day 4, and day 5). The ART test was used to compare the effects of type of platform used and day of the intervention and the interaction effect between platform and day on academic effort dependent variable (Fig. 4).

Our hypothesis was that students in the AR group will answer more multiple-choice questions earlier than those in the Web course.

Means and standard deviations of the academic effort measure can be found in Table 1.

Table 1. Mean scores and standard deviations of academic effort (AE) over days 1–5

	Day 1	Day 2	Day 3	Day 4	Day 5
AE AR group	1.03(2.69)	2.12(3.69)	5.20(5.21)	3.91(4.63)	1.00(2.46)
AE Web group	0.43(1.44)	1.30(2.24)	2.08(2.46)	2.34(2.51)	2.13(2.49)

Results of the ART test (see Table 2) show that the platform affected significantly the number of total answers answered (F (1,45) = 7.04, $p < 0.05$) and, there is a main effect on day (F (4,180) = 7.18, $p = < 0.001$). The detailed analysis shows that students using the AR-based platform tended to answer more questions than students using the Web-based platform (z = 2.65, $p < 0.01$). The analysis also suggests a difference on day with significant level between days 1 and 3 (z = −4.6, $p < 0.0001$) suggesting that students increased their activity on day 3 when comparing with day 1, and a weak significance between days 3 and 5 (z = 2.7, $p < 0.01$) which suggest that they did not wait, until the last moment to do their work. The results of the analysis suggest a significant interaction between the two experimental conditions and the day of the intervention when measured the number of total questions answered (F (4,180) = 3.64, $p < 0.01$) (see Table 2). Thus, our hypothesis is supported partially.

Table 2. Results of mixed-effect model analysis of variance for the academic effort

	Academic effort
Platform	F(1,45) = 7.04, $p < 0.05$
Day	F(4,180) = 7.18, $p < 0.001$
Platform × day	F(4,180) = 3.64, $p < 0.01$

4.3   Learning Outcomes

An independent-samples t-test was conducted to compare posttests marks in AR tool and Web tool. There was not a significant difference in the scores for experimental posttest scores (*M* = 3.67, *SD* = 2.21) and experimental mark posttest scores (*M* = 3.66, *SD* = 2.60), t (45)=0.012, p = 0.98. Thus, there was no difference in learning outcomes in the group that used the AR-tool and the one that used the Web-based tool.

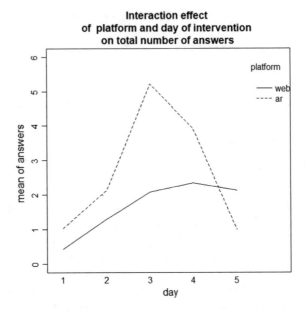

Fig. 4. Interaction effect of platform and day of intervention on total number of answers

5　Conclusion

This study presented and evaluated an augmented reality learning application to practice basic concepts of object-oriented programming. The study compared the AR-based application with a similar Web-based application in order to study its effectiveness on motivating students to avoid academic procrastination.

Results suggest that the AR-based tool was more effective than the Web-based tool in motivating students to answer the questions proposed before the deadline and in motivating students to answer more questions. Indeed, students in the experimental group finalize their assignment before the last day whereas the control group had still work to do on the fifth day of the intervention. However, results suggest that most of the students in both groups tended to wait 1 or 2 days before starting to work and there were no difference in learning outcomes between the two groups. Therefore, there still a room for further improvement.

Acknowledgements. This work was supported in part by Spanish Ministry of Economy and Competiveness: RESET project under grant no. CICYT (TIN2014-53199-C3-1-R), Smartlet project under grant no. TIN2017-85179-C3-1-R, funded by the Agencia Estatal de Investigación (AEI) and Fondo Europeo de Desarrollo Regional (FEDER); and Comunidad de Madrid (eMadrid project, S2013/ICE-2715).

References

1. Akinola, O. S., & Nosiru, K. A. (2014). Factors influencing students' performance in computer programming: A fuzzy set operations approach. *International Journal of Advances in Engineering Technology, 7*(4), 1141.
2. Shell, D. F., Hazley, M. P., Soh, L. K., Ingraham, E., & Ramsay, S. (2013). Associations of students' creativity, motivation, and self-regulation with learning and achievement in college computer science courses. In *Frontiers in Education Conference*, 2013 IEEE 1637-1643.
3. Michinov, N., Brunot, S., Le Bohec, O., Juhel, J., & Delaval, M. (2011). Procrastination, participation, and performance in online learning environments. *Computers & Education, 56* (1), 243–252.
4. Schraw, G., Wadkins, T., & Olafson, L. (2007). Doing the things we do: A grounded theory of academic procrastination. *Journal of Educational Psychology, 99*(1), 12.
5. Baumeister, R. F., Schmeichel, B. J., & Vohs, K. D. (2007). Self-regulation and the executive function: The self as controlling agent. In A. W. Kruglanski & E. T. Higgins (Eds.), *Social psychology: Handbook of basic principles.*
6. Lee, E. (2005). The relationship of motivation and flow experience to academic procrastination in university students. *The Journal of Genetic Psychology, 166*(1), 5–15.
7. Hernandez-Serrano, J., Choi, I., & Jonassen, D. H. (2000). Integrating constructivism and learning technologies. In *Integrated and holistic perspectives on learning, instruction and technology* (pp. 103–128). Dordrecht: Springer.
8. Bishop, J. L., & Verleger, M. A. (2013). The flipped classroom: A survey of the research. *ASEE National Conference Proceedings, Atlanta, GA., 30*(9), 1–18.
9. Lamberti, F., Hwang, G. J., Manjón, B. F., & Wang, W. (2018). Guest Editorial: Joint special issue on" innovation in technologies for educational computing". *IEEE Transactions on Learning Technologies, 11*(1), 2–4.
10. Di Serio, Á., Ibáñez, M. B., & Delgado-Kloos, C. (2013). Impact of an augmented reality system on students' motivation for a visual art course. *Computers & Education, 68,* 586–596.
11. Ibáñez, M. B., Di Serio, Á., Villarán, D., & Delgado-Kloos, C. (2014). Experimenting with electromagnetism using augmented reality: Impact on flow student experience and educational effectiveness. *Computers & Education, 71,* 1–13.
12. Tuckman, B., & Sexton, T. (1989). *Effects of relative feedback in overcoming procrastination on academic tasks.* Paper given at the meeting of the American Psychological Association, New Orleans, LA.
13. Sendcal, C., Koestner, R., & Vallerand, R. (1995). Self-regulation and academic procrastination. *The Journal of Social Psychology, 135,* 607–619.
14. Tuckman, B. (1998). Using tests as an incentive to motivate procrastinators to study. *Journal of Experimental Education, 66,* 141–147.
15. Azuma, R. T. (1997). A survey of augmented reality. *Presence: Teleoperators and Virtual Environments, 6,* 355–385.
16. Akçayır, M., & Akçayır, G. (2017). Advantages and challenges associated with augmented reality for education: A systematic review of the literature. *Educational Research Review, 20,* 1–11.
17. Bacca, J., Baldiris, S., Fabregat, R., & Graf, S. (2014). Augmented reality trends in education: A systematic review of research and applications. *Journal of Educational Technology & Society, 17*(4), 133–149.

18. Azuma, R., Baillot, Y., Behringer, R., Feiner, S., Julier, S., & MacIntyre, B. (2001). Recent advances in augmented reality. *IEEE Computer Graphics and Applications. IEEE Computer Society, 21*(6), 34–47.

19. Carmigniani, J., Furht, B., Anisetti, M., Ceravolo, P., Damiani, E., & Ivkovic, M. (2010). Augmented reality technologies, systems and applications. In *Multimedia Tools and Applications* (Vol. 51, no. 1, pp. 341–377). Springer Netherlands.

20. Huang, T. C., Chen, C. C., & Chou, Y. W. (2016). Animating eco-education: To see, feel, and discover in an augmented reality-based experiential learning environment. *Computers & Education, 96*, 72–82.

21. Kamarainen, A. M., Metcalf, S., Grotzer, T., Browne, A., Mazzuca, D., Tutwiler, M. S., et al. (2013). EcoMOBILE: Integrating augmented reality and probeware with environmental education field trips. *Computers & Education, 68*, 545–556.

22. Althoff, T., White, R. W., & Horvitz, E. (2016). Influence of Pokémon Go on physical activity: Study and implications. *Journal of Medical Internet Research, 18*(12). https://doi.org/10.2196/jmir.6759 (Published online 2016 Dec 6).

23. Carbonell Carrera, C., Saorín, J. L., & Hess Medler, S. (2018). Pokémon GO and improvement in spatial orientation skills. *Journal of Geography*, 1–9.

24. Cheng, K. H., & Tsai, C. C. (2014). Children and parents' reading of an augmented reality picture book: Analyses of behavioral patterns and cognitive attainment. *Computers & Education, 72*, 302–312.

25. Kaufmann, H., & Schmalstieg, D. (2002). Mathematics and geometry education with collaborative augmented reality. In *ACM SIGGRAPH 2002 Conference Abstracts and Applications* (pp. 37–41).

26. Leys, C., & Schumann, S. (2010). A nonparametric method to analyze interactions: The adjusted rank transform test. *Journal of Experimental Social Psychology, 46*(4), 684–688.

Mastering the MOOC: Exploring a Unique Approach to Online Course Development

Caitlin E. Mullarkey and Felicia Vulcu[✉]

Faculty of Health Sciences, Department of Biochemistry and Biomedical Sciences, McMaster University, Hamilton, ON, Canada
{mullarkc,vulcuf}@mcmaster.ca

Abstract. Since their initial launch, massive open online courses (MOOCs) have experienced explosive growth as a method of online distance learning. As of 2017, the MOOC landscape has ballooned to almost 10,000 courses offered by more than 800 universities worldwide. From an instructor's perspective, the biggest draws to such an online platform are the scalability and openness. The benefits are equally attractive from a learner's perspective as these courses lack prerequisites, are highly flexible, and provide access to education from top university professors. While the term MOOC was initially coined in Canada, Canadian universities have lagged behind their American counterparts in the market. On February 2018, in partnership with the MacPherson Institute and Labster virtual labs, we launched *DNA Decoded* on Coursera. Aimed at demystifying the complex topic of DNA to a lay audience, our course was just the fourth to be launched by McMaster University. In developing *DNA Decoded*, we incorporated several novel approaches including the use of gamified elements to promote engagement with content. In this paper, we explore the MOOC design process from an instructor's perspective and evaluate the performance of the course 6 months postlaunch.

Keywords: Massive open online course (MOOC) · Virtual lab simulations · Online videos · Open education resources

1 Introduction

The digital age has seen an explosion in the growth of e-learning and education technology platforms. Massive open online courses (MOOCs) offer, perhaps, the best example of a new type of online distance learning that has gained tremendous popularity in higher education in the past decade, grown at a furious pace, and yet at the same time, it has been the subject of great controversy. At the peak of their popularity in 2011–2012, MOOCs were often touted as a disruptive learning innovation that stood to alter current structures of higher education [1]. However, there is much debate as to whether this has actually come to fruition. Others have argued that the empirical evidence to suggest these open online courses that actually disturb higher education markets is weak [2]. Despite much discussion surrounding their utility and transformative abilities in higher education, the MOOC landscape has continued to demonstrate impressive growth. As of 2017, there were nearly 10,000 courses offered by more

© Springer Nature Switzerland AG 2019
M. E. Auer and T. Tsiatsos (eds.), *Mobile Technologies and Applications for the Internet of Things*, Advances in Intelligent Systems and Computing 909,
https://doi.org/10.1007/978-3-030-11434-3_22

than 800 universities worldwide [3]. Early classifications of MOOCs drew two broad categories of courses: cMOOCs or xMOOCs [4]. The former, based connectivist ideas [5], focused on collaborative learning between the student and instructor, as well as among the students in the course. On the other hand, xMOOCs highlighted individual learning and often employed a traditional course structure [6]. Yet, as the MOOC arena has evolved these categories have become overly restricted [7]. Several studies have advocated for the recognition of additional types of MOOCs [4, 8] based on their distinct goals. More specifically, Lane [7] and Beaven et al. [4] suggested three types of MOOCs: network-based, task-based, and content-based MOOCs.

Network-based MOOCs share many similar features to cMOOCs, where the primary goal is to develop a network of students. This type of MOOC is designed to engage a community of learners and promote peer learning. By providing a balanced mix of both instructivism and constructivism, task-based MOOCs emphasize the completion of assignments. These assignments can vary in format, but the overarching idea is that the execution of tasks promotes a particular skill set. Finally, content-based MOOCs focus on traditional content delivery which adopts a behaviorist pedagogical approach. The focus is on individual learning, typically through the use of videos to disseminate content, and quizzes to assess retention. The content-based MOOC design is extremely popular in platforms, such as Coursera, which aim to reach a large learner population.

Although MOOCs have been a global trend in higher education for nearly a decade, Canadian universities have lagged behind their American counterparts in the market. In collaboration with the MacPherson Institute for Leadership and Learning at McMaster University, we sought to develop and launch an innovative MOOC entitled DNA Decoded that relied on several novel approaches including gamified elements. Drawing on our own expertise and capitalizing on the growing public interest in DNA, we endeavored to demystify complex biochemical topics to a lay audience. In our design process, we created mainly content-based MOOC, which aimed at reaching a large base of learners. However, we also incorporated features of both network (emphasis on engaging learners) and task-based MOOCs (varied assignments and social learning activities).

Herein, we describe the development process, rationale, and learning objectives in creating this original MOOC. Furthermore, we assess preliminary data from the Coursera and Labster platforms on learner demographics, and quiz and course completion. Finally, we reflect on potential modifications to increase learner engagement that may be adopted and released in future versions of the course.

2   McMaster University Involvement in the MOOC Arena

The modern-day MOOC is evolved from earlier iterations of open distance education such as MIT Open Courseware (OCW) and Khan Academy, to name a few [9]. The original MOOC concept was created in 2008 by Canadians George Siemens and Stephen Downes. This MOOC type integrated key concepts—such as openness and flexibility—to engage and foster a community of learners [10]. The emergence of xMOOCs, a true departure from the original MOOC concept, paved the way for a true educational renaissance in open educational resources (OERs). The success of the xMOOC is due, in part, to Ivy League Universities which created and distributed numerous MOOCs on easily accessible online platforms like Coursera, Udacity, and EdX [10]. Courses such as Dr. Barbara Oakley's and Dr. Terry Sejnowski's "Learning how to Learn" have drawn millions of learners to the Coursera platform [11]. While Canada may have been the birthplace of the cMOOC, it is lagging behind in the xMOOC educational renaissance. There are currently a limited number of MOOC hubs spread throughout the country in places such as the University of Alberta and the University of Toronto. McMaster University, for a number of years, has had significant involvement in online learning through the eCampus Ontario portal. McMaster's footprint in online learning is continuing to grow and expand, especially in the MOOC arena. The first course offered by the University, and hosted on Coursera, entitled Experimentation for Improvement (Kevin Dunn) launched in 2014. This was followed by Finance for Everyone (Arshad Ahmad) in 2016 and the release of Mindshift (Barbara Oakley and Terence Sejnowski) in 2017. While all of these courses have been well-received and achieved various degrees of success, by in large, they all follow a more traditional didactic approach to content dissemination.

As we looked to design and execute DNA Decoded, we sought to differentiate our course from the myriad of other MOOCs by using humor, dialog, and narratives to explain science, and gamified interventions to bolster student engagement.

2.1 Rationale for Development of DNA Decoded and Integration of Gamified Elements

In science education, the laboratory setting is a valuable experience that can bridge the gap between theory and practice. However, there are obvious constraints to the completion of laboratory experiments in online distance education. Thankfully, a solution to this problem can be found in technology-enhanced learning (TEL) [12]. More specifically, virtual laboratory simulations represent a promising and innovative new tool to increase student engagement and enhance learning outcomes [13]. A number of studies have evaluated the utility of virtual simulations as compared to traditional teaching. Taken together, the results for these studies demonstrate that simulations can aid in the mastery of content, increase test scores, and enhance the overall learning experience [14–16].

Our department (Biochemistry and Biomedical Sciences) has spearheaded a novel initiative by partnering with Labster (a TEL-driven educational start-up) to introduce virtual laboratory simulations in two large Biochemistry courses. Recently, a large study of biology students in the Netherlands examined the use of Labster as compared to traditional teaching, and found a 76% increase in learning outcomes with the gamified laboratory simulation [17]. Moreover, the participants in this study also reported high levels of motivation and interest in completing the simulations [17]. Given our experience with the Labster platform, and ample evidence to support learner engagement and learning outcomes, we integrated three virtual laboratory simulations (Crime Scene Investigation, Molecular Cloning, and Next-Generation Sequencing) into DNA Decoded. These simulations were seamlessly incorporated onto the Coursera platform using a single-sign-on system that ensured ease of access to the software. We hypothesized that this would not only increase student engagement, but perhaps also encourage course completion (a chief concern faced by any MOOC).

2.2   MOOC Design Processes and Learning Objectives

We devised a 4-week course structure that began with basic biochemistry of DNA and slowly introduced increasingly complex topics with each subsequent week (Table 1). Strategically, in the fourth and final week of the course, we tackled trendier topics such as ancestry and health testing, genetically modified organisms, and ancient DNA. In this way, learners gained the tools to better understand this advanced material, and by withholding these more popular subjects, until the end, we hoped to promote learners' progression through all 4 weeks of the course. As part of the initial development process, we outlined overall learning objectives for the course as well as distinct learning objectives for each week (not shown). These objectives helped to define the content for each week and outline the expectations of acquired knowledge and skills for learners.

Table 1 Detailed outline of the DNA decoded course structure and a schematic of a typical weekly structure. DP—discussion prompt, Q—question(s), h—hours, min—minutes

DNA decoded MOOC learning objectives
✓ Understand the basic component and structural elements of DNA
✓ Identify the steps required to convert the genetic code to message and products
✓ Assess your knowledge through interactive and engaging exercises
✓ Explore a virtual lab setting and test your knowledge by manipulating DNA
✓ Reflect on and discuss how you may apply your knowledge of DNA to real-world applications
✓ Discover that science and learning about DNA can be easy and fun!

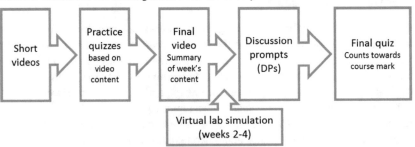

Week 1: cracking the genetic code (Estimated learning time: 2 h 35 m)

5 videos	5 practice quizzes	1 final video	3 DP	1 final quiz
Average: 5 min	Average: 3 Q			11 Q

Week 2: getting the message across (Estimated learning time: 2 h 36 m)

7 videos	8 practice quizzes	1 final video and 1 virtual lab	2 DP	1 final quiz 8 Q
Average: 7 min	Average: 2 Q			

Week 3: manipulating DNA (Estimated learning time: 2 h 50 m)

5 videos	6 practice quizzes	1 final video and 1 virtual lab	4 DP	1 final quiz 8 Q
Average: 7 min	Average: 3 Q			

Week 4: DNA and me (Estimated learning time: 2 h 53 m)

6 videos	4 practice quizzes	1 final video and 1 virtual lab	4 DP	1 final quiz 9 Q
Average: 6 min	Average: 2 Q			

We relied on a collaborative approach with the MacPherson Institute, Labster, and undergraduate students at McMaster University to execute and ultimately produce DNA Decoded as described below. Our team included digital media specialists, a professional writer, an instructional developer, a postdoctoral fellow, and ourselves as subject matter experts. In total, we filmed 27 videos that were on average 6 min long (Table 1). Virtual lab simulations were incorporated into Weeks 2, 3, and 4. We also implemented a number of sound pedagogical practices in our video construction to deeply engage learners in the content. These included storytelling, humor, and physical props to replace traditional lecture videos, as well as original animations throughout.

2.3 Structure and Execution

The structure of our MOOC follows a typical xMOOC layout with a 4-week time span with each week featuring short videos, quizzes, discussion prompts, and a list of resources. Table 1 depicts the layout of the course. Armed with foundational knowledge in Weeks 1 and 2, the learner can discover how DNA is manipulated using molecular cloning techniques in Week 3, and the personal and global implications of using DNA technology in Week 4. We looked to deviate from the standard content-based MOOC delivery style by incorporating a number of pedagogical interventions to engage our learners in the content presented. Discussion prompts were incorporated each week, as well as practice quizzes at the end of each video and virtual lab simulations in Weeks 2–4 (Table 1). To complete the course, learners must achieve a passing mark (50%) on the final weekly quizzes.

With a course structure in place and specific learning objectives identified, DNA Decoded was completed as a collective effort of our interdisciplinary team. We, as subject matter experts, wrote drafts of the scripts for each video with input from a professional writer. These scripts were edited and revised for clarity, continuity, and flow by both undergraduate students and an instructional developer. Filming took place over the period of several weeks under the direction of two digital media specialists. We chose the Biochemistry undergraduate teaching laboratories as the set for our videos, to provide learners with a firsthand view of a working laboratory setting. For each video, we served as "co-hosts" and not only did we utilize storytelling, humor, and physical props to replace traditional lecture videos, but we also relied heavily on dialog. Following filming, our digital media specialists were responsible for editing the videos and creating original animations. The content was ultimately uploaded on the Coursera platform, and the course was officially launched to the public on February 5, 2018.

3   Analyzing Measures of MOOC Performance and Success

3.1   Learner Demographics

Six months following our February 2018 launch on Coursera, DNA Decoded has had 6,372 total visitors and 2,251 total learners, with an average of approximately 10 new active learners enrolling every day. The demographics of our learners, as of August 2018, span 6 continents (Fig. 1) and represent 44% women and 56% men with a wide age range centered around 18–34 years of age.

Fig. 1 *DNA decoded enrolled user geographic representation.* The geographic location of total enrolled learners, expressed as a percent, was mapped on a world map. The cutoff for represented countries was 1.0% of total enrolled users (inclusive), which represents 75.7% of all users. This was based on data from 2,216 learners. Estimates are accurate to ±2.1 percentage points. Data was obtained as of August 16, 2018. World map and pin images obtained from Presenter Media, 2009–2018 Eclipse Digital Imaging, Inc

Of the learners surveyed, 37% are fully employed and 66.1% have a bachelor's degree or higher. To date, we have 51 paid learners and 88 course completers. The course reviews have been overwhelmingly positive (4.9 out of 5 stars), with 95% of learners answering in the affirmative when asked if they like or dislike their experience.

3.2   Assessment

Our initial rationale behind the structure of DNA Decoded was to build a knowledge base gradually from week to week, culminating in the exploration of topics we felt would be of most interest to a general audience in the final week of the course (Table 1). Our preliminary analysis of learner quiz completion rates reveals a number of interesting trends (Table 2).

If we examine each week as a stand-alone data segment, we can see a general increase in attrition rates with each weekly practice quiz, followed by a rather sharp attrition rate as learners move from practice quizzes to the final weekly quiz (Table 2). This trend was consistent for all 4 weeks. For example, for Week 1 quiz completion rates, we can see an average attrition rate of ∼24% between each of the practice quizzes. This rate spikes to 84% (compare Quiz 1: 703 to Final Quiz: 113) once we analyze the final quiz data for this week (Table 2, Week 1). This sudden spike in attrition rates between the practice quizzes (PQs) and the weekly final quiz (FQ) might be due to the fact that only the final quiz is counted toward the final assessment for the course and a large proportion of learners are enrolled for interest only. It could also indicate that learners are more willing to complete short quizzes, such as the PQs, which have on average 2.6 questions per quiz. The FQs, on the other hand, are much longer in length with an average of 9 questions per quiz. The observed decrease in completion rates is in line with published MOOC literature.

Table 2 *Total number of learners who completed each weekly quiz.* The highlighted rows indicate the number of learners who completed the final quiz, which counts toward the final course assessment. Each final quiz is weighted equally (25% of final course mark per final quiz). The weekly attrition was calculated independently for each week as follows: 100%—(weekly quiz completers/initial quiz completers (PQ 1) × 100). The final quiz attrition was calculated between weeks as follows: 100%—(weekly FQ completers/previous FQ completers × 100). PQ—practice quiz, FQ—final quiz. This data represents learner usage as of August 2018

Week	Quiz	Completers (#)	Weekly attrition (%)	Final attrition (%)
Week 1: cracking the genetic code	PQ 1	706	0	
	PQ 2	623	12	
	PQ 3	553	22	
	PQ 4	517	27	
	PQ 5	454	36	
	FQ	113	84	0
Week 2: getting the message across	PQ 1	320	0	
	PQ 2	307	4	
	PQ 3	289	10	
	PQ 4	268	16	
	PQ 5	204	36	
	PQ 6	136	58	
	PQ 7	223	30	
	PQ 8	183	43	
	FQ	99	69	12

Week	Quiz	Completers (#)	Weekly attrition (%)	Final attrition (%)
Week 3: manipulating DNA	PQ 1	198	0	
	PQ 2	184	7	
	PQ 3	173	13	
	PQ 4	167	16	
	PQ 5	165	17	
	PQ 6	147	26	
	FQ	92	54	7
Week 4: DNA and me	PQ 1	168	0	
	PQ 2	152	10	
	PQ 3	146	13	
	PQ 4	143	15	
	FQ	89	47	3

Another notable finding can be seen in Week 2, PQ 6 and 7 (Table 2). We detect a sharp decrease in the number of completers (58% attrition from PQ 1) for PQ 6, which could provide insight into the learner perseverance, as PQ 6 asks learners to apply learned content in order to transcribe and translate a given DNA segment. Interestingly, this severe decline is followed by a spike moving from PQ 6 to PQ 7, where the completion rate rebounds and even exceeds that of PQ 5 (compare PQ 5 203, PQ 6 136, and PQ 7 223). While it is difficult to fully account for this rebound, one explanation could be that the subject matter discussed in the video immediately preceding this quiz (polymerase chain reaction) has been mentioned in mainstream media with the popularization of forensic-based TV shows, and therefore may have been of great interest to learners. However, it is important to keep in mind that the total quiz completers represent a fraction of the total number of enrolled learners (2224 enrolled learners).

3.3   Labster Virtual Lab Simulations

We initially hypothesized that the inclusions of gamified interventions would enhance student engagement leading to higher course completion rates. While the previous studies have shown improved motivation, learning outcomes, and test scores in a classroom setting [17], overall, our data show that both course completion rates and simulation completion rates were low. Completion remains as one of the main challenges for any MOOC and ours was no exception, with rates hovering around 4%. Of the three virtual lab simulations, the crime scene investigation had the highest completion rate (3.4%), while the molecular cloning and next-generation sequencing simulations were modestly lower (2% and 1.4% respectively). With over 33 million users and 2,400 offered courses on Coursera, it is difficult to gauge average course completion rates. Some sources have suggested that it may be as high as 15% [18]. A 4-year review of MOOCs from HarvardX and MITx published in 2016 found that of courses offering free certificates, the median certification rate was 7.7%, but ranged from 0.4 to 34% [19]. An earlier study suggested that certification rates range from 2 to 10% [19]. Therefore, it appears our completion rates are in line with reported averages, when we anticipated virtual lab simulations might push these numbers higher. However, we should note that completion of simulations was not required for overall course completion.

4   Implications and Future Research Directions

The most critical outcome of this project to date is the creation of a rich trove of freely accessible online content that has the potential to be incorporated into existing science courses at the high school and undergraduate level. Although we anticipated that the employment of a gamified element would enhance the learner's experience and increase completion rates, our preliminary data does not support this hypothesis. Boredom and low student engagement have often been cited as explanations for MOOC course completion rates, and it is still our belief that gamification has the potential to combat these issues by enhancing learner experience [19]. In hindsight, the way in which the Labster simulations were integrated into DNA Decoded may not have been optimal for several reasons. First, as mentioned previously, the virtual laboratory simulations were an optional course element. In future versions of the course, it may be worthwhile to make the simulations part of the course completion requirement. Moreover, the simulations were placed at the end of weeks 2, 3, and 4. As evidenced by the quiz data, attrition rates increase as the week progresses. Therefore, shifting these simulations to the beginning or midway through the week may help us to realize the full potential of gamified elements in DNA Decoded. Finally, several methods could be employed to incentivize participation in gamified elements such as e-badges or leaderboards [19].

Initially, we envisioned our target audience to be the general public, however, 66.1% of our learners had some form of higher education (bachelor's degree or above). While this may reflect our overestimation of the complexity and difficulty of the content in DNA Decoded, in fact, these data are on par with the reported demographics of users who enroll in STEM MOOCs where the education level of 63% of learners is minimally a bachelor's degree [19]. DNA Decoded has continued to steadily grow its learner base since its launch 6 months ago. With an average of 10 new learners enrolling each day, the MOOC is on track to garner 4,000 students by its 1-year anniversary. While gamified interventions are not the sole answer to the completion rate dilemma posed by MOOCs, we believe there is still added value and scope for further incorporation of these elements into online learning. Due to the paucity of online courses to date which has utilized gamified interventions, future studies are needed to fully address their impact in the MOOC landscape.

References

1. Hyman, P. (2012). In the year of disruptive education. *Communications of the ACM, 55*(12), 20–22.
2. Al-Imarah, A. A. (2008). Shields, R. MOOCs, disruptive innovation and the future of higher education: A conceptual analysis. *Innovations in Education and Training International*, 1–12.
3. Shah, D. (2018). A product at every price: A review of MOOC stats and trends in 2017. *Class Central.*
4. Beaven, T., Hauck, M., Comas-Quinn, A., Lewis, T., & de los Arcos, B. (2014). MOOCs: Striking the right balance between facilitation and self-determination. *The MERLOT Journal of Online Learning and Teaching, 10*(1), 31–43.
5. Siemens, G. (2014). *Connectivism: A learning theory for the digital age.*
6. Siemens, G. (2012). *MOOCs are really a platform.*
7. Lane, L. M. (2015). Three kinds of MOOCs. 2012. http://www.lisahistory.net/wordpress/2012/08/threekindsmoocs.
8. Roberts, G., Waite, M., Lovegrove, E. J., & Mackness, J. (2013). Xvc: Hybridity in through and about MOOCs. *Creating a Virtuous Circle Proceedings OER13*, 1–8.
9. Gillani, N., & Eynon, R. (2014). Communication patterns in massively open online courses. *The Internet and Higher Education, 23*, 18–26.
10. Daniel, J. (2012). Making sense of MOOCs: Musings in a maze of myth, paradox and possibility. *Journal of Interactive Media in Education*, (3).

11. Schwartz, J. (2017). Learning to learn: You, too, can rewire your brain. *The New York Times.* https://www.nytimes.com/2017/08/04/education/edlif.

12. Scanlon, E., McAndrew, P., & O'Shea, T. (2015). Designing for educational technology to enhance the experience of learners in distance education: How open educational resources, learning design and MOOCs are influencing learning. *Journal of Interactive Media in Education,* (1).

13. Maldarelli, G. A., Hartmann, E. M., Cummings, P. J., Horner, R. D., Obom, K. M., Shingles, R., et al. (2009). Virtual lab demonstrations improve students' mastery of basic biology laboratory techniques. *Journal of Microbiology and Biology Education, JMBE, 10*(1), 51.

14. Gelbart, H., & Yarden, A. (2006). Learning genetics through an authentic research simulation in bioinformatics. *Journal of Biology Education, 40*(3), 107–112.

15. Kiboss, J. K., Ndirangu, M., & Wekesa, E. W. (2004). Effectiveness of a computer-mediated simulations program in school biology on pupils' learning outcomes in cell theory. *Journal of Science Education and Technology, 13*(2), 207–213.

16. Toth, E. E., Morrow, B. L., & Ludvico, L. R. (2009). Designing blended inquiry learning in a laboratory context: A study of incorporating hands-on and virtual laboratories. *Innovative Higher Education, 33*(5), 333–344.

17. Bonde, M. T., Makransky, G., Wandall, J., Larsen, M. V., Morsing, M., Jarmer, H., et al. (2014). Improving biotech education through gamified laboratory simulations. *Nature Biotechnology, 32*(7), 694.

18. Jordan, K. (2013). *MOOC completion rates: The data.*

19. Chuang, I., & Ho, A. (2016). HarvardX and MITx: Four Years of Open Online Courses–Fall 2012-Summer.

Using Gamification Technique to Increase Capacity in the Resolution of Problems During the Process Teaching and Learning Programming

Mónica Adriana Carreño-León[1(✉)],
Francisco Javier Rodríguez-Álvarez[2], and
Jesús Andrés Sandoval-Bringas[1]

[1] Universidad Autónoma de Baja California Sur, La Paz, B.C.S, Mexico
{mcarreno, sandoval}@uabcs.mx
[2] Universidad Autónoma de Aguascalientes, Aguascalientes, Mexico
fjalvar@correo.uaa.mx

Abstract. This research paper presents the results of an experience implemented in the Academic Department of Computational Systems (DASC, by its initials in Spanish) of the Autonomous University of Baja California Sur (UABCS, by its initials in Spanish) is presented, where a gamification technique was implemented in the classroom, in an introductory course of programming, as support to increase the capacity in the resolution from problems. An activity created with gamification principles was designed for development of the problem-solving competence. The technique was implemented with a group of 14 students from the second semester of Software Development Engineering (IDS, by its initials in Spanish).

Keywords: Gamification · Teaching programming · Teaching · Challenges · Motivation

1 Introduction

Learning to program is considered as basic for students who are studying a career in the areas of computer science and computing. The learning of the programming subjects is one of the most difficult and complex [1, 2]. Students have always shown difficulty in assimilating abstract notions, and therefore high rates of failure and dropout are shown in the courses [3–5]. To learn to program, it is necessary to know the programming structures and fundamentally to solve many exercises: "To learn to program, you have to program" [6]. The experience gained during the teaching of programming in the undergraduate level has been possible to detect how complicated it is to use the abstraction, that is, required for the design of an algorithm and obtain the desired results; for a student, it is difficult to determine if the algorithm design is correct. Additionally, the solution to a problem can be expressed in several ways and be correct.

© Springer Nature Switzerland AG 2019
M. E. Auer and T. Tsiatsos (eds.), *Mobile Technologies and Applications for the Internet of Things*, Advances in Intelligent Systems and Computing 909, https://doi.org/10.1007/978-3-030-11434-3_23

Many researchers, in their eagerness to improve the teaching of programming, have developed and experimented with different educational resources. Recently, the use of game elements and their design techniques, in a nongame context, is what is known as gamification [7].

Writing a computer program using a programming language requires several skills and abilities, which basically involve the ability to manipulate a set of interrelated abstractions to solve problems [8].

The resolution of problems usually involves an intellectual effort and the design of a set of steps that allow reaching the solution of the problem [9]. In [10], the four-step method for solving problems is proposed as a strategy: understanding a problem, designing a plan, executing a plan, and looking back.

Programming is an activity that requires the use of both sides of the brain. Logical-verbal reasoning is required to design and implement software correctly [11]. The research of several authors agrees that programming learning is complex, because it is necessary to use abstractions, the application of a logic of the paradigm of programming, and the construction of expressions, using the syntax and semantics of a programming language [12, 13]. In [14], it is stated that the greatest difficulties observed in the students, is the inability to develop a viable model that allows solving the problem, not being able to describe a comprehensible strategy for the computer or abstract the different behaviors of a task in a strategy that integrates them all.

The use of simulations through computers in the teaching-learning process has allowed favorable changes in the students, in terms of problem-solving, since they facilitate the possibility of access to teaching subjects that are difficult to understand and demonstrate [15]. The use of Information and Communication Technologies (ICT) are considered valuable tools to support teaching-learning processes [16]. Through them, the students can evaluate their knowledge, and thereby strengthen their training in a specific area. An important factor for the success or failure of the use of these is the pedagogical design of the tool.

On the other hand, software tools can also be found as didactic support to facilitate the teaching-learning of algorithms. In [17], it is mentioned that among these tools are those that exploit the use of microworlds such as Alice and JKarelRobot. Among the tools based on pseudocode representations or flow diagrams, the following stand out: PSeInt, RAPTOR, and DFD.

In more recent years, another technique, increasingly used is gamification, which is used as a method to encourage students to perform certain activities or tasks that generally would not. The gamification consists of the use of mechanics, elements, and game design techniques in a non-game context to engage users and solve problems [18, 19].

The incorporation of elements of games in contexts that are not games, such as teaching-learning processes, is a promising area in the field of education due to the possibilities of interaction and motivation that can generate in students. In [20], it is stated that the integration of gamification elements in education helps to achieve positive results.

The main argument is that the dynamics of the games can increase the attention of the students during the teaching-learning process also improving their satisfaction with this process [21]. In [22], a research is reported that makes use of gamification as an effective tool to improve the performance of students in different types of courses.

In a general sense, gamification is described as the process of game thinking and its mechanisms to attract users and make them solve problems [18]. This definition in the educational field refers to the use of elements of the game to involve students, motivate them to action, and promote learning and problem-solving [23].

In recent years, work has been done on the application of the basic principles of gamification to education [24], and in particular to the learning of programming [25–28].

According to [29], any process that fulfills the following premises can be transformed into a gamified game or being: (a) the activity can be learned; (b) the user's actions can be measured and (c) the feedback can be delivered in a timely manner to the user. Therefore, it is feasible that the training activities for the design of algorithms can be gamified.

2 Methodology

Gamification refers to the use of game design elements and principles to be used in non-gaming contexts [30]. That is, the theory and mechanics of games are used to involve, motivate, and engage people, to transform a routine and unattractive activity into a dynamic and motivating activity [31].

This learning proposal is not new, what is really important and challenging is to find ingenious and attractive ways to use it in the classroom. Thus, the research considered this methodological proposal of learning because gamification gives the possibility of being able to give a mechanic of interest, emotion, and fun to all the activities to be carried out [32].

As an example, and considering the interest of the project in favoring the development of problem-solving competence in students, the general design of an activity created with gamification principles is shown below:

(1) To apply the gamification in the classroom, the following steps were taken:

- Define the objective of algorithm resolution;
- Use cards with instructions in pseudocode;
- Solve an algorithm by ordering cards with instructions in pseudocode;
- Work in teams and in competition with the other teams, through the use of flags;
- Establish three levels: basic, intermediate, and advanced;

(2) It was discussed with the students, the activity was explained and what was expected of them.

(3) They were asked to form groups of three or four students, promoting a participatory and independent attitude.

(4) A multipurpose classroom was used to carry out the activities, which allowed for teacher mobility and greater comfort for the groups.

(5) During each session, each team was provided with a set of cards, each containing a pseudocode instruction, which together represents one of the possible solutions to the problem. The objective of the activity consisted in ordering the cards to obtain the sequence of steps that would allow the solution of the problem.

Figure 1 corresponds to the scheme of the activities or stages of the gamification technique from the perspective of the student. In level 1, the student is given the exact number of cards corresponding to the solution of an algorithm. At level 2, the student is given a larger number of cards with pseudocode fragments, but the number of cards that are part of the solution is indicated. At level 3, the student is given a larger number of cards, but is not told how many cards are part of the solution.

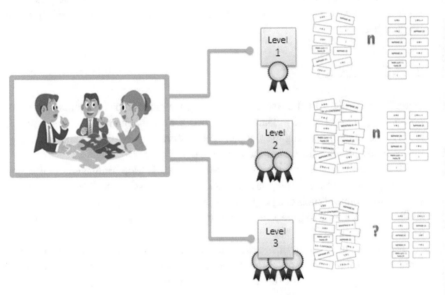

Fig. 1 Diagram of activities of the gamification technique (*Source* Authors)

3   Results

The study was implemented in the Academic Department of Computational Systems (DASC) of the Autonomous University of Baja California Sur (UABCS). To verify the effectiveness of the gamification technique, we had a group of university-level students of the Software Development Engineering (IDS) of the second semester. The gamification technique was used in a basic course of introduction to programming, by 14 students during 4 weeks in 2 sessions of 2 h each, to solve a total of 15 algorithms.

Four work teams were formed, two teams with three members and two teams with four members. The interaction between the teams that participated in the study consisted in showing permanently the progress in the development of the exercises, through the use of colored flags. Each exercise was assigned a specified color.

At the beginning of the experimentation phase, an examination of recognition of the level of each of the teams was applied, using the traditional algorithms resolution form: paper and pencil.

Figure 2 shows as evidence an image of the participation of students in the developed experiment, as well as the solution of some of the exercises solved by the gamification technique.

Fig. 2 Evidence of solution of some of the exercises solved by the gamification technique (*Source* Authors)

The review of the results obtained with the gamification technique was carried out through the evidence provided by each group of students of the algorithms they solved, as well as the time to achieve it.

Table 1 shows the number of algorithms solved by the teams during the eight sessions. Only one team solved all the algorithms in the time foreseen for the activities. All the teams solved the exercises of the basic and intermediate levels. And in the end,

Table 1 Number of algorithms solved by a team of students using the selected gamification technique in the classroom (*Source* Authors)

Team/Level	Basic (1)	Intermediate (2)	Advanced (3)	Total algorithms
Team 1	5	5	5	15
Team 2	5	5	3	13
Team 3	5	5	0	10
Team 4	5	5	2	12

a team could not solve any advanced level exercise.

In the graphs of Fig. 3, the behavior per team can be observed during each of the eight sessions that lasted the experimentation with the gamification technique. For team 1, their behavior was constant, solving two algorithms on an average per session, regardless of the level of complexity. Team 2, in the first session solved two algorithms, had some problems in the second session where they could only solve one algorithm, however, in the following sessions, they improved and could solve two algorithms on average, having a bit of problem when the complexity increased The exercise. The situation of team 4 was

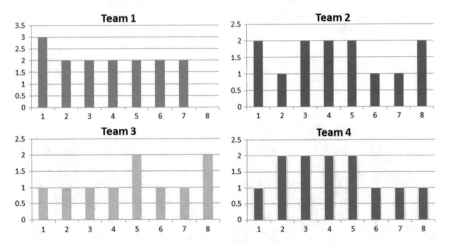

Fig. 3 Graph of the behavior of each team of students during each of the experimentation sessions with the gamification technique (*Source* Authors)

very similar to that of team 2. Team 3, on average, only solved one algorithm per session, and only two sessions solved more than one algorithm.

As a second revision of the results obtained with the gamification technique, a quiz was applied to the students. The quiz sought to measure different aspects of the gamification technique, according to the perspective of the students. For this, categories were established, covering both technical and perceptual aspects of the gamification technique used. For the final calculation, the average of the valuations was used for

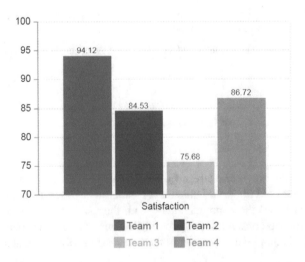

Fig. 4 Final results of the processing of the quiz (*Source* Authors)

each of the categories and with this, a satisfaction index was defined. The final results of the quiz processing can be seen in Fig. 4.

4 Conclusions

Gamification is the application of dynamics, mechanics, and game components in nongame environments. The paper evidences the creation of a gamification strategy in order to generate participation and commitment of students of introductory courses in teaching of programming at a higher level.

The preliminary results obtained with the use of the technique, in relation to the decrease in the rate of failure, show that designing a playful didactic that allows students to be more active, can be a very powerful tool, which is between a combination of serious learning and fun as he states [32].

In a study carried out in [33], it is affirmed that applying gamification techniques in subjects of difficult comprehension can be considered as a strategy of great value. This study has identified game elements that could be used in a learning material, which have already been validated by experts and students. Among them, to improve learning skills are considered: Results, Levels, and Skills' Percentages, which were incorporated into the developed environment.

According to the results, it is observed that a large percentage of students, improved capacity for problem-solving. On the other hand, from the perspective of the students, the analyzed results show that approximately 65% of the students are motivated to participate with the gamification classroom, and nearly 70% of the students would be satisfied if the gamification in other subjects will be implemented.

These results coincide with other research works [34, 35], which describe as a result of some advantages of gamification, such as increased participation and student interaction, collaborative learning and user fun as a motivation for realization of activities, and achievement of objectives.

References

1. Arévalo, C., & Solano, L. (2013). Patrones de Comportamiento de Estudiantes de Programación al Utilizar una Herramienta de Visualización de Protocolos Verbales. In *8va Conferencia Latinoamericana de Objetos de Aprendizaje y Tecnologías de Aprendizaje*, LACLO 2013. Valdivia, Chile.
2. Soler, Y., & Lezcano, M. (2009). Consideraciones sobre la tecnología educativa en el proceso de enseñanza-aprendizaje. Una experiencia en la asignatura Estructura de Datos. *Revista Iberoamericana de Educación, Organización de Estados Iberoamericanos para la Educación, la Ciencia y la Cultura*, 1–9.
3. Checa, R. (2011). La innovación metodológica en la enseñanza de la programación. Una aproximación pedagógica al aprendizaje activo en la asignatura Fundamentos de Programación. In *Interfases Revista Digital de la Facultad de Ingeniería de Sistemas* (pp. 67–87). Universidad de Lima.
4. Pullan, W., Drew, S., & Tucker, S. (2013). An integrated approach to teacning introductory programming. In *e-Learning and e-Technologies in Education (ICEEE) 2013* (pp. 81–86).

5. Niitsoo, M., Paales, M., Pedaste, M., Siiman, L., & Tõnisson, E. (2014). Predictors of informatics students' progress and graduation in university studies. In *International Technology, Education and Development Conference*. Valencia, Spain.

6. Sánchez-Ledesma, F., Ortiz, O., & Pastor, J. (2013). Aprendizaje de los lenguajes de programación en la educación universitaria a través de dispositivos móviles. *VI Jornadas de Introducción a la Investigación de la UPCT*, 100–102.

7. Deterding, S., Dixon, D., Khaled, R., & Nacke, L. (2011). From game design elements to gamefulness: Defining gamification. In *15th international academic MindTrek Conference: Envisioning future media environments* (pp. 9–15).

8. Chesñevar, C. (2000). Utilización de mapas conceptuales en la enseñanza de la programación. http://cs.uns.edu.ar/ ~ cic/2000/2000-jornadas-mapas/2000-jornadas-mapas. pdf.

9. Willing, P., Astudillo, G., & Bast, S. (2010). Aprender a programar (¿y a pensar?) jugando. In *Congreso de Tecnología en Educativa & Educación en Tecnología (TE&ET)* (pp. 167–176).

10. Polya, G. (1965). *Como plantear y resolver problemas*. México: Trillas.

11. Naps, T. (2002). Exploring the role of visualization and engagement in computer science education. *SIGCSE Bulletin ACM*.

12. Medina, A., & Chávez, A. (2011). Uso de herramientas informáticas como estrategia para la enseñanza de la programación de computadores. *Revista Unimar* (57), 23–32.

13. Rodriguez, J., Civit, A., Morgado, A., Jiménez, G., & Ferreiro, M. (2014). A game-based approach to the teaching of object-oriented programming languages. *Computers & Education*, 83–92.

14. Salgado, A., Berenguer, I., Sánchez, A., & Fernández, Y. (2013). Lógica Algorítmica para la resolución de problemas de programación computacional: una propuesta didáctica. *Didasc@lia: Didáctica y Educación*, 57–76.

15. Cataldi, Z. (2000). Metodología de diseño, desarrollo y evaluación de software educativo.

16. Rodrigez, L., García, M., & Landín, R. (2016). La apropiación de las TIC en los estudiantes universitarios: Una aproximación desde sus habitus y representaciones sociales. *Estudios λambda. Teoría y práctica de la didáctica en lengua y literatura*, 214–233.

17. Arellano, J., Nieva, O., Solar, R., & Arista, G. (2012). Software para la enseñanza-aprendizaje de algoritmos estructurados. *Revista Iberoamericana de Educación en Tecnología y Tecnología en Educación*, 23–33.

18. Zichermann, G., & Cunningham, C. (2011). *Gamification by design: Implementing game mechanics in web an mobile apps*. Canada: O'Reilly Media.

19. Werbach, K., & Hunter, D. (2012). For the win: How game thinking can revolutioniza your business. Wharton Digital Press.

20. Narasareddy, M., Singh, G., & Radermarcher, A. (2018). Gamification in computer science education: A systematic literature review. *American Society for Engineering Education*, 1–12.

21. Martí-Parreño, J., Queiro-Ameijeiras, C., Méndez-Ibañez, E., & Giménez-Fita, E. (2015). *El uso de la gamificación en la educación superior: el caso de Trade Ruler* (pp. 95–102). Aprendizaje experiencial: XII Jornadas Internacionales de Innovación Universitaria Educar para transformar.

22. Pineda-Corcho, A. (2014). Modelo tecno-pedagógico basado en ludificación y programación competitiva para el diseño de cursos de programación. Universidad Nacional de Colombia.

23. Kapp, K. (2012). *The gamification of learning and instruction*. Pfieer: Game-based methods and strategies for training and education.

24. Dicheva, D., Dichev, C., Agre, G., & Agelova, G. (2015). Gamification in education: A systematic mapping study. *Journal of Educational Technology & Society*, 1–15.

25. Kumar, B., & Khurana, P. (2012). Gamification in education-learn computer programming with fun. *International Journal of Computers and Distributed Systems, 2*(1), 46–53.
26. Bozorgmanesh, M., Sadighi, M., Nazarpour, M., & Branch, D. (2011). Increase the efficiency of adult education with the propoer use of learning styles. *Nature and Science, 9*(5), 140–145.
27. Swacha, J., & Baszuro, P. (2013). Gamification-based e-learning platform for computer programming education. *X World Conference on Computers in Education*, 122–130.
28. Azmi, S., Iahad, N., & Ahmad, N. (2015). Gamification in online collaborative learning for programming courses: A literature review. *ARPN Journal of Engineering and Applied Sciences*, 18087–18094.
29. Cook, W. (s.f.). Training Today: 5 Gamification Pitfalls. Training Magazine.
30. Contreras-Espinosa, R. S., & Eguia, J. L. (2016). *Gamificación en aulas universitarias.* Bellaterra: Institut de la Comunicació, Universitat Autònoma de Barcelona. ISBN 978-84-944171-6-0.
31. Kapp, K. (2012). The gamification of learning and instruction. Game-based methods and strategies for training and education. Pfier.
32. Villalustre, L., & Moral, E. (2015). Gamificación: Estrategia para optimizar el proceso de aprendizaje y la adquisición de competencias en contextos universitarios. *Digital Education Review*.
33. Khaleel, F., Wook, T., Ashaari, N., & Ismail, A. (2016). Gamification elements for learning applications. *International Journal on Advanced Science, Engineering and Information Technology*, 1–8.
34. Li, C., Dong, Z., Untch, R., Chasteen, M (2013). Engaging computer science students through gamification in an online social network based collaborative learning environment. *International Journal of Information and Education Technology*, 1–10.
35. Hamari, J., Koivisto, J., & Sarsa, H. (2014). Does gamification work?—A literature review of empirical studies on gamification. In *47th Hawaii international conference on system science*. IEEE.

Encouraging Student Motivation Through Gamification in Engineering Education

Santiago Criollo-C[1]([⊠]) and Sergio Luján-Mora[2]([⊠])

[1] Universidad de las Américas, Quito, Ecuador
Luis.criollo@udla.edu.ec
[2] University of Alicante, Alicante, Spain
sergio.lujan@ua.es

Abstract. At present, there is a wide variety of teaching methods and educational strategies mediated by technology: e-learning, m-learning, b-learning, gamification, flipped classroom, bring your own devices (BYOD), etc. This methodological variety creates an important student-participation space where challenges, problem solving and collaboration are put to the testing. Due to the need for communication, portability and ubiquity, several of these methods use mobile devices to achieve their goal. Laptops, smartphones and tablets, with access to wireless networks, can be used to improve teacher–student interaction and thereby stimulate students' motivation, participation and active learning. Gamification is a teaching-learning strategy, which incorporates knowledge based on games that can support and mostly motivate the work of students and teachers. This document presents the results of a study using the Kahoot Platform, which is a web service of social and gamified education learning, in order to create evaluations changed into games of knowledge and competition among students. When it is used as a questionnaire, it challenges users to answer the questions posed, with a controlled response time, accumulated points, badges and a podium with the first five places. This leads to an active and dynamic learning due to the strong playful component, competitive and teamwork. The results of the research indicate that students improved participation and motivation in classes, besides checking that they had better results when working with mobile devices. These data will allow to analyze future perspectives about the use of the learning platform based on games as an assistance in the teaching-learning process.

Keywords: Active learning · Dynamic learning · Gamification · Games · M-learning · Mobile learning

1 Introduction

Currently, information and communication technologies have produced significant changes in society, which must adapt to new ways of doing things. These changes affect the daily life of people, transform society, commerce, entertainment, communications, etc. Education systems are not unresponsive to the technological changes that happen every day and they must also respond with new teaching strategies. Students of the current generation are very passive since they are mostly accustomed to receiving

© Springer Nature Switzerland AG 2019
M. E. Auer and T. Tsiatsos (eds.), *Mobile Technologies and Applications*
for the Internet of Things, Advances in Intelligent Systems and Computing 909,
https://doi.org/10.1007/978-3-030-11434-3_24

multiple stimuli at once (Internet, music, television) and sometimes all at the same time. Students have grown up in the Internet era, they manage social networks and use mobile devices, all of which have changed the way they receive and process information. It means that being in the class with a teacher is boring for them; to solve this situation, strategies such as "learning by doing" can be used [1, 2].

Some researchers [3] found some obstacles to recent graduates compete with mid-level workers, a cause is poor quality learning. According to a study [4], the traditional lectures in the classrooms where the classes are received are not only boring, they are also inefficient and ineffective. The researchers suggest changing every 10 min with other teaching techniques that include some activity that encourages learning; the results indicate that in this way more students will be successful. It was found that undergraduate students who receive traditional lectures (stand-and-deliver) are 1.5 times more likely to fail than students who receive their classes using more stimulating active learning methods [4]. For this reason, new strategies based on current technologies are sought in order to educational institutions could change the traditional teaching model. The use of learning based on challenge and competition games has turned out to be a trend nowadays, revealing good results [5, 6].

Currently, there is a large number of previous works that show pedagogical tools that teachers can use to make their class more attractive to the student. These new forms of teaching should include mobile devices and gamification as a more natural way of learning for new generations [1]. For example [1, 7], they make a comparison between several ludic platforms and generate recommendations for the use of each of them. The study [8] was aimed at analyzing the perspectives of future teachers and teacher educators on the use of the game-based learning platform. Other studies [5, 9] analyze these platforms as learning strategies. Our work, through a survey analyzes the general impressions of the student, in terms of ease of use, utility, entertainment, interest, competence and motivation. The research question that guided this study was the following: "Can the attention, motivation, participation and learning experience of students be improved by introducing gamification in a classroom that uses traditional teaching"? The survey also addresses the issue of the advantages of the tendency to bring your own devices (BYOD) and the limitations and barriers that prevent it.

This work is structured in four sections. In this first section, we provide a brief description about the use of technology in the classroom, we refer several previous works that use gamification in support of education. The hypothesis and the problem addressed by the research are also established. The second section indicates the method used to find the students' perception regarding gamification and the use of mobile devices in the classroom. The discussion and results are presented in the third section. Finally, the fourth section contains the conclusions of the investigation.

2 Research Method

2.1 Design

The teaching method has a lot to do with motivating students to participate in activities within the classroom, it is important to explain what the purpose of using gamification and evaluate if they are effectively motivated with this teaching technique. It is possible to develop specific activities with gamification web tools, these tools will guide the students to challenge themselves and, possibly, their classmates. This investigation used the Kahoot platform. This tool allows to create educational environments using games to generate learning in a fun way [8]. The learning platforms based on games work on a questionnaire concept. Here, the participants compete with each other to obtain points if they can choose the correct answer among several proposals.

The use of questionnaires is adequate to verify the student's activity, analyze learning data, measure levels of knowledge and make decisions based on the information obtained [8]. This study was carried out in the classrooms, with different groups of students in different subjects. The scenario included a projector, the mobile devices belonging to the students and the competition game developed on the Kahoot platform.

2.2   Participants

The research question was answered using a sample of 86 students from the electronics and information networks career. For the activity, a subject was selected from each of the main thematic axes of the career: Infrastructure, Electronics and Management. The subjects chosen were: Networking, Electrical Engineering and Network Management, respectively. This activity took place in the period between September 2017 and February 2018. The participants in this study were undergraduate students of the Universidad de las Americas in Quito-Ecuador. All the participants were surveyed at the end of the instruction using the Microsoft collaboration tools. The age of the participants was between 18 and 25 years old, with 87% of male participation and 13% of female participation.

2.3   Measuring Instruments

We used a survey that contained of six questions. The survey was built to based Likert scale of three points: disagreement, neutral and agree [10]. The survey was designed in order to capture the general impressions of the student, in terms of ease of use, utility, entertainment, interest, competence, and motivation. In addition, this survey included questions about the benefits of bringing your own mobile devices to support the classroom (BYOD trend in the classroom), as well as the limitations and barriers that prevent it. In the first part of the research, the most relevant content of the selected theme is chosen to turn it into a playful activity. In the second part, the platform is used to carry out a proposed questionnaire in game form. In the third part, each student responds to a satisfaction survey that is carried out after the activity, finally the data are collected for analysis.

3   Results

The results of this research show that students agree with the advantages of the use of gamification in educational contexts. This methodology motivates the participation of students, being entertained generates greater interest and motivation, creates environments of participation, competition confidence, etc. Although the results of the satisfaction survey are positive, there are several limitations, for example, access to the Internet or mobile data plan, number of own mobile devices in the classroom, battery life, classroom infrastructure, design of the questionnaires in the Kahoot platform, etc.

At the end of the playful activity, all the participants carried out a survey. The analysis of each of the questions was made taking into account that the descriptive statistics have been summarized in the three categories: agreement, neutral and disagreement.

Table 1. Results of Question 1

Course	Students	Agree (%)	Neutral (%)	Disagree (%)
Course 1	35	43	29	29
Course 2	35	60	9	31
Course 3	16	50	6	44

Question 1. Was your participation in the activity using the Kahoot tool easy?

The values in Table 1 indicate the results perceived by the students regarding the ease of use of the tool used to carry out the activity. This first perception shows that there are some disadvantages and limitations when using the Kahoot tool. The main problems that the students identified were: The questions do not appear on mobile devices, students cannot choose more than one answer when answering the activity, not all students have access to a mobile data plan and the connection through the Wi-Fi network in the university it is not fast, not all the students have a smartphone and they have to do the activity on a desktop computer which is a disadvantage of time, etc.

Question 2. Was your participation in the activity useful for the development of the course?

Table 2. Results of Question 2

Course	Students	Agree (%)	Neutral (%)	Disagree (%)
Course 1	35	94	6	0
Course 2	35	91	6	3
Course 3	16	94	6	0

The values in Table 2 indicate the utility perceived by the students when completing the gamification activity in the classroom. This point is very important since the utility together with the ease of use of the play tool increases the satisfaction of learning. In the same way, the utility and satisfaction of learning are related to the creation of intentions of use. The data indicate that for the students, this activity was very useful.

Question 3. Was your participation in the activity satisfactory/entertaining?

Table 3. Results of Question 3

Course	Students	Agree (%)	Neutral (%)	Disagree (%)
Course 1	35	77	20	9
Course 2	35	71	20	3
Course 3	16	63	11	6

The values in Table 3 indicate the satisfaction perceived by the students after performing the recreational activity. This is related to the ease of use and the utility of the gamification tool. Satisfaction does not have an encouraging result, this may be related to the low level of perception that students had about the ease of use of Kahoot tool. Although the ease of use did not have the best reception, the students had fun with the activity.

Question 4. Did your participation in the activity increase your motivation and interest in the subject?

Table 4. Results of Question 4

Course	Students	Agree (%)	Neutral (%)	Disagree (%)
Course 1	35	86	8	6
Course 2	35	91	9	0
Course 3	16	88	6	6

The values in Table 4 indicate the perception of students' motivation after performing the gamification activity. The data show that the activities of play and challenge in the classroom motivated students and increased interest in the subject.

Question 5. Was the competition generated in the development of the activity motivating?

Table 5. Results of Question 5

Course	Students	Agree (%)	Neutral (%)	Disagree (%)
Course 1	35	97	3	0
Course 2	35	100	0	0
Course 3	16	94	0	6

The values in Table 5 indicate the students' perception of the motivation produced by the competence inherent in gamification. According to the results of the survey, the majority of students are motivated to compete, their interest and performance for learning increases when they are challenged in a competition by the rest of their classmates.

Question 6. Can you bring your own mobile device to the classroom?

Table 6. Results of Question 6

Course	Students	Agree (%)	Disagree (%)
Course 1	35	57	43
Course 2	35	43	57
Course 3	16	63	37

The values in Table 6 indicate the willingness of students to bring their own devices to the classroom. According to the results, there is a resistance on the part of the students towards the BYOD tendency. The majority surveyed agreed on the risks involved with BYOD, the main ones were: Risk of assault, risk of information security and, risk of virus infection. This last question was not evaluated with the Likert scale.

Finally, Fig. 1 shows the total results of the responses to the survey. This graph shows the acceptance by students of the use of gamification tools in the classroom. It

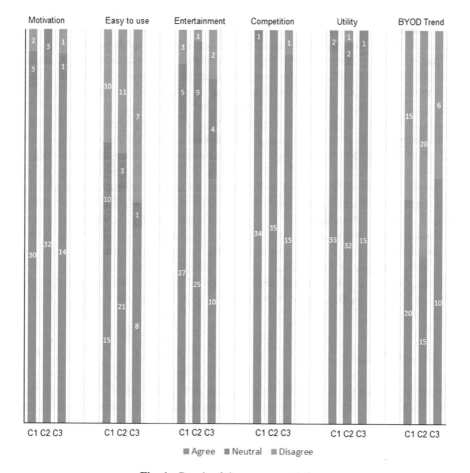

Fig. 1. Result of the survey carried out

can be seen that of the 86 students only 45 of them can to bring their own device to the classroom. Therefore, with this results, the low acceptance of the BYOD trend can be seen.

4 Conclusion

The current world is constantly changing, driven by information and communications technology. Technology is replaced by workers who perform routine functions, at the same time it complements workers with higher level skills, which allows them to be more productive and creative. Innovative companies, industries, and advanced economies offer many opportunities. They value more those who can adapt and contribute with communication skills, problem-solving and critical thinking [11]. For this reason, educational models must recognize the need to be able to adapt to new forms of teaching and learning, which allow them to prepare their students to successfully face the vertiginous change of technology of the twenty-first century.

The results of this research show that the use of gamification in educational contexts increases motivation, interest, and satisfaction. In addition, the usefulness of these activities creates fun environments for learning and participation. Although these platforms are free and available for the implementation of gamification, teacher training is necessary to successfully apply this methodology. Although the responses to the activity and the questionnaire are positive, there are some disadvantages and limitations that must be resolved before introducing this methodology into a higher education institution. These problems are identified by the students, examples of them are access to the Internet or data plan, number of own mobile devices in the classroom, battery life, classroom infrastructure, design of questionnaires and choice of digital platform, etc.

Another aspect of the activity with Kahoot is the relevant information about the learning that is obtained. If the questionnaires are well designed, the answers obtained from the students can measure the level of knowledge, identify the skills acquired and most importantly, evaluate the knowledge. As a cognitive tool, it can lead students to get involved and think more deeply about the subject under study, therefore, this methodology facilitates the construction of knowledge and reflection on the part of the students. At present, free digital platforms are suitable for the implementation of gamification in education, but it is necessary to train teachers for the effective use of this methodology.

5 Future Works

The use of Kahoot or other similar systems can pose problems of accessibility, these platforms may not be prepared for students with disabilities. It should be studied what are the main problems that people with disabilities could have when using these platforms. If we want to have an inclusive education, all these tools should be accessible. Otherwise, we should look for another tool or another method to include the gamification of education for all.

References

1. Chaiyo, Y. (2017, March 1–4). The effect of Kahoot, Quizizz and Google Forms on the student's perception in the classrooms response system. In *2nd International Conference on Digital Arts, Media and Technology (ICDAMT)*, Chiang Mai, Thailand (pp. 178–182).
2. Araújo, I. (2017, November 9–11). Empowering teachers to apply gamification. In *International Symposium on Computers in Education (SIIE)*, Lisbon, Portugal (pp. 1–5).
3. Prabawa H. (2017, October 25–26). Gamification development in attainment concept model learning for students' comprehension enhancement. In *3rd International Conference on Science in Information Technology (ICSITech)*, Bandung, Indonesia (pp. 685–689).
4. Freeman, S. (2014). Active learning increases student performance in science, engineering, and mathematics. *Proceedings of the National Academy of Sciences, 111*(23), 8410–8415.
5. Abidin, H. (2017, November 9–10). Students' perceptions on game-based classroom response system in a computer programming course. In *9th International Conference on Engineering Education (ICEED)*, Kanazahua, Japon (pp. 8410–8415).
6. Antonio, S. (2017, , October 18–21). Gamification in education: A methodology to identify student's profile. In *Frontiers in Education Conference (FIE)*, Indianapolis, USA (pp. 1–8).
7. Seralidou, E. (2018, April 17–20). Let's learn With Kahoot!. In *Global Engineering Education Conference (EDUCON)*, Sta Cruz de Tenerife, Canary Islands, Spain (pp. 683–691).
8. Correia, M. (2017, November 9–11). Game-based learning: The use of Kahoot in teacher education. In *International Symposium on Computers in Education (SIIE)*, Lisbon, Portugal (pp. 0–3).
9. Godinho, W. (2017, June 21–24). Quizzes as an active learning strategy: A study with students of pharmaceutical sciences. In *12th Iberian Conference on Information Systems and Technologies (CISTI)*, Lisboa, Portugal (pp. 1–6).
10. Wang, A. (2015). The wear out effect of a game-based student response system. *Computers and Education, 82,* 217–227.
11. James, B. (2010). *21st century skills rethinking how students learn.* Monroe, Indiana, USA: Solution Tree Press.

Design of Internet of Things Devices and Applications

A Survey of Mobile Learning Approaches for Teaching Internet of Things

Sigrid Schefer-Wenzl[✉], Igor Miladinovic, and Alice Ensor

University of Applied Science Campus Vienna, Favoritenstr. 226, 1100 Vienna,
Austria
{sigrid.schefer-wenzl,igor.miladinovic,alice.ensor}@fh-campuswien.ac.at

Abstract. Over the past few years, the interest in the Internet of Things
(IoT) as well as the integration of mobile technology into education—
called mobile learning—is rapidly rising. The purpose of this paper is to
review mobile learning approaches that have been applied to teach IoT
or related subjects and to derive lessons learned for designing mobile-
learning-based IoT courses. To achieve this, we performed a systematic interdisciplinary literature review. The contribution of this paper is
threefold. First, we present a survey of all identified publications in the
research area, providing insights into the development of this field and
showing its emerging importance. Second, we discuss different mobile
learning approaches, focusing on approaches concerned explicitly with
the special demands of these concepts in the context of IoT and related
Computer Science topics. Third, by deriving lessons learned we provide a
foundation for the informed selection of suitable mobile learning concepts
as well as for evaluating future research in this area.

Keywords: Computer Science · Internet of things ·
Literature review · Mobile learning

1 Introduction

Over the past three decades, teaching practices have increasingly shifted from
instructor-centered to student-centered learning concepts [5]. Additionally, in
the area of teaching and learning in higher education, mobile computing technologies have become a very important research topic. The use of mobile devices
for educational purposes has led to what is known as mobile learning, providing new opportunities for active learning by utilizing the advantages of mobility
and wireless technology. However, in the context of Computer Science education, traditional-style lectures are still much more common than mobile learning
settings [16,17,45].

© Springer Nature Switzerland AG 2019
M. E. Auer and T. Tsiatsos (eds.), *Mobile Technologies and Applications
for the Internet of Things*, Advances in Intelligent Systems and Computing 909,
https://doi.org/10.1007/978-3-030-11434-3_25

At our university, we are currently introducing two new courses into our Bachelor degree program "Computer Science and Digital Communications." These courses deal with topics in the area of the Internet of Things (IoT), focusing on relevant architectures, network concepts, and applications. The term IoT refers to the interconnection of billions of smart devices [19]. The steadily rising number of IoT devices with heterogeneous characteristics requires that future networks evolve to provide a new architecture to cope with the expected growth in data generation [47]. These developments have led to a major increase in demand from industry for students equipped with skills necessary in the IoT area. Thus, introducing IoT concepts in Computer Science studies is important to prepare today's students to be effective IoT developers and designers.

In the past, we successfully applied mobile learning for teaching software engineering, achieving strong improvements in learning outcomes [42,43]. As software engineering is also a part of IoT, we plan to extend our course concept to the two new IoT courses. As a first step toward this goal, we want to thoroughly investigate existing mobile learning approaches for teaching IoT. Leveraging these two activities will support us in selecting suitable didactic concepts for teaching IoT topics.

This paper presents the results of a literature review on mobile learning in the area of Computer Science studies, especially focussing on teaching IoT topics. In Sect. 2, we give an overview on the development of the research areas mobile learning, IoT, the combination of both topics, as well as the development of mobile learning in software engineering. Section 3 discusses selected mobile learning approaches and approaches for teaching IoT. Subsequently, Sect. 4 presents lessons learned from this study and Sect. 5 concludes the paper.

2   Development of the Research Area

In order to identify relevant mobile learning concepts for teaching IoT, we have carried out a literature review based on the guidelines presented in [6,24,50]. The search for relevant literature was conducted across four digital libraries that index scientific articles in Information Systems and Computer Science. In particular, our search included the following libraries: ACM Digital Library, IEEE Digital Library, Springer Link, and AIS Electronic Library.

To analyze the development of the research area in a first step, the databases were searched for contributions published until the end of 2017 containing in their full text at least one of the following search terms: "mobile learning", "internet of things", as well as "mobile learning," and "internet of things". The types of literature uncovered in this survey include conference papers, academic studies, book chapters, books, and articles from academic journals. The search results from all databases were combined and double entries were eliminated, leaving the following samples: 36,582 publications that included the term "internet of things", 5,119 publications dealing with "mobile learning", and 108 results that pertained to both IoT and mobile learning.

The first comprehensive result using the term "internet of things" was published in 2002 [4] citing a first appearance of this term in an MIT publication from

1997 [7]. The amount of publications per year in this research area increased consistently and this development is still ongoing, giving that there were 13,057 publications indexed in the examined scientific libraries solely in 2017 (see Fig. 1). In comparison, publications using the term "mobile learning" first appear in 1997 [39,44]. As also shown in Fig. 1, the number of publications on mobile learning has steadily increased, with a significant peak of 801 publications in 2015, remaining rather constant since then. A first approach applying mobile learning on IoT appeared in 2008 [18]. Figure 2 illustrates the development of publications using both terms. The number of relevant publications more than doubled from 2016 (20 publications) to 2017 (53 publications). Overall, the increasing total number of publications over the past few years in all three research areas demonstrates the increasing interest in these topics.

In a second step, we expanded the research by defining additional search terms with the goal of finding more mobile-learning-based approaches that can be applied for teaching IoT topics. These search terms were picked based on our previous knowledge of the research area as well as on screening searches. In particular, they were selected for their relevance to mobile learning in Computer Science, mobile learning at the university level, or teaching IoT. In this way, additional relevant papers could be included that dealt with, e.g., mobile learning in software engineering. This is because software engineering is also a part of IoT and thus corresponding teaching concepts might also be applicable to implementing a mobile learning course for IoT. Due to page restrictions, we only present the results for the search terms "mobile learning" and "software engineering" (see Fig. 2) in this paper. The numbers demonstrate that with the rising interest in mobile learning, researchers also increasingly dealt with applying these concepts in the area of software engineering. First corresponding teaching approaches were presented in 2002 (see, e.g., [31]).

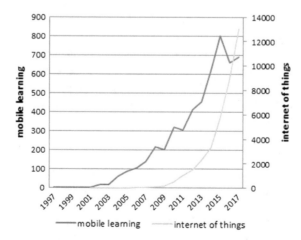

Fig. 1. Number of publications per year including the term "internet of things" or "mobile learning"

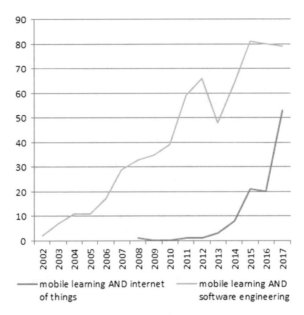

Fig. 2. Number of publications per year including the terms "mobile learning" and "internet of things" or "mobile learning" and "software engineering"

3   Discussion of Selected Approaches

To provide a better overview of the existing body of knowledge, we have thus further focused our study by selecting 168 articles that explicitly aim to evolve mobile learning approaches for IoT and related topics for an in-depth investigation. In particular, these articles all dealt with the presentation or active discussion of an implemented mobile learning program, learning theories that might improve the method, proposals for a mobile learning architecture or curriculum, or other e-learning-based approaches with interesting elements for mobile learning. Also included were articles covering mobile learning literature reviews because they can be further researched for additional sources and often contained proposals based on their results. In this section, we present 34 of these publications that were found to be particularly interesting for our goal to develop a mobile learning course to teach IoT topics.

3.1   Approaches for Teaching IoT

Ma et al. [32] use mobile devices to teach embedded systems, which are part of an IoT environment. The authors designed a mobile learning tool as a labware to engage students' active and hands-on learning in topics of embedded systems design and embedded security. Preliminary student feedback was encouraging and they showed further interest in learning about the software and hardware involved with embedded systems.

Another mobile-learning-based concept is introduced in [3]. It includes IoT models in the areas of home automatization and real-time sensor systems that are taught through a mobile application. Students showed an increased interest in these topics and it allowed them to gain practical knowledge. A flipped classroom approach in teaching an undergraduate level IoT development course is introduced in [27]. The authors used online video lectures so that more class time can be devoted to lab work. The proposed course design involved a large time investment on the instructors part, but overall they were able to better meet their students' needs compared to traditional lecture styles.

Samuel et al. [40] introduce the Microsoft Research Lab of Things and discuss what role it can play in education, specifically for teaching courses about IoT. They present an example course on smart home automation as well as student projects. In [21], positive experiences with applying this platform to teach a course on IoT application development at Kookmin University are reported. The team received positive feedback from students and felt they have gained practical knowledge in implementing IoT systems. Students found that the platform supported them especially in learning a new programming language (C#).

Förster et al. [15] describe the University of Bremen's approach to teaching IoT with hands-on labs and concept abstraction. It mainly used an e-learning platform to give assignments. The paper provides an overview of topics covered in their course weekly and also addresses different challenges such as dealing with new relevant platforms emerging every day. Overall, the students gave the course good evaluations. However, the instructors found the class material difficult to prepare, but worth it in the end.

Other not mobile-learning-based approaches also have elements that could be integrated into a mobile-learning-based IoT course. For example, in [33], the authors present a course about teaching IoT device prototyping that was held 3 years in a row. They used warm-up exercises, short teacher presentations and mainly project-based learning for teaching difficult topics. Despite different knowledge backgrounds, the students were able to produce versatile and complex student projects. Overall the courses were evaluated well. Ali [1] recounts the experiences, challenges, and reflections of teaching an IoT course. Students had a difficult time understanding some of the trickier concepts such as interrupts and wanted more opportunities to program. Dobrilovic and Stojanov [12] present an open-source IoT platform for use in university curricula. First, students' comments have also been positive.

3.2 Other Relevant Mobile Learning Approaches

There are many reports on applying mobile learning in the area of software engineering. For example, in [2], mobile learning was applied for teaching an introductory programming course. The authors found that the students were motivated and that the instructors were willing to try mobile devices. However, the paper does not provide a very detailed description of the study. Liang et al. [28] used a ubiquitous environment to teach students in C programming. This environment allowed students to practice their programming skills anywhere and at anytime. They were able to directly write, compile, execute, and debug their C programs on Android-based smartphones or tablets. Students found the concept interesting although the environment suffered from performance issues which hampered the learning experience. In [11], the authors created two apps and tested them in an introductory programming course. Results from surveys and tests showed a positive impact on student learning and engagement. An integrative approach for teaching embedded modeling and simulation in the context of software and telecommunication is presented in [34]. The authors present mobile and ubiquitous learning as an innovative way to teach these topics and describe the corresponding course structure, course offerings, and desired outcomes. However, the authors do not state if those outcomes were met. An evaluation study on mobile learning among 142 Computer Science undergraduates is presented in [35]. The authors found the results encouraging and that it improved learning outcomes.

Yu and Wang [48] describe a Massive Open Online Courses (MOOC) platform, especially designed for mobile learning that is used to provide 1000s of learners with online resources. It helps improve learner interaction and thus the quality of the experience. Similarly in [51], the authors evaluate differences in the performance, when a MOOC course is performed "on-the-go", i.e., via mobile learning or stationary. They found that users in the "on-the-go" mode performed less well than those in a stationary mode.

An approach for enhancing mobile learning via microlearning is introduced in [8]. The main goals were to overcome information overload experienced by users and to make repetition learning easier. They report on the results of three case studies with 100+ users. First results showed high usage and satisfaction among users. Fioravanti et al. [13] discuss the challenges in developing a mobile environment and present the prototypical implementation of two specific reference architectures for mobile learning. Experimental results show that users found their approach both satisfying and convenient. In [29], Socrative, which is a cloud-based audience response system, is used for educational purposes. It includes features such as in-class polling, multiple choice questions, and other interactive features. The paper describes a mobile-based interactive teaching model with in-class and off-class components aided by Socrative. Students felt their classroom experience was improved. However, attendance was not improved. In [23], a mobile learning artifact for students with a special focus on peer learning is presented. It is designed to connect with other peers in larger learning environments

Several papers discuss factors influencing the effectiveness of mobile learning. For example, a study on students' readiness for mobile learning is presented in [25]. The authors found a technological readiness and willingness to use applications on mobile devices. They found mobile learning to be especially suitable for teaching foreign languages, using game-based learning and also advocate for integrating audio-based learning resources. Poursaeed and Lee [38] investigate differences between mobile learning and ubiquitous learning, i.e., integrating the mobile learning application into a web portal. They evaluate efficiency, effectiveness, and acceptance of the two learning methods. Results showed that by adding a ubiquitous environment to a mobile learning curriculum, students showed an increase in motivation and learning efficiency. Overall, they reported good experiences. Krotov [26] introduces a qualitative case study to uncover several critical success factors for mobile learning, while Lu and Viehland [30] examine key factors determining the acceptance of mobile learning at six New Zealand universities. Similarly, a model that illustrates factors that may determine the effectiveness of a mobile learning system is introduced in [49]. A critical analysis of mobile learning, especially discussing the negative aspects of mobile learning and the interference of the technology in the classroom is presented in [36]. The authors especially stress that multitasking and orchestration need to be considered in a media-enriched mobile learning environment and it needs to be reflected in how far mobile learning affects the students' and teachers' roles.

Also, the usability of mobile learning environments is an important issue. A study with undergraduate students who were using mobile tablets to examine the effect it had on learning is presented in [22]. The results were compared to a control group. The authors found that the tablet group was more likely to read the assignments, but that it had a mixed result on their team dynamics and peer collaboration. However, this might be because the students did not quite know how to make use of related features. A systematic mapping study to evaluate the usability of mobile learning applications is presented in [9]. The authors found a surprising lack of usability tests and of frameworks and guidelines in this area. For this reason, the authors propose the creation of an evaluation framework.

In our study, we also came across several different literature reviews on mobile learning. We want to highlight some examples, such as the review of 889 studies concerning ubiquitous learning environments and thereby also including mobile learning environments presented in [46]. Cota et al. [10] present a research and systematic mapping study of mobile learning. It contains a literature review including the focus adopted by each publication. Furthermore, a model is proposed that is supposed to improve student experience and quality. Fong [14] presents a review of 105 articles about mobile learning and its implications. The author identified three trends, including the need for strategic planning and implementation of mobile learning as well as for effective instructional designs, and limitation of mobile technology. The main line of research for mobile learning as well as the necessary criteria for implementing mobile learning are described in [41]. The authors also try to connect the mobile learning methodology to different theories of education. In [20], a review on mobile learning and its differences from e-learning is described. It also includes a design for a mobile learning activity design model. Another comprehensive literature review on mobile learning projects is presented in [37].

4   Lessons Learned

One of the main advantages attributed to mobile learning is that it helps to address different levels of students' skills. This is a typical challenge in a Computer Science study, as especially programming skills of students are usually very diverse. Mobile learning may be applied in a way to ensure a certain entrance level of all students for a specific course. For example, mobile-ready warm-up exercises, i.e., quick exercises that students can do on their own before or during each lecture, have the benefit that students come to class having already read some material or having dealt with certain questions on a topic which increases their participation in-class discussions and problem-solving exercises.

Furthermore, mobile learning allows providing teaching resources in various formats to address different kinds of learning styles: video-, text-, and audio-based resources as well as interactive- or game-based exercises can be integrated into a mobile-learning-based course. Several approaches combine mobile learning with game-based teaching and report positive results because it can increase excitement and provides immediate rewards (see, e.g., [25]). However, it is frequently argued that designing resources for a mobile learning course requires a substantially high preparation effort for teachers.

Other advantages often mentioned are that mobile learning supports easy repetition of contents and different learning speeds. It is suitable to support peer learning, virtual collaboration, and teaching distributed and/or large groups. In such settings, the mobile learning environment needs to provide corresponding features, such as a forum, chat, or other collaboration tools.

In a typical university setting (i.e., not too large and not very distributed student groups), blended learning seems to provide a good learning setting as the benefits from both, distance learning and in-class hours, can be leveraged. When applying blended learning, in-class hours should be used for clarification of open questions, deepening knowledge of complex learning contents, or highly interactive exercises. It is also recommendable to provide ubiquitous functions to allow the full spectrum of advantages of mobile technologies and to enable students to learn anytime and anywhere.

Another lesson learned is to provide students with continuous feedback on their learning progress and to make the individual learning progress visible. This can be done, for example, by individual feedback from teachers, peer feedback from students, or by integrating game-based features, such as earning points for completed exercises or quizzes.

Important is a good structuring of the course and to ensure that people know about the features of the mobile learning environment. A short tutorial on features and how to use them might be a good idea. This leads to an important point as several papers report that performance and usability issues of the mobile learning environment used have contributed to students' frustration. Especially, usability is often neglected. Thus, prior usability tests should be performed.

Several publications cited provide recommendations on mobile/ e-learning/ blended-learning-based curricula for IoT courses (see, e.g., [1,3,15,27,32,33,40]). Specific challenges for courses in the IoT area are identified by different authors. One is to address the diverse levels of programming, networking and other Computer-Science-related skills of the participating students. As IoT is an interdisciplinary topic that combines the knowledge of highly complex areas, it is critical for the success of every IoT course to deal with these different skill levels. However, as stated above, mobile learning provides an ideal setting to deal with this challenge. In addition, IoT is a rapidly evolving research area. Thus, contents need to be updated constantly. New IoT platforms are emerging frequently, requiring that students are able to deal with many different software and hardware platforms. Thus, it is particularly important to teach the underlying concepts that enable students to transfer this knowledge to specific systems. Examples for such topics are given in [15]. Another frequently stated challenge is to enable students to apply their knowledge in realistic, industry-like projects. Again, mobile learning provides many features to support project-based learning. In a mobile learning setting, it is further necessary to provide sensors and actuators that can be accessed remotely or can be simulated via mobile phones. Connection to cloud services is also necessary to deploy IoT applications.

Finally, it should be noted that introducing innovative and unusual teaching concepts might always affect the results of an education process by acceptance or rejection of these innovations. Although students—and especially Computer Science students—today are usually very familiar with using mobile technology, mobile learning still is a teaching method that is uncommon in a university setting. Thus, it is recommendable to introduce mobile learning throughout the whole curriculum, to provide a slow start into this teaching method and to offer alternatives for students not ready for mobile learning.

5   Conclusion

Until now, there are only few publications that explicitly present mobile learning approaches for teaching IoT. However, our results demonstrate that research in the area has grown significantly over the past few years. Due to the recent boom of integrating mobile devices into teaching settings, we believe that the number of publications will continue to grow in the following years.

Furthermore, our results show that mobile learning has already been successfully applied to teach IoT and related topics repeatedly. Our main lessons learned for designing a mobile-learning-based course for teaching IoT are to leverage the advantages mobile learning provides to address different levels of skills by, e.g., warm-up exercises for complex subjects, to use the opportunity for providing teaching material for different learning styles also in audio- and video-format, to use in-class hours for interactive and deepening exercises, to provide continuous feedback, to consider usability issues, and optionally to integrate game-based elements and collaboration tools.

Specific lessons learned for IoT courses are to use a mobile learning setting to abstract from the rapidly changing IoT platforms, to support project-based learning, to deploy sensors and actuators that can be accessed remotely or can be simulated via smartphones and to provide cloud services.

In future work, we plan to expand our literature review to other scientific databases, to include further keywords that are related to mobile learning and IoT, to provide detailed mobile-learning-based designs for IoT courses, and to continuously evaluate results based on students' feedback and grades.

References

1. Ali, F. (2015). Teaching the internet of things concepts. In *Proceedings of the Workshop on Embedded and Cyber-Physical Systems Education*. New York, NY, USA.
2. Alsaggaf, W. (2012). Enhancement of learning programming experience by novices using mobile learning: Mobile learning in introductory programming lectures. In *Proceedings of the 9th Annual International Conference on International Computing Education Research*. New York, NY, USA (2012).
3. Babic, F., & Gaspar, V. (2017). Mobile technologies education based on smart laboratory models. In *Proceedings of the 15th International Conference on Emerging eLearning Technologies and Applications* (Oct 2017).
4. Boukraa, M., & Ando, S. (2002). Tag-rased vision: Assisting 3d scene analysis with radio-frequency tags. In *5th International Conference on Information Fusion* (July 2002).
5. Boyer, E. (1997). *Scholarship reconsidered: Priorities of the professoriate*. Hillsdale, NJ: Jossey-Bas.
6. Brereton, P., Kitchenham, B. A., Budgen, D., Turner, M., & Khalil, M. (2007). Lessons from applying the systematic literature review process within the software engineering domain. *Journal of Systems and Software, 80*(4), 571–583.
7. Brock, S. A. (1999). *The networked physical world*. MIT Auto-ID Center White Paper.
8. Bruck, P. A., Motiwalla, L., & Foerster, F. (2012). Mobile learning with micro-content: A framework and evaluation. In *Bled eConference*.
9. Cota, C. X. N., Díaz, A. I. M., & Duque, M. A. R. (2014). Developing a framework to evaluate usability in m-learning systems: Mapping study and proposal. In *Proceedings of the 2nd International Conference on Technological Ecosystems for Enhancing Multiculturality*. New York, NY, USA (2014).
10. Cota, C. X. N., Molina, A. I., & Redondo, M. A. (2014). Evaluation framework for m-learning systems: Current situation and proposal. In *Interacción*.
11. Dekhane, S., & Johnson, C. (2014). Using mobile apps to support novice programming students. In *Proceedings of the 15th Annual Conf. on Information Technology Education*. New York, NY, USA.
12. Dobrilovic, D., & Stojanov, Z. (2016). Design of open-source platform for introducing internet of things in university curricula. In *Proceedings of the 11th International Symposium on Applied Computational Intelligence and Informatics* (May 2016).
13. Fioravanti, M. L., Filho, N. F. D., Fronza, L. B., & Barbosa, E. F. (2017). Towards a mobile learning environment using reference architectures. In *Hawaii International Conference on System Sciences 2017 (HICSS-50)*. Hawaii (January 2017).
14. Fong, W. W. (2013). The trends in mobile learning. In S. K. S. Cheung, J. Fong, W. Fong, F. L. Wang, & L. F. Kwok (Eds.), *Hybrid learning and continuing education* (pp. 301–312). Berlin, Heidelberg: Springer.
15. Förster, A., Dede, J., Könsgen, A., Udugama, A., & Zaman, I. (2017). Teaching the internet of things. *GetMobile: Mobile Computing and Communications, 20*(3), 24–28.
16. Gary, K., Lindquist, T., Bansal, S., & Ghazarian, A. (2013). A project spine for software engineering curricular design. In *Proceedings of the 26th International Conference on Software Engineering Education and Training* (May 2013).

17. Ghezzi, C., & Mandrioli, D. (2006). *The challenges of software engineering education* (pp. 115–127). Berlin, Heidelberg: Springer.
18. Gonzalez, G. R., Organero, M. M., & Kloos, C. D. (2008). Early infrastructure of an internet of things in spaces for learning. In *Proceedings of the 8th International Conference on Advanced Learning Technologies* (July 2008).
19. Hanes, D., Salgueiro, G., Grossetete, P., Barton, R., & Henry, J. (2017). *IoT fundamentals: Networking technologies, protocols, and use cases for the internet of things*. Pearson Education: Fundamentals.
20. Huang, R., Zhang, H., Li, Y., & Yang, J. (2012). A framework of designing learning activities for mobile learning. In S. K. S. Cheung, J. Fong, L. F. Kwok, K. Li, & R. Kwan (Eds.), *Hybrid learning* (pp. 9–22). Berlin, Heidelberg: Springer.
21. Jeong, G. M., Truong, P. H., Lee, T. Y., Choi, J. W., & Lee, M. (2016). Course design for internet of things using lab of things of microsoft research. In *IEEE Frontiers in Education Conference (FIE)* (pp. 1–6).
22. Kaganer, E. A., Giordano, G. A., Brion, S., & Tortoriello, M. (November 2013). Media tablets for mobile learning. *Communications of the ACM, 56*(11), 68–75.
23. Kallookaran, M., & Siemon, D. (2017). Using mobile learning to create a reciprocal peer learning environment. In *AMCIS 2017*. Boston, Massachusetts, USA.
24. Kitchenham, B., Brereton, O. P., Budgen, D., Turner, M., Bailey, J., & Linkman, S. (2009). Systematic literature reviews in software engineering—A systematic literature review. *Information and Software Technology, 51*(1).
25. Kopackova, H., & Bilkova, R. (2014). Mobile devices in learning—Are students ready for the change? In *Proceedings of the 12th IEEE International Conference on Emerging eLearning Technologies and Applications*.
26. Krotov, V. (2015). Critical success factors in m-learning: A socio-technical perspective. *Communications of the Association for Information Systems, 36*(6).
27. Lei, C., Yau, C., Lui, K.., Yum, P., Tam, V., Yuen, A.H., et al. (2017). Teaching internet of things: Enhancing learning efficiency via full-semester flipped classroom. In *Proceedings of the 6th International Conference on Teaching, Assessment, and Learning for Engineering*.
28. Liang, T. Y., Li, H. F., & Chen, Y. C. (2014). A ubiquitous integrated development environment for c programming on mobile devices. In *Proceedings of the 12th International Conference on Dependable, Autonomic and Secure Computing*.
29. Lim, W. N. (2017). Improving student engagement in higher education through mobile-based interactive teaching model using socrative. In *IEEE Global Engineering Education Conference (EDUCON)* (pp. 404–412).
30. Lu, X., & Viehland, D. (2008). Factors influencing the adoption of mobile learning. In *ACIS 2008 Proceedings*. New Zealand.
31. Luchini, K., Bobrowsky, W., Curtis, M., Quintana, C., & Soloway, E. (2002). Supporting learning in context: Extending learner-centered design to the development of handheld educational software. In: *Proceedings. IEEE International Workshop on Wireless and Mobile Technologies in Education* (pp. 107–111).
32. Ma, K., Ma, Y., Wang, Y., Qian, K., Zheng, Q., & Hong, L. (2015). Innovative mobile tool for engineering embedded design and security educations. In *IEEE Frontiers in Education Conference (FIE)* (pp. 1–4).
33. Mäenpää, H., Varjonen, S., Hellas, A., Tarkoma, S., & Männistö, T. (2017). Assessing IOT projects in university education: A framework for problem-based learning. In *Proceedings of the 39th International Conference on Software Engineering*. Piscataway, NJ, USA.

34. Möller, D. P. F., Haas, R., & Vakilzadian, H. (2013). Ubiquitous learning: Teaching modeling and simulation with technology. In *Proceedings of the Grand Challenges on Modeling and Simulation Conference*. Vista, CA.

35. Oyelere, S. S., Suhonen, J., Wajiga, G. M., & Sutinen, E. (2016). Evaluating mobileedu: Third-year undergraduate computer science students' mobile learning achievements. In *Proceedings of the 16th Koli Calling International Conference on Computing Education Research*. New York, NY, USA.

36. Pedro, L. F. M. G., Barbosa, C. M. M. D. O., & Santos, C. M. D.N. (2018). A critical review of mobile learning integration in formal educational contexts. *International Journal of Educational Technology in Higher Education ,15*(1), 10.

37. Pereira, O. R. E., & Rodrigues, J. J. P. C. (2013). Survey and analysis of current mobile learning applications and technologies. *ACM Computing Surveys, 46*(2), 27:1–27:35.

38. Poursaeed, B., & Lee, C. (2010). Self-initiated curriculum planning, visualization and assessment in improving meaningful learning: A comparison between mobile and ubiquitous learning. In *International Conference on Technology for Education*.

39. Rieger, R., & Gay, G. (1997). Using mobile computing to enhance field study. In *Proceedings of the 2nd International Conference on Computer Support for Collaborative Learning*.

40. Samuel, A., Mohamedally, D., Banerjee, N., Brush, A. J., & Mahajan, R. (2015). Lab of things in education. *GetMobile: Mobile Computing and Communications, 19*(1), 18–24.

41. Sánchez Prieto, J. C., Migueláñez, S. O., & García-Peñalvo, F. J. (2013). Mobile learning: Tendencies and lines of research. In *Proceedings of the 1st International Conference on Technological Ecosystem for Enhancing Multiculturality*. New York, NY, USA.

42. Schefer-Wenzl, S., & Miladinovic, I. (2017). Game changing mobile learning based method mix for teaching software development. In *mLearn 2017, 16th World Conference on Mobile and Contextual Learning*.

43. Schefer-Wenzl, S., & Miladinovic, I. (2018). Leveraging collaborative mobile learning for sustained software development skills. In *Proceedings of the 21 International Conference on Interactive Collaborative Learning*.

44. Sendov, B. (1997). Towards global wisdom in the era of digitalization and communication. *Prospects, 27*(3), 415–426.

45. Tillmann, N., Moskal, M., de Halleux, J., Fahndrich, M., Bishop, J., Samuel, A., et al. (2012). The future of teaching programming is on mobile devices. In *Proceedings of the 17th ACM Annual Conference on Innovation and Technology in Computer Science Education*. New York, NY, USA (2012)

46. Virtanen, M. A., Haavisto, E., Liikanen, E., & Kääriäinen, M. (2018). Ubiquitous learning environments in higher education: A scoping literature review. *Education and Information Technologies, 23*(2), 985–998.

47. Weldon, M. K. (2016). *The Future X Network*. A Bell Labs Perspective: CRC Press.

48. Yu, M., Wang, J.: The design and implementation of a mobile massive open online courses platform. In: Proc. of Intern. Conf. on Information Integration and Web-based Applications & Services. New York, NY, USA (2013)
49. Yunis, M.M., Liu, L.C., Koong, K.S.: Towards a framework for perceived effectiveness of mobile learning. In: AMCIS 2011 Proceedings. Detroit, Michigan, USA (August 2011)
50. Zhang, H., & Babar, M. A. (2013). Systematic reviews in software engineering: An empirical investigation. *Information and Software Technology, 55*(7),
51. Zhao, Y., Robal, T., Lofi, C., Hauff, C.: Stationary vs. non-stationary mobile learning in moocs. In: Proc. of the 26th International Conference on User Modeling, Adaptation and Personalization (June 2018)

The Design and Implementation of a Low-Cost Demo Tool to Teach Dynamics in the IOT Era

Timber K. M. Yuen[✉], Lucian Balan, and Moein Mehrtash

W. Booth School of Engineering Practice and Technology, McMaster University,
1280 Main Street West, ETB 509, Hamilton, ON L8S 0A3, Canada
timber@mcmaster.ca

Abstract. A primary concern in the teaching of an engineering curriculum is to deliver interesting contents to keep the students engaged during a lecture. In this IOT era, students rely on resources posted online by their professors to preview the lectures. The extensive use of online postings, on one hand, provides much convenience to the students; however, it also creates problems to the professors on how to exceed the students' expectations such that they would choose to attend the lectures. This paper describes a solution to the challenge of creating interesting contents for a machine dynamics/mechanical vibrations course. The approach includes the use of a live demonstration platform to facilitate in-class participation and discussion. The use of demos not only will make the explanation of complex concepts much easier, it will also help the students retain concepts discussed in class much easier.

Keywords: Hand tremor · Vibration absorber · Demo tool · Teaching resources design · Hands-on learning

1 How to Encourage Students to Go to Lectures?

One of the major challenges for professors today is to make their lecture material interesting. Students often decide whether to go to a lecture based on whether they think the professor will have something interesting to say in class. If a professor simply reads off the presentation slides, chances are that most students will not show up to the next class. This often puts the course in a downward spiral and the professor would think that since the students did not care to come to class anyway, there is no point putting in the effort in preparing the lecture.

As educators, we often face the challenges of posting just enough of lecture notes online to tease the students so that they would come to class to listen to the whole story. Engineers, in general, like to solve problems. Therefore, being able to identify the problem at hand and learn to come up with a solution is often considered an essential ingredient for a successful lecture.

© Springer Nature Switzerland AG 2019
M. E. Auer and T. Tsiatsos (eds.), *Mobile Technologies and Applications
for the Internet of Things*, Advances in Intelligent Systems and Computing 909,
https://doi.org/10.1007/978-3-030-11434-3_26

2 Making a Vibration Problem Interesting

Essential tremor is a common involuntary movement disorder that affects the lives of many people in the world. Patients with this disability will see their hands oscillating in the range of 4–12 Hz [1]. It is very difficult for the people suffering from hand tremor to perform daily tasks such as eating and drinking. Medical treatments and brain surgery can be applied to tremor patients. However, for people not responding well to medical treatments, mechanical vibration absorbers can be a good alternative [2–4].

In order to study the effects of vibration absorbers on hand tremor patients, a model of such a patient has been designed and constructed (see Fig. 1). A model, although incomplete, is very useful in the studying of a system and in predicting the effects of different components. To narrow the scope of the project, we designed the model to simulate a patient with hand tremor sitting at a dining table trying to eat with a spoon. A DC fan with rotating unbalance is mounted on the model to create the tremor. The benefit of using such a model to study a vibration problem is that very repeatable results can be expected. Also, since the model is made out of wood, it is very lightweight and portable. Therefore, it could be easily moved around and used as a demo tool during lectures.

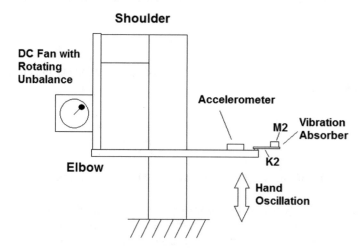

Fig. 1. Hand tremor patient model

3 Vibration Absorber Design

When a single degree of freedom system is subjected to a forcing function input $f(t) = Fo \sin \omega t$ along the x-axis, the following differential equation can be used to describe the system [5].

$$M\ddot{x} + C\dot{x} + Kx = F_o \sin \omega t \qquad (1)$$

For a rotating unbalance with mass m and eccentricity R, the amplitude of the input forcing function is

$$F_o = mR\omega^2 \tag{2}$$

Since the transient response of the system disappears quickly over time, it is common to consider just the steady-state response of the system. The steady-state response can be expressed as [5]

$$x\_ss(t) = \frac{F_o}{(K - \omega^2)^2 + (C\omega)^2} \left[(K - M\omega^2) \sin \omega t - C\omega \cos \omega t\right] \tag{3}$$

The steady-state response can also be expressed in an alternative form [5]:

$$x\_ss(t) \frac{F_o}{K} \frac{1}{\sqrt{(1 - r^2)^2 + (2\zeta r)^2}} \sin (\omega t + \phi) \tag{4}$$

where r is the ratio between the input frequency and the system natural frequency.

$$r = \frac{\omega}{\omega_n} \tag{5}$$

And the steady-state magnitude ratio, M, is often used to study the effects of r on the vibration amplitude.

$$M = \frac{|x\_ss(t)|}{F_o/K} = \frac{1}{\sqrt{(1 - r^2)^2 + (2\zeta r)^2}} \tag{6}$$

A typical graph of magnitude versus frequency ratio is shown below (with $\zeta = 0.1$) in Fig. 2a.

One can see clearly that as the frequency ratio increases, the magnitude ratio, M, goes through a peak at r = 1. When r = 1, the input frequency is identical to one of the system's resonance frequencies, the steady-state amplitude is at its maximum.

Figure 2b shows how the phase angle (between the input and output oscillation) versus r looks like when $\zeta = 0.1$. Note regardless what ζ is, when r = 1, the phase angle is always = $-90°$.

Typically, vibration absorbers can be designed to control system oscillations if the following criteria are met.

(1) When the excitation frequency is constant.
(2) When the excitation frequency is close to the system natural frequency.

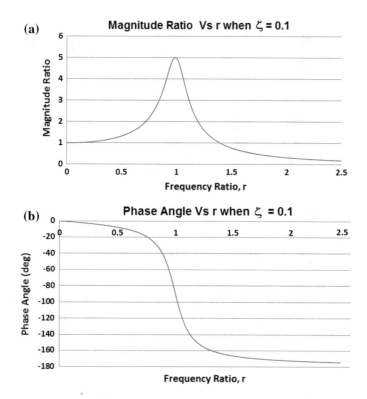

Fig. 2. Magnitude and phase angle versus frequency ratio graphs

A common design of a vibration absorber for an exhaust pipe of a vehicle is shown below in Fig. 3. The absorber consists of a mass element, M2 mounted at the end of a stiffness element, K2, and a clamp to mount the absorber onto the system for vibration control.

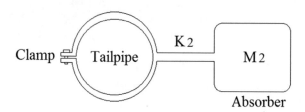

Fig. 3. Absorber design for a tailpipe of a vehicle

To understand why the excitation frequency needs to be close to the system natural frequency to work, it is useful to consider the phase angle information between the input and the absorber provided in Fig. 4. When the absorber mass is −180° out of phase with the input, cancelation will occur.

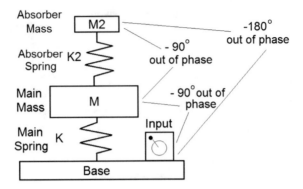

Fig. 4. Phase angle between excitation input and absorber

4   Hand Tremor—Without Vibration Absorber

Figure 5 shows the hand tremor displacement versus time graph for the patient model shown in Fig. 1 with the DC fan turned on. The speed of the fan was adjusted to 11 Hz to excite one of the vibration modes of the system. The amplitude of oscillation was recorded by an accelerometer and converted into displacement. As shown below, the displacement amplitude at steady state was recorded as 2.8 mm (or 5.9 mm peak to peak).

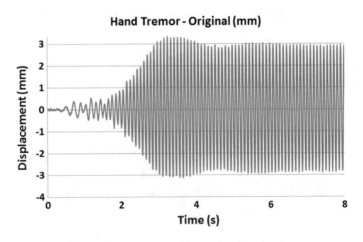

Fig. 5. Hand tremor without vibration absorber

It is apparent that a hand tremor patient with 5.9 mm peak to peak vibration amplitude at 11 Hz would not be able to eat with a spoon. A vibration absorber with natural frequency at 11 Hz was designed with a mass of M2 = 0.128 kg. Using $\omega n = \sqrt{(K2/M2)}$, the absorber stiffness K2 was calculated as 611.1 N/m (Table 1).

Table 1. Vibration absorber design parameters

Vibration absorber linear natural frequency, fn	11 Hz
Vibration absorber angular natural frequency, ωn	69.1 rad/s
Vibration absorber mass M2	0.128 kg
Vibration absorber stiffness K2	611.1 N/m

5 Hand Tremor Control with Vibration Absorber

Once the vibration absorber was mounted onto the model, the DC fan was turned on again at 11 Hz. Figure 6 shows the arm tremor versus time graph for the model with the unbalanced fan turned on. The amplitude of oscillation was reduced to 0.7 mm (or 1.4 mm peak to peak). A reduction of over 76% in vibration amplitude has been achieved by using this vibration absorber.

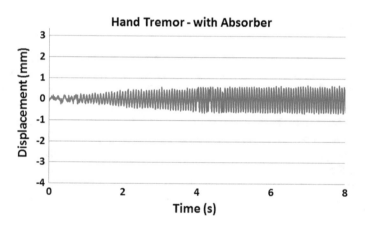

Fig. 6. Hand tremor with vibration absorber

6 Hand Tremor—With Poorly Designed Vibration Absorber

To illustrate the importance of designing the absorber correctly, the same absorber mass M2 was incorrectly mounted at a larger cantilever location to create a much smaller stiffness K2 value for the absorber. Figure 7 below was obtained with this poorly designed absorber. The vibration amplitude increased by 10% to a 6.6 mm peak-to-peak amplitude when compared with the original case without any absorbers.

Hand Tremor with Poorly Designed Absorber

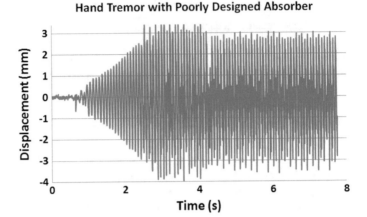

Fig. 7. Hand tremor with a poorly designed absorber

7 Conclusions

Mechanical vibration is mathematically intense and it could be a very dry subject to the students. By using a demo tool at the beginning of a lecture, lifeless math in the textbook is turned into a powerful tool to help solve a problem. After seeing the demo, the students would try to figure out a way to help this patient with hand tremor issues to stabilize his hand so the patient can eat properly. By designing a vibration absorber, the hand vibration amplitude could be reduced significantly. However, a poorly designed absorber could make the hand tremor get worst. The empathy of the students is used throughout the demo to encourage them to stay engaged in class to learn the theories behind vibration absorbers. In addition, the visual image created by seeing a vibration absorber actually worked (and failed) in front of them will help the students retain the concepts studied in class much easier.

References

1. Louis, E. D. (2001). Essential tremor. *New England Journal of Medicine, 345*, 887–891.
2. Hashemi, S. M., Golnaraghi, M. F., & Patla, A. E. (2004). Tuned vibration absorber for suppression of rest tremor in Parkinson's disease. *Medical and Biological Engineering and Computing, 42*, 61–70.
3. Quan, R. & Stech, D. (1996, January). Time varying passive vibration absorption for flexible structures. *Transaction of ASME, 118*.
4. Gebai, S., Hammoud, M., Hallal, A. & Khachfe, H. (2016, September). Tremor reduction at the palm of a Parkinson's patient using dynamic vibration absorber. *Bioengineering (Basel), 3*(3), 18.
5. Palm, W. (2007). *Mechanical vibration*. Wiley.

Interactive and Adaptable Mobile-Friendly e-Learning Environments for K-12 and Higher STEM Education and Skills Training

Yakov Cherner[1(✉)], Gary Witus[2], James Uhomoibhi[3],
Tatyana Cherner[1], Bruce Van Dyke[4], Irina Popova[5], and Hui Wang[3]

[1] ATeL-Advanced Tools for e-Learning, Swampscott, MA, USA
ycherner@ATeLearning.com
[2] College of Engineering, Wayne State University, Detroit, MI, USA
[3] Ulster University, Jordanstown, Northern Ireland, UK
[4] Quincy College, Quincy, MA, USA
[5] Federal Education Development Institute, Moscow, Russia

Abstract. This paper presents the use of highly interactive simulation-based virtual equipment, customizable online laboratories, and integrated e-learning environments for STEM education and professional training. The covered subject areas include X-ray equipment and technologies and their applications for investigating crystal structure and microstructure, various biomanufacturing technologies and equipment, energy sources, energy efficiency, responsible energy consumption, and other topics. The activity-centered approach and gamified pedagogy are consistent with the learning habits of millennials. Adaptable and immersive virtual activities facilitate personalized cross-disciplinary learning STEM principles and laws in the realistic contexts familiar to students from their everyday life and workplace experience. Mastering practical skills are combined with studying theoretical knowledge. The described e-learning tools have been used for performing authentic laboratory and workplace tasks online, preparing students and company personnel for conventional hands-on training, lesson demonstrations, performance-based assessment of gained knowledge and skills, and much more. The browser-based applications are mobile-friendly and can run on Windows, iOS and, Android platforms. They can also be incorporated into online courses available on MOOC platforms.

Keywords: Virtual laboratories · Interactive STEM learning · Mobile learning · Online experimentation · Mobile-friendly apps · Biomanufacturing education and training · Virtual house

1 Introduction

Online STEM courses and programs are experiencing an explosive growth and becoming indispensable parts of today's K-12 and higher education. However, one of the major weaknesses of distance STEM education and technical training is the lack of hands-on practice and experimentation that are integral parts of any conventional

© Springer Nature Switzerland AG 2019
M. E. Auer and T. Tsiatsos (eds.), *Mobile Technologies and Applications for the Internet of Things*, Advances in Intelligent Systems and Computing 909,
https://doi.org/10.1007/978-3-030-11434-3_27

science or engineering curricula or technical skills training programs. Self-guided online practice activities, employing virtual equipment, and simulation-based laboratories (v-Labs) that accurately represents actual instruments and technological processes, can substitute, supplement, or even partially replace traditional laboratory practice.

v-Labs have great potential for transforming traditional "e-learning by reading and watching" into more effective "e-learning by doing" and promoting efficient active-based pedagogy and contextual learning. Gamified interactive online experimentation meets the learning habits of today's students and engages them in STEM learning.

Today, when most students own smartphones and tablets, teachers often unsuccessfully compete with mobile devices that distract student attention from social media and websites. Gamified mobile versions partially address this challenge by engaging students with the interactive STEM content in fun and exciting ways.

For many years, our research and development [1–5] have been focused on addressing the challenges in today's K12 and undergraduate STEM education, including:

- Absence of authentic hands-on practice in distance learning STEM courses and programs.
- Limited availability of advanced research equipment for educational purposes.
- Shortage of mentally and emotionally engaging interactive online resources capable of bridging the gap between theoretical knowledge of STEM principles and methods with students' ability to apply them to solving daily living and workplace problems.
- Low reusability of expensive simulation-based interactive educational resources and limited ability to personalize and adapt them to specific courses, learning objectives, and student levels and backgrounds.
- Poor understanding by undergraduate and graduate students of operational principles of research and production equipment and factors which affect data accuracy, method reliability, and limitations—often caused by the use of fully computerized equipment that executes most tasks automatically without student participation or needs for their understanding of the principles involved.
- Long duration of many conventional hands-on experiments that does not allow students to explore alternatives or even complete them within the time allocated for lab practice.

2   Adaptable Browser-Based Virtual Laboratories and e-Learning Environments

A key instructional challenge is to keep students engaged in the learning or practicing task. This is especially difficult to achieve and verify in online self-directed learning. A way to address this problem is to provide immersive virtual activities around objects, situations, or natural processes, familiar to students from their everyday lives or workplace experience—coupled with STEM education objectives. With this in mind, golf game, cooking, home appliances and systems, actual devices, or systems (e.g., cars, solar water heating or photovoltaic (PV) panels), and research and production equipment (e.g., bioreactors, chromatographs) are used as a context for STEM learning and skills development. To keep learner's attention, all self-guided learning or practice assignments are short enough or are split into small parts.

Highly realistic and interactive simulations form the core of virtual labs and environments. They incorporate solid science/math models that accurately imitate natural phenomena and equipment such as home appliances, laboratory or industrial equipment or production processes. Simulations facilitate online experimentation and exploration and enable learners to explore links between fundamental principles and their practical applications.

To achieve particular learning or training objectives, simulations are integrated with instructional, assessment and educational resources into self-directed online or blended student assignments. In addition to the main simulation, each assignment includes a defined scenario, step-by-step performance guide, an optional worksheet, and task-related documentation, as well as embedded assessment and associated educational multimedia resources for just-in-time learning. Auxiliary simulations extend the functionality of the core simulations (e.g., visualize hidden processes or explore the studying phenomenon from different perspectives). Third-party resources can be integrated as well.

In traditional settings, learning and practice activities have to focus on the classroom average and are not aligned with the needs of either advanced students or those requiring extra help. Too often, there is inadequate attention to tailoring educational content to the individual needs and learning styles of students. According to NMC Horizon Report [6], "the demand for personalized learning is not adequately supported by current technology or practices". Implementation of simulation-based and activity-focused integrated learning/training units along with the open unit framework and easy to use authoring tool enable instructors and curriculum developers to create new assignments and tailor existing ones to specific educational needs, student levels and backgrounds, course contents, pedagogical models, etc. This technology and approach assure high flexibility and customization and unparalleled reusability of interactive learning resources. A wide range of student assignments can be built based on each simulation.

2.1 Virtual Equipment and Laboratories

Today's advanced research equipment is typically fully computerized with most tasks executed routinely with limited user participation. Procedures such as equipment calibration, data collection, data handling, and interpretation are performed automatically. While this provides enormous benefits for researchers, it also creates a number of substantial educational drawbacks and limitations. Except for sample preparation and installation, procedures related to pattern analysis, measurements, peak indexing, etc., are executed by a computer without student involvement. This is one of the reasons why many students experience difficulties in assessing applicability and limitations of methods, and factors affecting data accuracy. Hence, they cannot correctly estimate the reliability of the results. In addition, many students, especially those enrolled in online or large-class science and engineering courses, lack opportunities for hands-on practice and experimentation that involve expensive contemporary equipment due to limited availability and accessibility of it for educational purposes.

To overcome some of these problems and address the demands of distance and blended education the customizable virtual X-ray diffractometer (v-Diffractometer) and online laboratory (e-XRLab) has been developed [1]. The e-XRLab enables students to become familiar with nondestructive research and testing methods widely used in science and industry by performing authentic experiments online using virtual replicas of a fully functional X-ray powder diffractometer. It expands the spectrum of knowledge and practical skills that could be acquired using a real computerized X-ray equipment.

In contrast with other XRD simulations [7, 8] that focus on specific models of commercial diffractometer and enable users to execute only predefined scenarios with a single or prefixed set of samples, the v-Diffractometer, described below, is much more flexible and customizable. It enables students, to explore the design and operation of X-ray equipment and its major parts, and to learn underlying scientific and engineering principles and laws. Through online and blended activities, students gain the practical skills required for conducting actual experiments, mastering to collect, analyze, and interpret experimental data, and examining the instrumental factors affecting data accuracy.

The e-XRLab includes simulations, an open library of samples, and expandable sets of online experiments. It also contains supplementary educational resources, authoring tools, learning and content management system (LCMS), and cyberinfrastructure. Highly interactive simulations authentically reproduce equipment design and operation, and realistically model relevant physical processes. e-XRLab includes educational analytical and modeling software as well (Fig. 1).

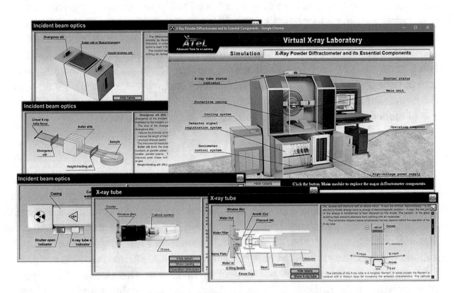

Fig. 1. The *Virtual Diffractometer* simulation enables students to explore the design and function of an XRD powder diffractometer. A student can select any essential part (e.g., incident beam optics—top left) and virtually inspect it inside out (middle left) to better understand its operation. Screenshots on the bottom illustrate how students can examine design, major components, and basic parameters of an X-ray tube and learn the principles on which the tube's operation is based.

The efficiency and usefulness of students' laboratory practice depend on how well student assignments match specific educational objectives, course content, learners level, etc. The same is true, to an even greater degree, for professional training. The framework of the e-XRLab and a complimentary authoring tool enables instructors to adjust existing virtual experiments and create new ones. The experiments developed by the authors of this paper and their colleagues were substantially different due to the great diversity of course subjects, student majors, educational goals, and settings.

The e-XRLab contains an associated open collection of virtual samples available for online experimentation that includes alloys, minerals, ceramics, polymers, nanostructured materials, and thin films. Instructors are able to add their own virtual samples to the collection. The collection may include measured digital patterns obtained using actual diffractometer or theoretical X-ray diffraction spectra calculated by employing the pattern simulation tools associated with the e-XRLab and using available crystal structure data of substances or materials (e.g., from the Cambridge Crystallographic Data Centre (CCDC) [9]) or d-spacing and intensity (d, I) data sets (Fig. 2).

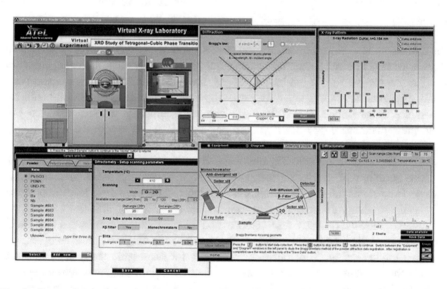

Fig. 2. Screenshots of the virtual experiment "XRD Study of Tetragonal–Cubic Transition in Ferroelectrics". A sample selection form (bottom left) presents the list of available samples. The sample should be installed in a sample holder (left). The setup form (bottom center) is used to specify parameters of scanning. The left panel of the data collection interface (bottom right) dynamically displays either the device (goniometer) status or Bragg–Brentano focusing geometry diagram (shown in figure). The right panel shows the scattering X-ray pattern being recorded. The instructional panel below the simulation displays step-by-step experiment instructions for the student. During the experiment, the auxiliary simulation modeling X-ray scattering from a crystal lattice (Bragg's law) can be invoked (top right).

Similar to a conventional hands-on experiment, a student following an instruction has to select a sample to be investigated, virtually open the safety doors and install the sample into the holder, specify scanning parameters, turn on the X-ray radiation, open the shutter, and press the Start scanning button. One of the main goals of such detailed guidelines is to help students build a mental algorithm for executing the necessary steps and required safety procedures. After the scanning is completed, the generated X-ray pattern can be saved for further examination and obtained data are used for solving various relevant problems. Data obtained over the course of virtual experiments can be collected and handled manually or automatically. Virtual data can be exported to popular third-party software as well.

2.2 e-Learning Tools for Biomanufacturing Education and Professional Training

e-Learning tools for biotechnology education and biomanufacturing training are designed to address the shortage of skilled technical professionals for the biopharmaceutical industry, that currently experiences an explosive growth. Conventional hands-on training of biopharmaceutical personnel is very expensive and slow as it requires costly equipment, materials, and supplies. It also has severe constraints due to the fact that most biomanufacturing processes occur extremely slowly (often lasting days or weeks) in clean rooms which are required rigorously controlled contamination.

The interactive simulation-based training has successfully overcome many drawbacks of traditional biomanufacturing workforce development. By executing authentic workplace tasks online, the college students and company workers enhance their conceptual understanding prior to performing tasks in a real workplace.

Our e-learning tools are focused on single-use upstream processing [2], low-pressure (LP) chromatography [5] and upstream processing using stainless bioreactors.

The adoption of single-use technologies is an increasing trend in today's biopharmaceutical manufacturing. Disposable systems, which are easy to use and maintain, and are therefore cost-efficient, are replacing relatively inflexible, hard-piped equipment. However, many life science and biotechnology companies are facing shortages in skilled personnel skilled with experience in single-use technologies.

Two modules (Fig. 3) of our e-learning tool for disposable technology aimed to expose the student to industrial scale production equipment, and allow them to understand what procedures and how it should be executed to complete this stage of bioprocessing, as well as what equipment should be employed. The "Big picture" of upstream processing helps learners to develop transferable knowledge and skills that will enable them to keep pace with fast and continuous technological advances in biomanufacturing.

The set of customizable virtual experiments allow students to perform authentic biomanufacturing tasks such as initial *Inoculation, Cell Growth* in a large industrial scale bioreactor and *Cell Counting*.

During the "Inoculation" experiment, students execute all procedures of an initial cell expansion: retrieving cells from a frozen vial kept in a cell bank, thawing them in a warm water bath, cells counting and the determination of an appropriate volume media needed for an initial cell expansion in a T flask or spinner flask, labeling the flask with increased number of cells and placing it into an incubator. The student also learns necessary aseptic techniques that must be applied to avoid any source of contamination.

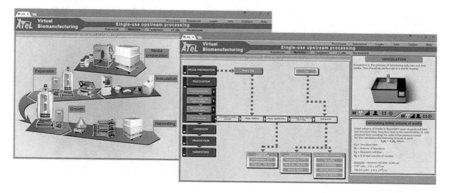

Fig. 3. The left screenshot shows an interactive diagram introducing an entire chain of disposable equipment used in upstream processing. The right screenshot presents an expandable overall flowchart of upstream processing using disposable technology. A description of the selected procedure is shown in the bottom right panel. A video clip that demonstrates how the procedure is performed by an expert is displayed in the top right panel.

In the "Cell Growth" experiment, the student has to assemble the system, perform calculations for the amount of reagents, media, and cells. After assembling the system and setting up the required process parameters, he or she has to operate the reactor and keep the process on track. When some of the parameters are beyond the preset permissible limits, the system generates messages and the student ought to decide how to fix the problem. Based on regular cells counting data, the student decides when to complete the process.

The virtual exercise "Cell Counting" helps students develop skills for accurate and consistent cell counts. The student should aseptically put inside a biological safety cabinet necessary instruments, reagents, accessories, and supplies and organize his/her workplace, prepare a sample, load a hemocytometer, count viable and nonviable cells under a microscope (Fig. 4, bottom right) and determine the concentration of viable cells. The v-Labs "Cell Counting" is connected with other v-Labs that require to perform this procedure regularly.

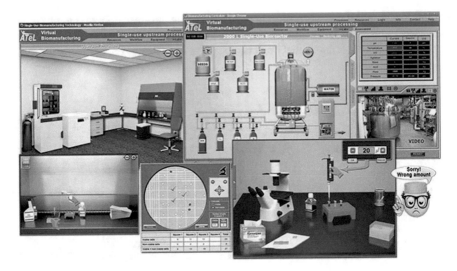

Fig. 4. The v-Labs "*Inoculation*" (top left) helps students better understand the process and gain the practical skills necessary to perform the initial cell growth task. One of emphasizes is placed on aseptic techniques and culture transfer procedures (bottom right). A series of "Cell Growth" experiments (top right) requires students to assemble the system, calculate and set up process parameter and keep the process on track. The virtual exercise "*Cell Counting*" (bottom middle and right) helps students develop skills for accurate and consistent cell counts. An animated assistant (bottom far right) provides immediate feedback, tips, and comments.

2.3   Energy Efficient House—Cross-Disciplinary e-Learning Environment

Interactive Energy Efficient House (EEH) is an integrated simulation-based reconfigurable educational environment designed to enable a wide range of secondary and higher education students to explore energy consumption by major residential systems and home appliances and get familiar with the domestic use of renewable energy sources. Also, the EEH is aimed to educate the general public and help them acquire more energy-responsible behavior.

The EHH associated learning activities are focused on cross-disciplinary conceptual studying underlying fundamental principles and related STEM subjects and are designed to enhance the understanding of technical concepts and the economic aspects of using energy efficient devices and techniques.

This Virtual Kitchen helps learners study (in highly interactive and gamified manner) and better understands STEM principles and laws relevant to kitchen appliances, cooking, and cookware. The primary focus is on energy-efficient appliances and cooking technologies. Students can replace conventional appliances with certified energy efficient ones and observe how these replacements affect energy use and the utility bill. Simulations make it possible to visualize the cooking processes and processes responsible for appliances operation and to study related physical, chemical and biochemical theories (Fig. 5).

The user can customize the virtual house by selecting various home appliances, lights, solar panels, heaters or air conditioners, etc. (Fig. 6).

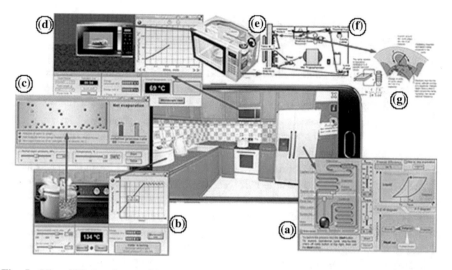

Fig. 5. Virtual Kitchen (a part of the EEH system) running on a smartphone (in the middle). Tapping on an appliance opens the simulation or virtual experiments related to this appliance. For example, refrigeration system and underlying principles (**a**); pressure cooker and processes occurring in it (**b, c**); microwave oven (**d**); its parts arrangement and interconnections (**e, f**); and principles underlying generation of high-frequency electromagnetic waves by a cavity magnetron (**g**).

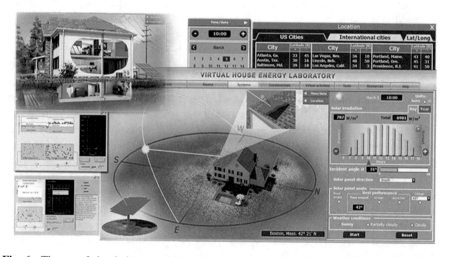

Fig. 6. The set of simulations modeling solar heating and photovoltaic (PV) systems. Students can change many parameters of simulations and explore how solar irradiation and efficiency of the systems depends on seasons and daytime (**a**), geographical location (**b**), house and panel orientation, weather conditions, and some other factors. It is possible to switch between fixed tilt and sun-tracking PV systems. Auxiliary simulations (shown in the left) imitating the basic processes in semiconductors responsible for converting that convert light energy into electricity. Enable students to study photovoltaic effect in semiconductors and explore factors affecting the efficiency of PV panels.

3   The Use of Activity-Based e-Learning Tools in Secondary and Higher Education and Professional Training

Today, when virtually every student owns a smartphone or tablet, mobile-friendly content is critical and online courses should be optimized for mobiles. All applications described above are entirely browser-based and cross-platform. They include a user agent that automatically detects a device type, browser, and operating system and tunes up the functionality and interface design. respectively. User interfaces support mouse, touch, and gesture-based interaction. Accordingly, simulation interfaces and pages of the student assignments are optimized to fit mobile device screen format and resolution.

The e-learning system design is based on contemporary constructivist and situated cognitive theories [10], which suggest that effective learning occurs in a situational context, drawing on each student's prior knowledge and meaningful experiences in their own frame of reference. It employs effective "learning-by-doing" and problem-based learning methodologies [10, 11] and facilitates dual-impact learning experiences. Students participate in contextual cross-disciplinary learning of STEM theoretical principles and laws while mastering practical skills in such a way that it makes sense to them in their own frame of reference. According to contextual learning theory, learning skills and acquiring knowledge "in context" is the most efficient learning strategy [12, 13].

The modules have been used as parts of different K-12 STEM curricula, undergraduate courses, for out of class activities and for professional training in several countries including US, UK, Russia, Ghana, and some others. They have proved to be useful for both self-directed online learning and as a supplement to traditional instructor-led learning.

Performance-Based Assessment

Embedded digital assessment is an important and almost mandatory component of each virtual experiment. In addition to conventual testing formats, contextual performance-based (PB) evaluation of theoretical knowledge and practical skills is widely used across our e-learning modules. In contrast with the guided practice activity, in the PB assessment mode student or trainee has to accomplish the required task without detailed performance instructions. When professional skills are evaluated, employees perform authentic workplace tasks and may require to resolve a troubleshooting situation.

The system keeps track of detailed trainee/student actions including time spent on each procedure, number of incorrect steps, and number of requests for help. It is possible to rate each performance step and/or action and at the end of the activity, the instructor (and the trainee as well) will be provided with a detailed report and grade.

The described virtual laboratories, e-learning environments, and associated online activities have been used as follows:

- as the only practice on the relevant subjects for students enrolled in large-scale lecture classes or in distance learning courses;
- for online experimentation as a part of online or traditional courses;
- as online training tools for preparing students and employees to perform hands-on assignments at college labs;

- for preparing company new hires and incumbent employees for conventional training using actual equipment workplaces;
- for performance-based assessment of students' ability to operate the equipment and apply gained knowledge to workplace tasks;
- for in-class lecture demonstrations;
- for blended experimentation in combination with actual equipment and instruments;
- as interactive manuals for hands-on laboratory experiments or workplace tasks using real equipment.

At Massachusetts Institute of Technology (MIT), the virtual X-ray laboratory was integrated into the undergraduate course "Introduction to Solid State Chemistry" available on the MOOC edX platform (see Fig. 7). As a part of a homework assignment, students had to determinate the crystal structure, lattice constant, and atomic radius of an unknown cubic metal. Before the integration of e-XRLab, the students had only just to make calculations using ready-to-be used data, After the incorporating e-XRLab, the students had to become familiar with design and functionality of an X-ray powder diffractometer and understand its underlying operating principles, then scan a sample and collect data, and perform all procedures required in the actual hands-on experiment. Only after that, they were able to calculate the required parameters. Students' feedback and test results revealed that virtual experimentation substantially enhanced their understanding of the method and the impact of various factors on the result accuracy.

Fig. 7. On the left—students and faculty of three Russian and two U.S. universities use the e-XRLab to collaboratively perform online X-ray experiment via WebEx video conferencing. On the right—the screenshots of assessment pages of the undergraduate MIT course "Introduction to Solid State Chemistry" available on the MOOC **edX** platform.

According to surveys of Quincy College biotechnology faculty, the use of virtual biomanufacturing laboratories for training 2-year college students and biopharma employees enrolled in professional development made them comfortable to perform actual laboratory tasks and enabled the faculty to cut conventional hands-on training time on up to 40%.

The same approach has been employed over the years at Ulster University with postgraduate students in engineering specifically in Product Innovation and Work-Based Learning as well as with undergraduate students in Control Engineering, Digital Electronics. Feedback from students has been very positive where they specify that availability and use of these virtual resources have helped them to understand from the practical perspective the concepts being covered but also it has provided them with an enriched positive learning experience. This is confirmed by the performance of students in these modules where pass rate is greater than 90% compared to the period when these facilities were not available. In the present era where there are the challenges of falling student numbers in STEM subjects and the problem of retention of students in these areas in higher education, such approaches have proven to be a solution which could be gainfully employed.

Students are appreciated for being able to perform laboratory assignments from home at their own pace at a convenient time. They also pointed out that the combination of interactive simulations with synchronized online multimedia resources associated with experiments enabled them to perform experiment tasks faster and more meaningfully. Students were interested in the opportunity provided by the cyberinfrastructure to communicate with their peers from other schools and even from different countries.

Faculty and students give high praise to real-time collaborative online X-ray experiment via WebEx involved three Russian and two U.S. universities.

References

1. Cherner, Y. E., Kuklja, M. M., Cima, M. J., Rusakov, A. I., Sigov, A. S., & Settens, C. (2017). The use of web-based virtual X-ray diffraction laboratory for teaching materials science and engineering. *MRS Advances, 2*(31-32), 1687–1692.
2. Cherner, Y. E., & Van Dyke, B. (2014) Interactive web-based virtual environment for learning single-use biomanufacturing technologies. In *Proceedings of the 121st ASEE Annual Conference*.
3. Cherner, Y. E., & Darkwa, O. K. (2013, December 8–12). Adjustable simulation-based virtual laboratories for learning telecommunication, alternative energy, and energy conservation. In *Proceedings of the ICEE/ICIT-2013 (Joint International Conference on Engineering Education and Research)*, Cape Town, SA (pp. 276–281).
4. Cherner, Y. E., Mullett, G., Khan, A., & Karim A. (2011). Use of adaptable simulation-based virtual laboratories for teaching alternative energy and energy conservation in engineering & technology programs. In *Proceedings of the 2011 American Society for Engineering Education Annual Conference & Exposition*.
5. Cherner, Y. E., & Wallman, S. (2011). Virtual and blended liquid chromatograph for chemical and biological engineering education. In *Proceedings of the 2011 American Society for Engineering Education Annual Conference & Exposition*.

6. NMC Horizon Report. (2013). "2013 K-12 Edition", New Media Consortium, the Consortium for School Networking, and the International Society for Technology in Education.
7. MyScope Outreach, Virtual X-Ray Diffractometer. http://www.ammrf.org.au/myscope/xrd/practice/virtualxrd/.
8. Virtual Powder X-Ray Diffraction online learning module. http://web.mse.ntu.edu.sg/omni/xrd/.
9. The Cambridge Crystallographic Data Centre (CCDC). (2018). CSD 2018—The world's repository for small-molecule crystal structures (over 900,000 curated entries). https://www.ccdc.cam.ac.uk/.
10. Bransford, J. D., Brown, A. L., & Cocking, R. R. (Eds.). (1999). *How people learn: Brain, mind, experience, and school.* Washington, DC: National Academy Press.
11. Lee, M. E. (1999). Distance learning as "learning by doing". *Educational Technology & Society, 2*(3). http://ifets.ieee.org/periodical/vol_3_99/mary_e_lee.html.
12. Contextual Teaching and Learning, Center for Occupational Research and Development (CORD Inc.), Waco, TX. (2008). http://www.cord.org/contextual-teaching-and-learning/.
13. Martin, L. (2009). Learning in Context, ASTC—Resource Center. http://www.astc.org/resource/education/learning_martin.htm.

A Novel Approach for Secure In-class Delivery of Educational Content via Mobile Routers with Functionally Enhanced Firmware

Kamen Kanev[1], Federico Gelsomini[1,2(✉)], Paolo Bottoni[2],
Francesco Ficarola[2], Massimiliano Pedone[2], Domenico Vitali[2],
and Patrick C. K. Hung[3]

[1] Shizuoka University, Hamamatsu, Japan
kanev@inf.shizuoka.ac.jp, federico.
gelsomini@uniroma1.it
[2] Sapienza University of Rome, Rome, Italy
{bottoni,vitali}@di.uniroma1.it,
francesco.ficarola@gmail.com, {federico.gelsomini,
massimiliano.pedone}@uniroma1.it
[3] University of Ontario Institute of Technology, Oshawa, Canada
Patrick.Hung@uoit.ca

Abstract. In this work, we identify some of the important security issues arising from the rapidly expanding use of privately owned communication devices such as smartphones in business environments following the Bring Your Own Device (BYOD) concept. Given the limitations of the existing software-only approaches, we introduce a novel hybrid solution that involves inexpensive mobile routes with functionally enhanced firmware based on open source software. We then implement experimental software components as functional enhancements for integration into the router firmware and generation of new customized versions suitable for installation on different hardware platforms. Finally, in-class delivery of educational content stored on a functionally enhanced mobile router controlled by the instructor is employed for testing of the embedded security features.

Keywords: Customized routers · Open firmware · Bring Your Own Device (BYOD) · Connection security · Mobile learning (m-learning)

1 Introduction

1.1 Context

Advancements in communication technologies have made wireless connectivity so ubiquitous that online information is always at the fingertips of the users. Nowadays, many learning environments are equipped with a variety of technical and pedagogical facilities that allow instructors to carry out advanced teaching activities [1]. Furthermore, as students can be reached through a number of communication channels, they can be engaged in different learning styles [2]. This is particularly true and important in

© Springer Nature Switzerland AG 2019
M. E. Auer and T. Tsiatsos (eds.), *Mobile Technologies and Applications for the Internet of Things*, Advances in Intelligent Systems and Computing 909,
https://doi.org/10.1007/978-3-030-11434-3_28

contexts where advanced technologies could better support students in developing their practical and functional skills [3]. In such educational settings, along with the traditional oral/verbal channel, forms of advanced visuals and interactive experience are employed to stimulate the processing and understanding of learning material, thus improving acquisition. In particular, collaboration and interaction among students stimulate all the perceptive channels that are involved in the development of communication skills and, hence, are strongly encouraged.

However, while smartphones may give instant access to a wide range of global information resources anywhere in the world, access to institutional resources such as work- and study-related ones is often restricted. Banks and military establishments, for example, would limit the use of employees' private communication devices on their premises for security reasons, hospitals would do so to protect patients, and airlines will ask passengers to turn off their devices to prevent interference with aircraft equipment.

With respect to security, if a private smartphone is already 3G or LTE connected, another simultaneous connection to an institutional network, for example, may result in a rogue bypass of security policies and thus lead to a security breach. These risks are of particular concern in specialized business environments dealing with secret and confidential information and require huge infrastructural investments to prevent potential information leakage [4]. Such elevated equipment cost, however, is often hard to sustain for typical businesses that put the stress on the accessibility and wide availability of their communication services. A more affordable security solution, although highly desirable, seems not immediately available for a number of reasons as discussed below.

Secure protocols, such as 802.1x and more advanced cryptographic functionalities have higher hardware and software demands, so many "legacy" devices cannot be equipped with such support. Old versions of operating systems, for example, would often require the installation of add-ons or custom software in order to get access to the network. Similar considerations are applicable to Small Office, Home Office (SOHO) secure approaches. For example, the well-known family of WPA secure protocols is considered a good solution for small environments but they require management of security keys for user network access. In large distributed networks, e.g., universities with several campuses, etc., such key management (i.e., dissemination, refresh, and so on) is not a viable network administration policy.

Most of the higher security protocols require substantial configuration effort by the clients. In more detail, the 802.1x protocols define the Extensible Authentication Protocol (EAP), namely an authentication framework used to implement port-based Network Access Control functions. EAP supports a multiplicity of secure methods, e.g., EAP-MD5, EAP-TLS, EAP-TTLS, EAP-IKEv2, and so on, some of which rely on the use of client certificates rather than simple settings. All users must then be informed of the network settings and they must be in a position to properly set up their communication devices, which in turn must properly support such settings, or otherwise, no connection could be established. As some EAP security settings are quite exigent, institutions are often forced to choose custom settings based on the skills of the available network administration staff and the awareness of the users and their communication devices specifications rather than in accordance with the real security requirements.

In order to minimize security risks, approaches based on antivirus, firewall, and other security software installed on user devices and maintained by users themselves are also employed. In this context, however, simple user mistakes can often be detrimental to the overall security of the network. User involvement in security management can be decreased by centralized installation and management of security software on client computers, but this is only viable for institution-owned devices lent to customers. Hardware ownership appears to play a central role in this: with no way to enforce proper software maintenance on user-owned devices, businesses often have no other option but to ban the use of such devices on their premises altogether. On the other hand, providing institution-owned devices to all users may be very expensive in terms of cost and management.

The main question that arises from the above discussion is, therefore, how to minimize the security risks associated with the wider use of private communication devices in different business environments, which appears to be a nontrivial task.

1.2 Research Purpose

The fast evolution and high penetration rate of mobile technology in many aspects of everyday life leads to its employment in a wide range of educational environments and stimulates the adoption of various learner-centered solutions [5]. In education, security requirements are not as stringent as in finance and military, so less centralized management methods could be employed. Some businesses, for example, have begun encouraging customers to bring and use their devices, adhering to the Bring Your Own Device (BYOD) concept [6–8]. This approach is becoming increasingly popular and is spreading quickly to educational institutions and universities in particular. Educational content is, nowadays, being redesigned and optimized for smooth presentation on user devices with various screen sizes and physical characteristics that employ different operating systems and software. By allowing students to use their own devices directly in the educational process, universities can significantly reduce the equipment installed on the premises and thus bring down both the capital investments and the maintenance cost. Mobile learning (m-learning) is distinguished by many scholars for being: (i) location-independent, (ii) time-independent, and (iii) providing meaningful content [9], which clearly differentiates it from both e-learning and web-based learning [10]. The notion of m-learning is thus in a position to offer significant educational benefits and improvements by allowing students to access materials and conduct activities without time and/or place restrictions, thanks to the mobility, flexibility, and seamless connectivity of the devices employed [11, 12].

However, since the current BYOD model does not address the security issues discussed above in a comprehensive way, universities adhering to it are often facing elevated levels of security incidents [13]. While it may still be economically viable to employ BYOD as it is and allocate some of the accumulated revenue for recovery from security incidents, just one high impact security incident with severe consequences may quickly reverse the situation.

The purpose of our research is, therefore, to consider the different security risks associated with the expanding use of private communication devices in business environments, identify possible approaches to address them, conduct experiments, and implement practical solutions compatible with the BYOD concept. This paper is organized as follows: Sect. 2 describes the proposed hybrid approach, Sect. 3 introduces the target institutional network environments, Sect. 4 discusses the implementation and experimental work, and Sect. 5 concludes the paper outlining our future plans.

2 Adopted Approach

As we consider addressing the above security concerns an important research target, and given the limitations of software-only approaches, we propose a novel hybrid approach that relies on both hardware and software components. The idea is to employ inexpensive mobile routers as communication access points that act as secure gateways to an institutional network. Such devices are an order of magnitude cheaper than standard smartphones (e.g., less than US$20 vs. US$200 and up) and will cost even less if purchased in large quantities (e.g., by a university for all enrolled students and employees). A university can distribute such devices to students for free while retaining the ownership of the devices (i.e., the model employed by credit card companies that never release the ownership of the physical cards). Since the university will own the devices, it can impose and enforce the policies needed to ensure the proper installation and maintenance of the software on the devices. Access to the institutional network will then be available only through these devices, so it will be both highly individualized and traceable due to the security software and private keys embedded into the devices.

Adopting this approach opens a range of interesting possibilities for enhancing and augmenting the educational process. Instructors, for example, can use their devices for storage and online delivery of educational content specific to each of their classes and sessions. In such a case, the instructor's mobile router will act as a localized Web server that is only accessible by the students attending the current session. Note that, this is different from placing the content on a centralized web server for shared (albeit controlled) access that may raise security concerns. The instructor's portable device will effectively restrict access only to students that are physically in the lecture room and will block all external connections so that security risks, if any, stay confined to the local class environment. If particular class activities should require access to external online resources, however, the instructor's mobile router can act as a proxy server for the students already connected to it.

In general, by introducing a custom hardware component, the resulting architecture allows instructors to define new, controlled, and manageable environments isolated from the underlying network infrastructure. These features bring distinct advantages to the instructor: (i) the class network environment is independent of the network (with respect to the protocols and settings imposed by the network administration staff) and (ii) the instructor is free to configure the custom hardware according to the needs of the class work.

The practical implementation of the approach briefly described above requires careful planning and selection of necessary hardware and software components. Mobile routers are manufactured nowadays by many companies and a wide range of models is readily available. Such models, however, are based on different, often incompatible chipsets and mostly employ proprietary firmware. For a wider router model coverage, we adopted an open source approach, employing DD-WRT and OpenWrt-based firmware [14] in our experimental work. So far, we have customized and installed the community version of the DD-WRT firmware on a number of Buffalo home routers such as WZR-300HP, WZR-HP-AG300H, and WZR-600DHP (Fig. 1).

Fig. 1. Buffalo wireless home routers suitable for DD-WRT firmware installation.

Those routers are not mobile but they are fairly compact and we have successfully moved them to different lecture rooms after each class. A standard Apache server has been included in the firmware to serve educational content for Italian language classes at A2, B1, and B2 levels. In this field, software supporting computer-assisted language learning with respect to different aspects of linguistic abilities has been developed [15] and mobile-based interactive learning setting facilitating and enhancing student learning outcomes and motivation have been implemented [16]. Other studies focus on QR-code based interactions and Augmented Reality (AR) applications for enhanced m-learning in language settings through linking specific information and activities in a context-aware manner [17].

We have extended the above approaches with specialized support for in-class collaborative learning activities through scripts installed on a mobile router. In particular, we have conducted experiments with configuring and testing some very compact and truly mobile router models from GL Technologies such as GL-MiFi4G, GL-AR150, GL-AR300 M, and GL-MT300 N-V2 (Fig. 2).

Fig. 2. GL technologies mobile routers suitable for OpenWrt firmware installation.

3 Captive Portal Networks and the Sapienza Wi-Fi

Captive portal networks are widely employed in different contexts: airports, campuses, and public offices often use this technology since it does not require specific configuration for the clients and it does not impose any encryption overhead. In fact, captive portal networks are highly suitable for legacy devices as they are compatible with any, even low-spec device, as long as it is wireless compliant. Other wireless technologies, e.g., the 802.1x standard, provide more secure connections, but often require significant configuration efforts on the client side and, due to hardware limitations, some legacy devices may not perform well enough computationally.

The Sapienza wireless network is a hot-spot-like Wi-Fi network. It is centrally managed and serves connections to more than 30,000 distinct users each month. The

Fig. 3. The web interface for connecting to Sapienza Wi-Fi.

technology behind its infrastructure is simple but effective: the wireless network requires no encryption so any client can connect and get into the network without restrictions. However, as soon as a client (usually a browser) tries to reach the internet, the Sapienza captive portal shows a login page (Fig. 3). If the provided credentials are correct, the portal enables the firewall for the host, and the connection can be correctly established. No connection is provided otherwise.

Based on the above considerations we deemed captive portal networks an attractive development target and thus decided to employ the Sapienza campus network as a testbed environment for conducting connectivity, authentication, management, and other communication security related experiments with the mobile routers.

4 Experimental Work

We have conducted a range of experiments with employing different approaches that: (i) restrict the distribution of the e-learning materials prepared by the instructor to students physically present in the classroom, (ii) enable such students to access the said materials through the authorized mobile devices in their possession in a transparent way, and (iii) effectively isolate the classroom e-learning environment from the university network while providing controlled access to specific resources identified by the instructor. The experiments were conducted in different Sapienza University classrooms employing OpenWrt-based routers with a number of enhancements.

4.1 Settings and Procedures for Access to Local Resources

The mobile router used in class by the instructor was equipped with an enhanced content delivery system specifically designed for collaborative language learning. The student interface of the system as shown in Fig. 4, is adjusted for direct visualization on a wide range of mobile devices with different screen resolutions and performance. The Wi-Fi password of the mobile router was set differently for each class and was written on the blackboard so that only students physically present in the classroom could see it. Forwarding/leaking the password to students not in the classroom was meaningless as the router Wi-Fi was barely reachable from behind the classroom walls and no access through its remote interface was allowed. For more transparent access and also as an additional precaution, the URL address of the current assignment page was not directly conveyed to the students. Instead the access link was encoded in a 2D-barcode that was presented to the students to scan with their mobile devices in order to automatically open the link in a browser.

While the above-described approach enabled us to provide location-based transparent access to e-learning materials for students in class, it was limited to a mobile router with no external connections. In some cases, however, instructors needed to provide students with access to globally available resources that could not be replicated on the mobile router for technical, legal, or other reasons. To address this issue we have developed and conducted experiments with different methods for connectivity automation and management of router access to external networks and the internet. The employed mobile router models could be configured for wired WAN connections, for wireless Wi-Fi connections, and even for 3G/LTE connections through embedded or attachable wireless modems.

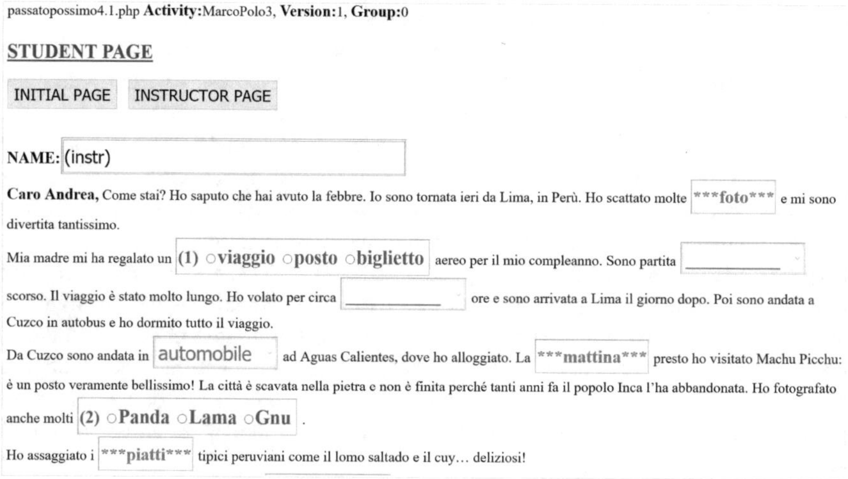

passatopossimo4.1.php **Activity:**MarcoPolo3, **Version:**1, **Group:**0

STUDENT PAGE

INITIAL PAGE   INSTRUCTOR PAGE

NAME: (instr)

Caro Andrea, Come stai? Ho saputo che hai avuto la febbre. Io sono tornata ieri da Lima, in Perù. Ho scattato molte ***foto*** e mi sono divertita tantissimo.

Mia madre mi ha regalato un (1) ○**viaggio** ○**posto** ○**biglietto** aereo per il mio compleanno. Sono partita _____

scorso. Il viaggio è stato molto lungo. Ho volato per circa _____ ore e sono arrivata a Lima il giorno dopo. Poi sono andata a Cuzco in autobus e ho dormito tutto il viaggio.

Da Cuzco sono andata in automobile ad Aguas Calientes, dove ho alloggiato. La ***mattina*** presto ho visitato Machu Picchu: è un posto veramente bellissimo! La città è scavata nella pietra e non è finita perché tanti anni fa il popolo Inca l'ha abbandonata. Ho fotografato anche molti (2) ○**Panda** ○**Lama** ○**Gnu** .

Ho assaggiato i ***piatti*** tipici peruviani come il lomo saltado e il cuy... deliziosi!

Fig. 4. The student interface of the system that is accessible through a variety of student-owned mobile devices.

Although the experimental work, reported in the following subsections, is confined to Wi-Fi connectivity, we are also exploring the wired and 3G/LTE communication modes and planning to report the obtained results in forthcoming publications.

4.2 Sapienza Wi-Fi Connectivity and Access to Global Resources

A snapshot of the Wi-Fi configuration interface of the mobile router is shown in Fig. 5. To provide student access to the content installed on the wireless router as described in the previous sections, a Wi-Fi interface was set up in Master mode (the first entry in Fig. 5).

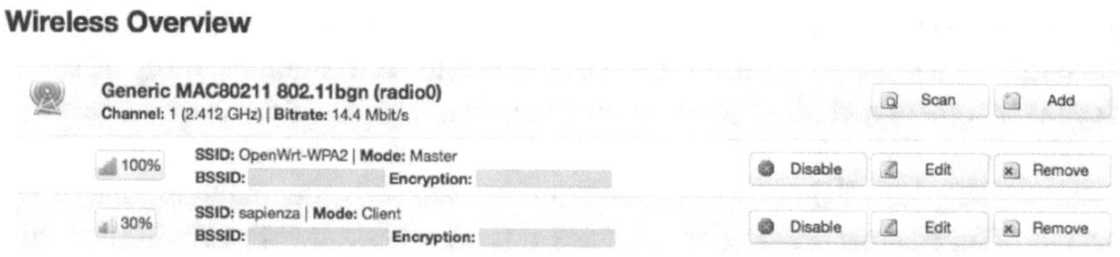

Wireless Overview

Generic MAC80211 802.11bgn (radio0)
Channel: 1 (2.412 GHz) | Bitrate: 14.4 Mbit/s   Scan   Add

100%   SSID: OpenWrt-WPA2 | **Mode:** Master
       BSSID: Encryption: Disable   Edit   Remove

30%   SSID: sapienza | **Mode:** Client
       BSSID: Encryption: Disable   Edit   Remove

Fig. 5. Establishing the settings

For global connectivity, another Wi-Fi interface was set up in client mode (the second entry in Fig. 5) and configured to connect to a wireless access point of the

Sapienza Wi-Fi. With this setup, the mobile router could accept requests for access to global resources from the student devices connected to its master interface and route them through its client interface for processing. However, as discussed in the previous sections, the Sapienza Wi-Fi would not allow access to the internet without a prior authentication through the captive portal interface (Fig. 3).

Our educational content delivery model assumes centralized access to global resources controlled by the instructor. This implies that access to such resources is managed by the instructor who is responsible for providing the necessary credentials. Students, on the other hand, may or may not have credentials to access the global resource directly. It becomes, therefore, clear that captive portal authentication should not be passed over to the connected student devices, but rather be carried out by the instructor or delegated to the router as an automated procedure.

The mobile router itself is not equipped with a web browser, so opening the captive portal page on it and letting the instructor authenticate manually is not an option. We have therefore adopted the second approach and created a streamlined procedure for transparent authentication at request. For this, browser connection functionality was emulated by preparing a `curl` command to POST the necessary data to the server. Invoking the `curl` command could be done at any time by establishing an SSH connection to the mobile router.

We automated this process even further by putting the curl command in a shell script and setting it up for automatic invocation when the router connects to the Sapienza Wi-Fi for the first time after a startup. Unfortunately, connection drops followed by automatic reconnection attempts by the router is not properly handled by this approach. This seems to be partially due to some bugs in the implemented reconnection protocol so we are planning to conduct further experiments with newer versions of the OpenWrt firmware and, as well, explore other automation means.

4.3   Internet Connectivity Access Control

One of the objectives of our experiments was to provide transparent access to the entire educational content installed on the router, while effectively restricting access to the internet through the established Wi-Fi connections only to resources endorsed by the instructor. This clearly provides better information security and helps the students stay focused on the educational materials pertinent to the learning objectives of the course. Note the difference from the standard access model where the educational content is placed on an institutional or a public server, and students use a fully-fledged Internet connection to access it. In the latter case, students can simultaneously access the educational content and other services such as e-mail and social networks, while in our implementation, the router will block access to unwanted services.

Another aspect of Internet connectivity is that laws, regulations, and privacy concerns often impose requirements for logging of some network activities. With the setup discussed in the previous subsection, the router is connected to the internet through a specific account determined by the credentials provided to the Sapienza Captive Portal at startup or manually at a later time. Consequently, all connections that are made through the router will be attributed to the same account making the individual students practically indistinguishable. To avoid possible complications we set up on the router an internal Captive Portal, so that connecting to Internet would require additional authentication. The captive portal was implemented through `nodogs-plash`,[1] a hot spot management software package in the OpenWrt project.

In general, when browsing educational content installed directly in the mobile router, such additional authentication is not required. However, for the downloading of the teaching materials, we have decided to make the authentication obligatory.

5 Conclusion and Future Work

In this work, we have identified some important security issues arising from the rapidly expanding use of privately owned communication devices such as smartphones in business environments, following the BYOD concept. Given the limitations of existing software-only approaches, we have adopted a hybrid approach involving inexpensive mobile routes with functionally enhanced firmware based on open source. We have thus developed experimental software components as functional enhancements for integration into router firmware and generation of new customized version.

The related experimental work was carried out in a university-level educational environment involving students from Sapienza University of Rome, Italy and Shizuoka University, Japan. As most of the mobile router content delivery and BYOD-assisted collaborative work was conducted in Italian language classes, many foreign students from EU and other countries participated in the experiments. We see this as a well-diversified pool of students that ensures the validity of the obtained results over a wide range of countries and regions worldwide.

We are planning to continue this line of research by further software development and functional enhancement improvements of the customized mobile rooter firmware. Continuing experimental work in classroom environments is also envisaged aiming for a thorough testing and more extensive data collection.

Acknowledgements. We thank Keisuke Inoue for the implementations of the tests during his stay at Sapienza University of Rome.

This work was partially supported by the 2018 Cooperative Research Projects at Research Center for Biomedical Engineering with RIE Shizuoka University.

References

1. Lever-Duffy, J., McDonald, J., & Mizell, A. (2002). *The 21st-century classroom: Teaching and learning with technology*. Boston, MA, USA: Addison-Wesley Longman Publishing Co., Inc.
2. Hein, T. L., & Budny, D. D. (1999). Teaching to students' learning styles: Approaches that work. In *Frontiers in Education Conference, 1999 FIE '99 29th Annual*. ieeexplore.ieee.org (Vol. 2, pp. 12C1/7–12C114).

---

[1] https://github.com/nodogsplash/nodogsplash.

3. Levy, M. (2009). Technologies in use for second language learning. In *The Modern Language Journal, 93,* 769–782. Wiley Online Library.

4. Loughry, J., & Umphress, D. A. (2002). Information leakage from optical emanations. *ACM Transactions on Information and System Security (TISSEC).*

5. Nezarat, A., & Mosavi Miangah, T. (2012). Mobile-assisted language learning. *International Journal of Parallel, Emergent and Distributed Systems, 3,* 309–319.

6. French, A. M, Guo, C., & Shim, J. P. (2014). Current status, issues, and future of bring your own device (BYOD). *CAIS, 35,* 10. researchgate.net.

7. Disterer, G., & Kleiner, C. (2013). BYOD bring your own device. *Procedia Technology, 9,* 43–53. Elsevier.

8. Gelsomini, F., Kanev, K., Hung, P., Kapralos, B., Jenkin, M., Barneva, R. P., et al. (2017). BYOD collaborative Kanji learning in tangible augmented reality settings. In D. Luca, L. Sirghi, & C. Costin (Eds.), *Recent advances in technology research and education* (pp. 315–325). Cham: Springer.

9. So, S. (2008). A study on the acceptance of mobile phones for teaching and learning with a group of pre-service teachers in Hong Kong. *Journal of Educational Technology Development and Exchange.* aquila.usm.edu.

10. Law, C.-Y., & So, S. (2010). QR codes in education. *Journal of Educational Technology Development and Exchange (JETDE), 3,* 7. aquila.usm.edu.

11. Ogata, H., Yano, Y. (2004). Context-aware support for computer supported ubiquitous learning. *Wireless and Mobile Technologies in Education,* 27–34.

12. Ogata, H., & Yano, Y. (2004). Knowledge awareness for a computer-assisted language learning using handhelds. *International Journal of Continuing Engineering Education & Lifelong Learning, 14,* 435–449.

13. Pedone, M., Kanev, K., Bottoni, P., Vitali, D., & Mei. A. (2018). Firmware enhancements for BYOD-aware network security. In *Recent advances in technology research and education.* Springer (pp. 273–280).

14. Kanev, K., Mei, A., & Bottoni, P. (2015). Home communications and services with enhanced security: Augmented embedded systems for communication appliances as an educational platform. *The Japan Society of Applied Physics.*

15. Uther, M., Zipitria, I., Singh, P., & Uther, J. (2005). Mobile adaptive CALL (MAC): A case-study in developing a mobile learning application for speech/audio language training. IEEE.

16. Tan, T.-H., & Liu, T.-Y. (2004). The mobile-based interactive learning environment (MOBILE) and a case study for assisting elementary school English learning. In *2004 Proceedings IEEE International Conference on Advanced Learning Technologies* (pp. 530–534). ieeexplore.ieee.org.

17. Liu, T.-Y., Tan, T.-H., & Chu, Y.-L. (2010). QR code and augmented reality-supported mobile English learning system. pdfs.semanticscholar.org.

Leveraging Low-Power Wide Area Networks for Precision Farming: Limabora—A Smart Farming Case Using LoRa Modules, Gateway, TTN and Firebase in Kenya

Leonard Mabele$^{(\boxtimes)}$ and Lorna Mutegi

@iLabAfrica Research Centre, Strathmore University, Nairobi, Kenya
{lmabele, lmutegi}@strathmore.edu

Abstract. Over the last couple of years, the Internet of things (IoT) technology has dominated the headlines globally. A number of forums, conferences and seminars have been organised locally to inform and educate people, specifically the C-suite, about IoT and the opportunities it brings concerning digital transformation for better business. In this regard, @iLabAfrica, an ICT and Innovation Research Centre based in Strathmore University, Nairobi, Kenya, set up a lab in 2016 to foster industry-led research and innovation in the area of IoT. The lab has implemented a number of IoT projects that address various sectors including agriculture. A noteworthy project is the Limabora—a remote farm monitoring system project, an ongoing collaborative effort between @iLabAfrica, IBM Kenya, Oregon State University and Trans-African Hydro-Meteorological Observatory (TAHMO), which leverages on IoT technology and data analytics to facilitate precision farming. This is meant to address the fear of food insecurity that has been raised by the interchanging flood and drought plagues that have subsequently affected the remote regions in Kenya. 'Lima' and 'Bora' are both Swahili words that mean 'to farm' and 'well/good/better' in English, respectively. The real value of IoT lies in the data and the insights that can be derived from it through various analytics algorithms. The big data revolution has provided novel ways that large amounts of data can be analysed to derive meaningful insights in various fields, including agriculture. This paper provides a detailed account of smart farming—a remote farm monitoring system project, which has proven reproducible outcomes towards ensuring food security and the realisation of development goals in Africa.

Keywords: Precision farming · Agriculture · Internet of things (IoT) ·
LoRa · Machine-to-machine (M2M) · Low-power wide area networks
(LPWAN) · Big data

1 Introduction

The 3rd Green Revolution has transformed agriculture which is often seen as a very traditional sector, into a technology reliant venture to provide for efficient, innovative, dynamic and eco-friendly farms. Farmers are now embracing information communication technology (ICT) solutions to monitor and manage their farms remotely.

© Springer Nature Switzerland AG 2019
M. E. Auer and T. Tsiatsos (eds.), *Mobile Technologies and Applications
for the Internet of Things*, Advances in Intelligent Systems and Computing 909,
https://doi.org/10.1007/978-3-030-11434-3_29

Specifically, farmers are looking to monitor and manage resources such as water and energy usage and supply in their farms, in addition to parameters concerning the weather (humidity, temperature) and soil (soil moisture). However, in the rural regions of Africa, specifically Kenya, the benefits of smart farming are not accessible to farmers. The IoT lab of @iLabAfrica has embarked on a research project that seeks to provide an innovative solution for remote farm monitoring, which is built on IoT and big data analytics technologies. The goal of this project is to build a usable solution for farmers in rural Kenya and to inform on the implementation and sustainability approaches of similar systems in other developing countries towards the realisation of national and global development goals, especially the sustainable development goals (SGDs).

2   Background Research

The research of the technology presented in this study was motivated by our passion to bridge the gap of connectivity and nurture the adoption of the Internet of things (IoT) ecosystem that hit the Kenyan tech scene with a buzz. With the challenges of food security for the last four subsequent years and the vibrant tech ecosystem in our market, we shifted our focus into developing a low-cost, very relevant solution that would fit most of our rural areas. We therefore set out to experiment the LoRa technology to properly understand its reliability if implemented to improve yield through intelligent techniques of collecting and analysing data from the rural farms.

The first test we carried out at our small Internet of things (IoT), laboratory together with Reha Yurdakul (IBM Kenya) and Kelvin Mwega (IBM Kenya) involved the use of two notebooks and two RN2483 modules. One module was connected to one notebook and left stationary at the lab on our fifth floor while the other module was connected to the second notebook whose position was being shifted in three locations per floor to the ground floor. Our focus at this juncture was to establish the stability of the signal from both transmitters in order to select a case. Table 1 shows the finding of this experiment. These findings inspired us to leverage this technology with a full stack implementation to address farming challenges. In this context, the Arduino Serial command/response interface was used alongside the LoRa module's mac and radio commands. Some of these commands with their functions include:

- radio get mod—to output the modulation scheme;
- radio get freq—reads back the current frequency the transceiver communicates on;
- radio get pwr—reads back the current transmit output power;
- radio get sf—reads back the spreading factor settings;
- radio get afcbw—reads back the current automatic frequency correction bandwidth;
- radio get rxbw—reads back the receive signal bandwidth;
- radio get wdt—reads back the current time-out value applied to the watchdog;
- radio get bw—gets the operational bandwidth;
- radio get snr—gets the signal-to-noise ratio for the previous packet reception.

Most of the commands for the two modules gave similar output and on testing the transmission of data, two major commands were used. These are radio tx and radio rx which configures the modules to transmit and receive packets, respectively.

3   Implementation Approach

The remote farm monitoring system leverages on Internet of things (IoT) architecture, through remote sensing and low-power wide area networks, and big data analytics to provide precision farming to farmers located in rural areas. The data collected so far include temperature, humidity, water level (rate of water evaporation also monitored through the use of evaporimeters), battery level and the signal strength of the data transmitting devices. Due to poor signal strength experienced in rural areas, GSM/GPRS and long-range (LoRa) connectivity have been used extensively to achieve the desired machine-to-machine data exchange between the farms and the cloud. The data is transmitted in real time to an Android mobile application through Google's Firebase—a mobile application development platform.

The infrastructural setup implemented for the remote farm monitoring system constitutes usage of open-source electronics and software. The edge devices encompass an 8-bit AVR microcontroller (MCU). In this case, an ATmega328P MCU which operates at a power voltage of 5 V and packages a 16 MHz crystal oscillator to handle the processing of the sensor data is used. The soil moisture and the water level sensors use the analogue interface while the humidity and temperature sensors utilise the digital interface of the MCU. An RN2483 LoRa module is also integrated to the MCU through the receiver–transmitter (UART) interface for the transmission of the data on one set of the nodes while a GSM/GPRS module from Adafruit is connected to the other set of the nodes using the same UART interface.

The LoRa module transmits data through an LoRa-based gateway which has been implemented with a Raspberry Pi 3, an open-source system-on-chip. It hence utilises the LoRaWAN protocol as described in [1]. The gateway leverages the Wi-Fi interface of the Raspberry Pi to relay the sensor data to the Firebase cloud through The Things Network (TTN). TTN is an open-source LoRa-based platform. The data currently streams from all the nodes in real time since the nodes are still under study. This implementation also makes use of mini solar panels which power the sensor nodes. The gateway is currently located indoors and is hence connected to the main power supply backed up with an uninterruptible power supply (UPS) unit.

3.1   Overview of Low-Power Wide Area Networks (LPWAN)

Low-power wide area networks (LPWAN) have become a popular low-rate long-range radio communication technology for adoption globally. Sigfox, LoRa and NB-IoT are the three leading LPWAN technologies that compete for large-scale IoT deployment. This paper focuses on the usage of the LoRa technology within the domain of agriculture. A brief comparative study of these technologies, which serve as efficient solutions to connect smart, autonomous and heterogeneous devices is also discussed here. We show that Sigfox and LoRa are advantageous in terms of battery lifetime, capacity and cost. Meanwhile, narrowband-IoT (NB-IoT) offers benefits in terms of latency and quality of service [2].

LPWAN is increasingly gaining popularity in industrial and research communities because of its low-power, long-range and low-cost communication characteristics. It provides long-range communication up to 10–40 km in rural zones and 1–5 km in urban zones [3]. In addition, it is highly energy efficient (i.e. 10+ years of battery lifetime) and inexpensive, with the cost of a radio chipset being less than two dollars and an operating cost of one dollar per device per year. These promising aspects of LPWAN have prompted recent experimental studies on the performance of LPWAN in outdoor and indoor environments [4].

LoRa

LoRa is a spread spectrum modulation scheme that uses wideband linear frequency modulated pulses whose frequency increases or decreases over a certain amount of time to encode information. The main advantages of this approach are twofold: a substantial increase in receiver sensitivity due to the processing gain of the spread spectrum technique and a high tolerance to frequency misalignment between receiver and transmitter. Just Like Sigfox, LoRa uses unlicensed ISM bands, i.e. 868 MHz in Europe, 915 MHz in North America and 433 MHz in Asia. The bidirectional communication is provided by the chirp spread spectrum (CSS) modulation that spreads a narrowband signal over a wider channel bandwidth. The resulting signal has low noise levels, enabling high interference resilience, and is difficult to detect or jam.

LoRa uses six spreading factors (SF7 to SF12) to adapt the data rate and range trade-off. Higher spreading factor allows longer range at the expense of lower data rate and vice versa. The LoRa data rate is between 300 bps and 50 kbps depending on spreading factor and channel bandwidth. Further, messages transmitted using different spreading factors can be received simultaneously by LoRa base stations. The maximum payload length is 243 bytes. An LoRa-based communication protocol called LoRaWAN was standardised by LoRa-Alliance (first version in 2015). Using LoRaWAN, each message transmitted by an end device is received by all the base stations in the range. By exploiting this redundant reception, LoRaWAN improves the successfully received messages ratio [5].

Sigfox

Sigfox is an LPWAN network operator that offers an end-to-end IoT connectivity solution based on its patented technologies. Sigfox deploys its proprietary base stations equipped with cognitive software-defined radios and connects them to the back-end servers using an IP-based network. The end devices connected to these base stations using binary phase-shift keying (BPSK) modulation in an ultra-narrowband (100 Hz) sub-GHZ ISM band carrier. Sigfox uses unlicensed ISM bands, for example, 868 MHz in Europe, 915 MHz in North America, and 433 MHz in Asia. By employing the ultra-narrowband, Sigfox uses the frequency bandwidth efficiently and experiences very low noise levels, leading to very low power consumption, high receiver sensitivity and low-cost antenna design at the expense of maximum throughput of only 100 bps. Sigfox initially supported only uplink communication, but later evolved to bidirectional technology with a significant link asymmetry. The downlink communication, i.e. data from the base stations to the end devices can only occur following an uplink communication. The number of messages over the uplink is limited to 140 messages per day. The maximum payload length for each uplink message is 12 bytes. However, the number of messages over the downlink is limited to four messages per day, which means that the acknowledgment of every uplink message is not supported. The maximum payload length for each downlink message is eight bytes.

Narrowband-IoT (NB-IoT)

NB-IoT is a narrowband-IoT technology specified in Release 13 of the 3GPP in June 2016. NB-IoT can coexist with global system for mobile communications (GSM) and long-term evolution (LTE) under licensed frequency bands (e.g. 700, 800 and 900 MHz). NB-IoT occupies a frequency bandwidth of 200 kHz, which corresponds to one resource block in GSM and LTE transmission [6]. In fact, the 3GPP recommends the integration of NB-IoT in conjunction with the LTE cellular networks. NB-IoT can be supported with only a software upgrade in addition to the existing LTE infrastructure. The NB-IoT communication protocol is based on the LTE protocol. In fact, NB-IoT reduces LTE protocol functionalities to the minimum and enhances them as required for IoT applications. For example, the LTE backend system is used to broadcast information that is valid for all end devices within a cell. As the broadcasting back-end system obtains resources and consumes battery power from each end device, it is kept to a minimum, in size as well as in its occurrence. It was optimised to small and infrequent data messages and avoids the features not required for the IoT purpose, e.g. measurements to monitor the channel quality, carrier aggregation and dual connectivity. Therefore, the end devices require only a small amount of battery, thus making it cost-efficient [7].

3.2 Development of the Sensor Node Environment

The devices used in the development of the prototype to collect data from their farms under this study include:

1. aTmega328P MCU,
2. soil moisture sensor from Sparkfun,
3. DHT11 sensor module for temperature and humidity sensing, and
4. an RN2483 LoRa module.

The soil moisture sensor connects to the ATmega328P [2] microcontroller unit (MCU) using one of the analogue pins of this MCU chip while the DHT11 uses any of the digital pins. In this case, pins A5 and D2 were used. The RN2483 module uses the Universal Asynchronous Receiver/Transmitter (UART) [7] interface to connect to the MCU [8]. One can use hardware serial or software serial to make this connection. In this case, software serial has been used. The diagram of the connection is as shown below. Both sensors use 5 V power input while the RN2483 communication module uses 3.3 V. Since the ATmega328P MCU used is the 16 MHz, 5 V input, the LD1117 regulator is used to get the 3.3 V for the RN2483 LoRa module.

The ATmega328P was programmed with platformio as described in the platformio documentation using the C++ language. Platformio is added as a plugin to your preferred text editor such as vim, visual studio code, eclipse or atom. Visual studio code is the preferred environment in this case although the code is uploaded on the terminal interface. The diagram of this connection is as shown in Fig. 1. The soil moisture and the RN2483 connections have only been described on how they could be interacted to the circuit.

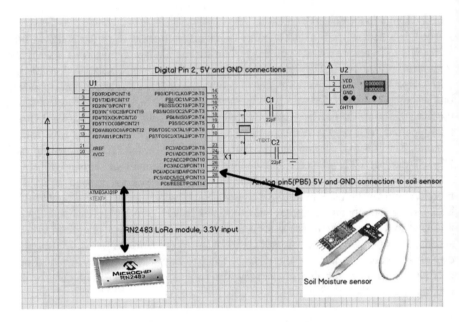

Fig. 1. Connection diagram on ISIS Proteus

Figure 2 shows a serial image of the windows cmd on transmission of data from the sensors to The Things Network (TTN) platform. The diagram shows how the data gets to be transmitted on the LoRa network both on fail and successful transmission.

3.3   The Communication Architecture

Obtaining sensor data to an interface of use by the farmers is a major challenge within the context of this study since the farmers targeted in our development live in the rural areas of Kenya. Most of these areas occasionally experience poor signal strength on the cellular band. Internet connectivity has not penetrated in these places as it has done in the major cities and towns in Kenya. Hence, most farmers only rely on the use of feature phones through Unstructured Supplementary Service Data (USSD) and Short Message Services (SMS) or mobile data for the ones using the smartphone [9]. With this picture in mind, our approach of implementation focused on the use of the GSM/GPRS infrastructure which has also had its cons overriding the pros to deliver a precision farming test case in terms of signal stability, power consumption and management of data. We therefore opted to combine that with long-range (LoRa) technology in our development to deliver a cheaper, reliable and energy efficient solution. The implementation of the LoRa communication architecture leverages the low-power wide area network technology that taps into an unlicensed band of 868 MHz in the ITU region 1. Based on the LoRa radio modulation technology, invented in 2010 by the French start-up Cycleo and then acquired by Semtech, a media access control (MAC) layer has been added to standardise and extend the LoRa physical layer onto Internet networks [8]. Its low power performance capability provides a longer battery

```
-- LOOP
Temperature: 1600
Humidity: 5800
Moisture: 36564
Sending: mac tx uncnf 1 064016A88ED4
Response is not OK: no_free_ch
Send command failed
-- LOOP
Temperature: 1600
Humidity: 5800
Moisture: 36464
Sending: mac tx uncnf 1 064016A88E70
Response is not OK: no_free_ch
Send command failed
-- LOOP
Temperature: 1600
Humidity: 5800
Moisture: 36564
Sending: mac tx uncnf 1 064016A88ED4
Successful transmission
-- LOOP
Temperature: 1600
Humidity: 5700
Moisture: 36564
Sending: mac tx uncnf 1 064016448ED4
Successful transmission
```

Fig. 2. CMD screenshot on transmission of data over the LoRa network

life enhancing the ability of monitoring farms that are not closely linked to grid power. Our implementation elucidated our agreement to the following advantages coming from a number of signal strength and power tests we carried within our University on five floors as shown in Table 1.

The advantages of the LoRa technology are as follows:

1. It uses 868 MHz/915 MHz Industrial, Scientific and Medical (ISM) bands which is available worldwide.
2. It has very wide coverage range about 5 km in urban areas and 15 km in suburban areas.
3. It consumes less power and hence battery will last for longer duration.
4. Single long-range (LoRa) gateway device is designed to take care of 1000s of end devices or nodes.
5. It is easy to deploy due to its simple architecture.
6. It uses adaptive data rate technique to vary output data rate/radio frequency (R.F.) output of end devices. This helps in maximising battery life as well as overall capacity of the LoRaWAN network. The data rate can be varied from 0.3 kbps to 27 Kbps for 125 kHz bandwidth.

7. The physical layer uses robust CSS modulation. CSS stands for Chirp Spread Spectrum. It uses 6 spreading factor (S.F.) from SF 7 to 12. This delivers orthogonal transmissions at different data rates. Additionally, it provides processing gain and hence transmitter output power can be reduced with the same RF link budget and hence will increase battery life.
8. It uses LoRa modulation which has constant envelope modulation similar to frequency shift keying (FSK) modulation type and hence available PA (power amplifier) stages having low cost and low power with high efficiency can be used.

In the development of Limabora, an LoRa gateway was constructed using a Raspberry Pi 3 and the iC880a (purchased from IMST) concentrator. The iC880a was able to receive packets of different end devices sent with different spreading factors on up to eight channels in parallel. The LoRa nodes with larger distances from the concentrator must use higher spreading factors while the nodes closer to the concentrator must use lower spreading factors eliciting the concept of dynamic data rate. The larger the distance, the lower the data rate and vice versa. The usage of this technology in this study is to help bridge the IoT connectivity gap and present a case of LPWAN that is well suited to support services which need long-range communication (dozens of kilometres) to reach devices which must have a low power consumption budget in order to operate several years on a battery pack. The trade-off is a low data rate delivered by the low-power wide area network technologies, from 300 bps up to 5 kbps (with 125 kHz bandwidth) in LoRa modulation as outlined in [10].

The LoRa node connected to the ATmega328P MCU packages the payload of the two sensors and transmits them to the Internet through the iC880a concentrator sitting on the Raspberry Pi 3 to The Things Network (TTN) platform. The TTN platform has also been used to decode the payload from hexadecimal to float. The payload is packaged as bytes in the microcontroller code.

Figure 3 shows how the communication architecture is developed in trying to transmit the data to the cloud. The RN2483 packages the data and sends to the iC880a concentrator on the Raspberry Pi. This is what is called uplink transmission. The concentrator has LoRa on one end and Internet on the other end through the Raspberry Pi. In this, an Ethernet cable has been used on the Pi although Wi-Fi has also been implemented on the other Raspberry Pi devices used in the project.

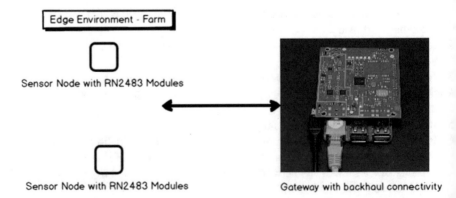

Fig. 3. Transmission of data from the nodes to the gateway with an Internet backhaul to cloud

3.4   Transmitting the Data to Firebase

The implementation of the data store in our case is important for the development of the desired analytics and capability of A.I. to extract the farm trends in terms of the monitored parameters such as the temperature, humidity, pH and soil moisture. This should help in the development of more informative insights in terms of the future climatic changes to anticipate the best crops to grow, the best fertiliser to put to use among other relevant details. In our case, the database selected to handle both user data and the sensor data is Firebase. Although previous test developments had already been done with MySQL and MongoDB, our research approach within this context explored the use of Firebase for two reasons: The first one being the development of a new knowledge locally in regard to the use of Firebase and the second one relating to the benefit farmers can derive from a real-time implementation of a NoSQL approach of the database including real-time analytics.

Firebase is a real-time database designed to accelerate the integration of cloud-based feature into mobile and web applications. In 2014, Google completed the acquisition of a San Francisco-based company named Firebase Inc. It combined the services initially packaged with Firebase with a number of complementary features previously included as part of the Google Cloud Platform [11]. These features include functionality like analytics, databases, messaging and automatic scalability which are really useful in handling IoT-based data.

The Things Network under the integrations tab on the application interface has provided an easy way of integrating with Firebase. Our project leverages this on HTTP integration pointing to our Firebase url where the database is created. The payload has to be in json format.

The result is as shown in Fig. 4.

Fig. 4. JSON format of the farm sensor data

The app_id, counter, device_id and the downlink url are all fetched from The Things Network. The payload is received in json which is the format used by firebase. One can decode the payload_raw to obtain the sensor values for viewing it on a web interface or mobile App. The payload_raw is in base64 format as described in [11].

4   Results

This section gives an overview of the research findings that are related on the use of long-range, low-power wide area network technology to transmit data from a sensor node on the farm to a mobile application made available to the farmer in an effort to achieve precision farming through usage of IoT technology. The capability of transmitting data from the device environment to the online LoRa platform is also presented here. Table 1 is a representation of the tests carried between two nodes to gauge the signal quality of transmitting data using the LoRa technology.

Table 1. Signal power tests on the LoRa RN2483 modules

Floor number	Position	Output power (in decibels)														
		0	1	2	3	4	5	6	7	8	9	10	11	12	13	14
Fifth	1	✓														
	2	✓														
	3	✓														
Fourth	1	✓														
	2	✓														
	3	✓														
Third	1	×	×	×	–	–	✓									
	2	×	×	×	×	×	✓									
	3	×	×	×	×	×	✓									
Second	1	×	×	×	×	×	✓									
	2	×	×	×	×	×	–	–	✓							
	3	×	×	×	×	×	×	–	✓							
First	1							✓								
	2							✓								
	3							✓								
GF	1	×	×	×	×	×	×	×	✓	–	–	✓				
	2	×	×	×	×	×	×	×	–	✓		✓				
	3	×	×	×	×	×	×	×	×	×	×	×	×	×	×	✓

From this table:

- A tick (✓) illustrates successful transmission of data between the modules.
- A dash (–) shows an uncertainty in the communication between the modules; a huge probability of successful packet transmission exists.
- A cross (✗) shows nonsuccessful communication.

The two interfaces in Fig. 5 show how one can view the real-time data from the sensor node environment as gateway traffic and on the application registered on The Things Network. It is critical for one to activate the node device before adding it to The Things Network. The gateway traffic also indicates the frequency used, the modulation scheme used which in this case is LoRa, the spreading factor, bandwidth and the payload size.

The snippet of the programmed code for the transmission of data is shown here.

```
void loop()
{
  debugSerial.println("-- LOOP");

  // Read sensor values and multiply by 100 to effectively have 2 decimals
  uint16_t humidity = dht.readHumidity(false) * 100;

  // false: Celsius (default)
  // true: Farenheit
  uint16_t temperature = dht.readTemperature(false) * 100;

  //soilSensor reading
  uint16_t moistureValue = analogRead(soilSensor) *100 ;

  // Split both words (16 bits) into 2 bytes of 8
  byte payload[6];
  payload[0] = highByte(temperature);
  payload[1] = lowByte(temperature);
  payload[2] = highByte(humidity);
  payload[3] = lowByte(humidity);
  payload[4] = highByte(moistureValue);
  payload[5] = lowByte(moistureValue);

  debugSerial.print("Temperature: ");
  debugSerial.println(temperature);
  debugSerial.print("Humidity: ");
  debugSerial.println(humidity);
  debugSerial.print("Moisture: ");
  debugSerial.println(moistureValue);

  ttn.sendBytes(payload, sizeof(payload));

  delay(20000);
}
```

Fig. 5. The application and gateway traffic on TTN

4.1 Presenting the Data on the Android Mobile Application

In this project, we have developed an Android application that presents the data from the farm to the farmer in real time. The application enables the farmer to view the real-time changes of soil moisture, temperature and humidity on his farm leveraging the LPWAN infrastructure. More features are currently being added to the application to improve its user interface and possibly integrate to the actuators. Historical graphs of the data and the predictive capability are also being integrated to the application to derive the value of the data analytics.

The Android interface that presents this data is shown in Fig. 6.

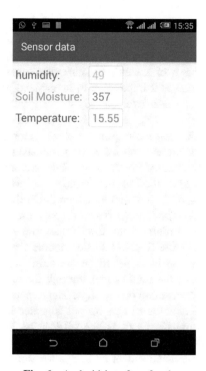

Fig. 6. Android interface for data

5 Anticipated Outcomes

The development tests done for this project informs on the implementation of a smart farming case that leverages on LoRa low-power wide area network (LPWAN) connectivity in Kenya, with reproducible outcomes in other developing economies. The project envisions a huge breakthrough to overcome connectivity challenges in the rural farms by leveraging the LPWAN technology to enable thousands of farmers 'get connected' with their farms. Through the remote farm monitoring system, farmers in rural Kenya will be able to view the overall state of their farm in terms of soil moisture, humidity, pH, temperature among other parameters to realise precision farming. The overall goal is to provide a data-driven approach to farming through low-cost remote sensing that can help farmers make better decisions such as when to plant, what to plant, where to plant and how to plant.

6 Conclusions and Implications

Farming is one of the most critical blocs of the Kenyan economy. The current government of Kenya headed by President Uhuru Kenyatta has included food security as part of its 'big four' agenda which has manufacturing, universal health care, affordable housing alongside the food security. Usage of innovative technologies to help address some of the problems faced by farmers in our rural regions will certainly provide a boost of the yield that would strengthen our food security. Some of the problems faced include waterlogging, under and overuse of fertilisers, and lack of understanding of soil health. There is also a dearth of accurate baseline data regarding different fields and different soil types. This makes it impossible to understand how the yield of the land has either increased or decreased over the years. Even with the addressing of these problems, technology challenges do exist such as connectivity that can help provide data that can be used to develop more problem-solving strategies.

In this case, this work presents an interesting implementation of the LoRa technology especially with the use of the ATmega328P microcontroller. In Kenya, this implementation is unique since the LoRa technology has not gained huge recognition or technical understanding on how it can be adopted. On the other hand, the existing implementation studied on The Things Network only present usage of the already assembled LoRa chip with Arduino Uno. In this case, the usage of the ATmega328P (not on Arduino Uno) and the RN2483 LoRa module from Semtech Corporations shows that miniaturisation can be achieved on the farm through low-cost electronics and reliably transmit data to the rural farmer through the cloud services of TTN and Firebase without Internet connection on the edge. The remote farm monitoring systems demonstrate how these technologies can be put together to help address the food insecurity problem that rural Kenya faces annually. In addition, the system provides historical farm data to inform on planting seasons in relation to drought or flood trends, which in turn facilitates the achievement of Kenya's vision 2030 and the sustainable development goals (SGDs).

References

1. Blum, J. (2013). *Exploring Arduino: Tools and techniques for engineering wizardsry.* Indianapolis: Wiley.
2. Centenaro, L., Vangelista, L., Zanella, A., & Zorzi, M. (2016). The rising star in the IoT and smart city scenarios. *IEEE*, 60–67.
3. Ducrot, N., & Hersent, O. (2016). *LoRa device developer guide.* Orange.
4. Evans, B. (2007). *Beginning Arduino programming.* Apress.
5. Heine, G. (1999). *GSM networks: Protocols, terminology and implementation.* Norwood: Artech House Mobile Communications Library.
6. Liang, D. (2013). *Introduction to programming using Python.* Pearson.
7. Wang, Y. E., Grovlen, L. A., Sui, Y., & Bergman, J. (2016). A primer on 3GPP narrowband internet of things. *IEEE Communication Magazine*, 117–123.
8. Mekki, K., Bajic, E., Chaxel, F., & Meyer, F. (2017). A comparative study of LPWAN technologies for large scale IoT deployment. *Science Direct*, 7.
9. Vermessan, O., & Fries, P. (2017). IoT from research and innovation to market deployment *IERC*, 1–400.
10. Robyns, P., Marin, E., Lamotte, W., Quax, P., Singelee, D., & Preneel, B. (2017). Physical-layer fingerprinting of LoRa devices using supervised and zero-shot learning. In *Proceedings of Wisec ' 17 Boston* (p. 6). Boston, MA: ACM.
11. Smyth, N. (2017). FIrebase essentials—Android edition. In N. Smyth (Ed.), *Firebase essentials* (p. 53). Payload Media.

Mobile Application as a Teaching Strategy to Learn English as a Second Language for Preschool Children

Rocío Rodríguez Guerrero[(✉)], Miguel Andrés Gomez, and Carlos Alberto Vanegas

Universidad Distrital FJC, Bogotá, Colombia
{rrodriguezg, cavanegas}@udistrital.edu.co,
miagomezd@correo.udistrital.edu.co

Abstract. This article shows the process of design to create a mobile solution to reinforce learning in order to teach English as a second language to children under 5 years of age who attend preschool. Initially, planning is done identifying the generalities of the problem. The information about the requirements that clarify the scope of the software is collected. The pertinent analysis of a previously realized design is carried out, a design that includes previsualization of the application. The implementation is completed and the process is concluded with a phase of different tests that allow to identify flaws, for later correction of errors. The development was realized at the Universidad Distrital Francisco Jose de Caldas and the tests in the Localidad Ciudad Bolivar.

Keywords: Android · Application · Learning · Technology in education

1 Introduction

The impact of Information and Communications Technology (ICT) on educational processes has generated a change in the current traditional education model, creating an alternative pattern of more pedagogical teaching in the development of contents in the classrooms. In this way, its use promises a new change in the learning models that were previously used, more didactic and precise, becoming a support tool for students and teachers.

When applying the technology of a mobile system into education is possible to have the potential to improve the learning environments in which children are educated, therefore, it will be given a better use of this technology to support learning, providing the child with a change in the current education system they receive.

On the other hand, the creation of a mobile environment that integrates the development of thematic contents of the areas of knowledge of a second language will involve the development of new skills and the learning process in students.

© Springer Nature Switzerland AG 2019
M. E. Auer and T. Tsiatsos (eds.), *Mobile Technologies and Applications for the Internet of Things*, Advances in Intelligent Systems and Computing 909,
https://doi.org/10.1007/978-3-030-11434-3_30

2 Development Mobile Application as a Teaching Strategy for Learning

2.1 Analysis

The children that to learn a second language from an early age develop a linguistic superiority to the rest of their age, to better ability to learn and in the future, it will be easier to relate in that new language in a successful way. First, understand the language, and second, you will get many opportunities, either professionally or personally [1]. It is for this reason that their study and understanding are of utmost importance for the development of this knowledge.

The importance of teaching a second language to children like the English language will help them in their development as professionals. To this is added the latest discoveries of educators, who believe that the smaller the children are introduced to a foreign language, the better it is for them. Why? They believe their school results generally improve when they are exposed to another culture or way of thinking [2].

The different stages of knowledge, such as elementary and high school, show clearly how in the first stage of learning that is elementary and is shaped the attitude that the student will have in the future with respect to one or several subjects, for this particular case the English language [3].

By centralizing and identifying the problem, it is important to gather certain information for the process of acquiring the second language. In general, this process has its evolutionary development:

A. Pronunciation,
B. Grammar,
C. Vocabulary,
D. Fluency, and
E. Understanding.

This article was focused on the first three stages of evolutionary development in which was created the solid foundation for developing the following two stages. The topics of the class were taken from the childhood education in Colombia, which are:

- The colors: Red, yellow, blue, coffee, pink, green, orange, black, and white. The numbers of the 1–9.
- The alphabet.
- The animals: Dog, cat, lion, fish, bird, monkey, mouse, rabbit, and cow.
- The fruits: Grapes, cherry, watermelon, banana, pear, pineapple, apple, strawberry, and orange.
- The family: Grandmother, grandfather, mom, dad, brother, and sister.

The previous topics are the basic content of study of children under 5 years, because it is not suitable to put topics they have not seen in their regular classes. If they don't know the word, it will make learning difficult, and having in mind the meaning of the words in your native language will help them to remember grammar and pronunciation in English.

In general, children enter school with a good knowledge of their native language, which serves as a natural basis to learn English. The process to acquire English as a second language is progressive and follows a similar process model of how they acquired their native language. For example, produce simple sentences before the more complex ones.

The lifting of requirements was carried out at the Jardin Infantil Universo del Saber (located in the south-west zone of Bogota–Colombia). With the help of the teachers were identified certain critical points in which the children have more disadvantages. These were classified with respect to the already mentioned evolutionary development.

A. *Pronunciation*

In a traditional English class, the words are repeated several times to be able to hold them and to be able to pronounce them properly. In this part, the first critical point is evidenced, the infant does not have the confidence to ask the teacher how to pronounce the words already reviewed and the words they do not understand.

B. *Grammar and Vocabulary*

When assessing the students, the second critical point is evident, which is that they do not retain the syntax of the words studied. Some of the reasons are the lack of practice in class hours and the lack of dedication in their respective homes.

C. *Development Tools*

For the development were applied different tools that are used to create a user interface of well functioning and of easy handling. The software used was:

- Android Studio,
- Adobe Photoshop,
- SQLite, and
- Adobe Illustrator.

One of the most important pillars to learn a second language is constant practice and when referring to children under 5 years, it is necessary to have to think of a flashy and didactic solution, which is why this project was proposed.

D. *Teaching Methodology*

The best method (to teach reading and writing) is one in which the children learn to read and write in a way in which both things are in a playful situation. [...] In the same way that children learn to speak, they should learn to read and write (Vygotsky 1978).

Vygotsky expresses his conviction that written language develops, like discourse, in the context of its use. It indicates their holistic inclination and their awareness of the need for learners to be immersed in language so that learning literacy is easy [4].

The processes of teaching a second language in preschool children have been characterized by playful methods, either the child with an object, or the child with the teacher, but with a didactic approach. The most pleasant methods used to be taught are shown in Table 1.

Table 1. Methodology whole language

Playful activities	Description
Association with photos or drawings	Children tend to be visual, so use books and illustrations to help those associate words in English with illustrations
Treasure hunt	Give each child a list of items written in English that they must find and collect

2.2   Development

The pedagogical tools to use to learn a second language have a fundamental role, in this case, in the process of learning English.

It is important to emphasize the importance of the work that the teacher performs, when teaching the content of the subject in the classroom, where the didactic skills used are a fundamental piece so that the student can learn the subjects taught in the subject. The reason of this project is intend to offer a technological tool through a mobile interactive system, which aims to reinforce the study of English and also to support the methodological strategies for future undergraduate teachers in basic education with an emphasis on English for the development of preschool classes.

The importance of developing the application on mobile technology has to do with the fact that it allows, independently of the place or time, to perform different processing activities on the devices through the different communication options.

The mobile application was designed with two roles: a child and a girl, with their respective colors and their respective voices in terms of phonetics of the vocabulary as can see in Fig. 1.

Fig. 1. Roles respective

The design and development of an interactive mobile system to support the learning of English to children of preschool were integrated by the following modules:

E. History

First of all, the user can select the topic to start. The initial state of the roles is shown in black and white and according to those that complete, the child will change their state to color.

When completing the themes, the achievements already completed in the story are unlocked. When all six accomplishments are completed, it is assumed that the child saw and understood the themes.

Accessing the respective themes, within each one will find the respective vocabulary and pronunciation of this subject as shown in Fig. 2, pressing each image will reproduce the sound of pronunciation and can be pressed as many times as desired children.

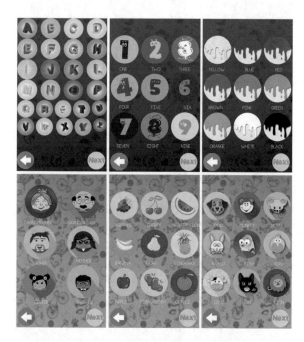

Fig. 2. Themes with their respective vocabulary

The mobile app can realize an evaluation and comprises three levels which the user has to overcome, as can be seen in Fig. 3.

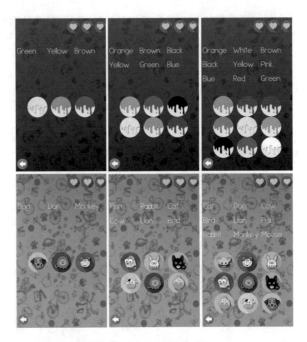

Fig. 3. Assessment levels

The evaluation is handled with a system of opportunities, each level has three lives; if for some reason all three lives are lost in one level of the evaluation, the user has to start again from the beginning, see Fig. 4.

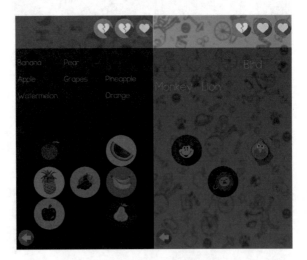

Fig. 4. Lives of the evaluation

F. Challenge

Already back to the main menu the child has this option of challenge, see Fig. 5.

Fig. 5. Challenge module

The score is used to evaluate and to practice what is learned through play words as shown in Fig. 6, when performing the challenge this is can save.

Fig. 6. Word game challenge

G. Score

The mobile app has control of the challenges and record of the practice of students in their free time.

H. Help

This option will allow the members of the community to have a tool that provides helpful information so that users know how to properly handle the application as shown in Fig. 7.

Fig. 7. Help module

The methodology used for the development of the project was Rational Unified Process (RUP), since it allowed to estimate risks early. The project is especially aimed at preschool children, this methodology allowed to realize several tests, and this could correct those areas immediately for an improvement of the project in terms of code, modules, and libraries [5].

The development of this methodology was carried through four phases:

Start phase:

It is one of the most important phases, since it is described in an organized way how the software will be developed. Users who interact with the system and the classes that comprise it are the most important aspects to take into account in the start-up phase.

Processing phase:

In this phase, we continue to work with analysis flows and design. On the other hand, they propose and analyze the risks and threats that can affect the project throughout its development.

Construction phase:

It is an independent and transient phase, since the focus of work is in the operational development of the final product, that is, it is the moment of greater intervention of the programmers.

Deployment phase:

The interaction with the customer is greater in this phase, since the test cycle begins and the possible lifting of new requirements according to the client's request [6].

2.3   Implementation

Before implementation in the garden mentioned above, the application was reviewed and tested by a professor of the department of languages of the *Universidad Distrital Francisco Jose de Caldas* to be completely sure that the application in question has no error in terms of pronunciation and grammar.

Tests were carried out in the *kinder garden universe of knowledge* and the mobile app was installed on the following devices:

1. Motorola Moto g (Android 4.4.4 KitKat),
2. Motorola Moto x (Android 5.0 Lollipop), and
3. LG Magna C90 (Android 5.1 Lollipop).

To evaluate the application took into account features such as graphics and game play, during the test was evidenced how some students who did not solve the games had to use the help of the vocabulary, read them, and repeat them several times to solve the exercises. It was also possible to analyze the behavior of the application in several devices and their operation was normal, only one device was slow to execute, due to the physical limitations of the device as processor and RAM. As previously mentioned the application was tested in the *kinder garden universe of knowledge* in children with an average age of 5 years. The application was used during the subject of the class, 1 week for 30 min.

Table 2 shows the assessment given by the students who interacted with the application. The valuation is given by numbers from 1 to 5, where 1 is deficient and 5 is excellent.

Table 2. Assessment of students

Characteristics	Student rating										
Graphic	5	5	5	5	5	5	4	5	4	5	5
Playability	4	5	5	5	5	5	5	4	5	4	5
New	5	5	5	5	5	5	5	5	5	5	5
Continue	5	5	5	5	5	5	5	5	5	5	5
History	5	4	5	4	5	5	5	5	5	5	5
Challenge	4	5	5	2	5	5	4	5	4	3	5
Help	4	5	3	4	4	3	4	3	4	5	2
Score	5	5	5	5	5	5	5	5	5	5	5
Vocabulary	5	5	5	5	5	5	5	5	5	5	5
Total average	4.6 approximately										

Figures 8 and 9 show that mobile application can reinforce learning in order to teach English as a second language to children under 5 years of age; the concentration level increases with the use of different elements.

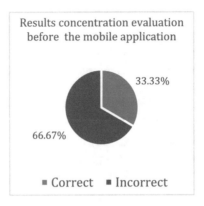

Fig. 8. Results before the mobile application

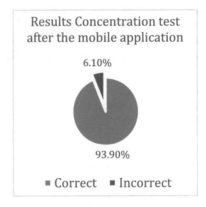

Fig. 9. Results after the mobile application

3 Conclusion

It was concluded that using thematic that gives priority to the interaction through a roll with characters that identify the individual in this case to the children generated a greater interest and facilitated the understanding of thematic of the application.

With the use of a simple introductory module, the user easily locates all the application tools.

The tool of a dictionary that contains the meaning of a basic and simple understandable language offers the possibility of achieving a better learning and with better results.

Appendix

Classrooms are more valuable than any application, so it is not a replacement for traditional classes but a support for children's learning.

References

1. Pavesi, M., Bertocchi, D., Hofmannová, M., & Kazianka, M. (2016). Enseñar en una lengua extranjera. Retrieved from http://www.ub.edu/filoan/CLIL/profesores.pdf.
2. Evelio Jesús, I. P., & Solís, M. (2016). minerva Raquel departamento de educación preescolar. https://goo.gl/HzQ2mp.
3. Carrie Pérez, M. (2006). módulos por competencias para el aprendizaje del inglés como lengua extranjera para niños de edad preescolar. https://goo.gl/2HjlGx.
4. Odman, Y. M., & Goodman, K. S. (1978). Vygotsky desde la perspeciva del lenguaje (p. 118).
5. Francisco Apodaca, German Encinas (2011). https://goo.gl/pDwfzG.
6. María de Lourdes. (2010). Desarrollando aplicaciones informáticas con el Proceso de Desarrollo Unificado. https://goo.gl/Bur3pn.

Poster: Carris Application for Public Transportation

João Fernandes[2], Micaela Esteves[1,2(✉)], and Sónia Luz[2]

[1] CIIC - Computer Science and Communication Research – ESTG, Leiria,
Portugal
micaela.dinis@ipleiria.pt
[2] Polytechnic Institute of Leiria, Leiria, Portugal
2162439@my.ipleiria.pt, {micaela.dinis,sonia.luz}
@ipleiria.pt

Abstract. The Carris application is an application for mobile devices (Android and iOS) personalized to the needs of the client, which is as intuitive as possible, respecting the age diversification of Carris users. This application, in addition to promoting the image of Carris, encourages the use of public transport in the Lisbon region. The application provides detailed information about stops and routes belonging to Carris such as the presentation of the stop's time estimations. The application responds to the geographic context of the user by presenting the closest stop next to him or by planning a trip. The work was developed according to the good rules of software development, following the Apple and the Google guidelines. It has been putting into production based on the requirements, the tests carried out and after all, got the approval of the client. The Android and iOS apps have been deployed in their stores with many uses and installations on both platforms.

Keywords: Mobile applications · Public transport · Real time · Mobile development

1 Introduction

Today, with the presence of sophisticated technology and the diversity of mobile devices, it is possible to simplify the tasks that involve the user's daily life. For this, mobile applications are created to guide and inform the user about the environment in which he is. The use of this technology has become increasingly common and allows to spend the time used on waiting for the bus more productively. The *Companhia de Carris de Ferro de Lisboa* (Lisbon Railroad Company), or just Carris, is a public passenger transport company in the city of Lisbon, in Portugal and it was founded in 1872. The Carris application comes with the objective of showing information in real time, improving the user interaction on the Carris services. In addition to encouraging the use of public transport, this application allows to promote the image of Carris itself and eliminate the barriers between its own services.

© Springer Nature Switzerland AG 2019
M. E. Auer and T. Tsiatsos (eds.), *Mobile Technologies and Applications*
for the Internet of Things, Advances in Intelligent Systems and Computing 909,
https://doi.org/10.1007/978-3-030-11434-3_31

2   Related Work

With the evolution of electronic technology, there is an exponential increase of features and functionalities that a mobile device has been gaining. This technology has not ceased to upgrade and has so vastly spread that facilitates the daily activities of the user. With this, it allowed to exist some variety of models and platforms, being the most known and used: Android and iOS. This section is used to present the main applications for public transport passengers, namely Citymapper, Google Maps, and Move-Me.

The Citymapper is an application developed by Citymapper Limited in 2012 and is widely used in urban public transports, namely on bus, tram, metro, and train. It includes a wide range of features, namely the travel planner that uses real-time information and allows the user to select where will be his departure and destination. In addition to these features, there are variations of paths to the user that involve bus routes, subway, routes without rain, among others, including information about the current context. The application indicates crucial issues to the frequent user such as closure of a line or congestion of the route [1] and is useful for this same reason [2]. This application is adapted to several cities in the world such as Lisbon, London, Paris, Brussels, Madrid, among others.

Google Maps is a smartphone application developed by Google for the Android operating system in 2008 and iOS in 2012. This application provides direction, indication, real-time traffic status, satellite view, street view, offline maps, and local information for users of public transport applications to orient/guide themselves. The application is supported for iOS, Android, Windows Mobile, BlackBerry OS, and more.

The Move-Me is a location-based service, distributed in the mobile markets (Android and iOS) in Lisbon (Lisbon MOVE-ME Application), Porto (MOVE-ME. AMP application), and Coimbra (MOVE-ME Application Coimbra), developed in 2012. These applications add value to the local transport network, allowing the users to search for the buses daily schedule, therefore, permitting the possibility of defining a route through the network and obtaining stops near to the user or any geographical location. The system, however, calls for the involvement of transmission system operators and its storing data that, if analyzed, can result in an important metrics to increase the efficiency of the transport network and user satisfaction [3].

3   Methodology

The software development methodologies aim to structure, plan, and control the development process and the software life cycle. There are two main types of methodologies: the traditional and the agile [4]. The traditional methodologies are characterized by having well-defined phases in the software life cycle, more documentation on product development phases, and greater formality in communication among team members. For example the waterfall model, the prototyping and the spiral model [5].

The agile methodologies present several incremental cycles in the realization of the software. Unlike the traditional methodologies mentioned previously, these have a lower documentation and a greater focus on the development of a product. In addition, there is a more informal communication between the members, thus making it easier to structure the team. There are some methodologies such as Agile, Extreme Programming (XP), Scrum, Kanban, and others [6].

To support the development of the project, it was necessary to define a development support methodology. The methodology used by the company that developed the Carris application, Tecmic, is based on waterfall, but with adaptations that allow bidirectional navigation between the various stages of development [7, 8].

However, it was considered that the methodology most suitable for the realization of the project would be a methodology based on a combination between Scrum and waterfall. This option was based not only on the involvement and constant participation of the client during the development of the project, but also on the transition state of the processes in which the company was in. This methodology was always adapted to the needs of the project, and several iterations of functional software were presented through several meetings with the client. At each meeting, the adjustments to the project were discussed with the client. These adjustments could never affect the principal requirements originally documented.

4   Carris Application

The Carris application comes with the main purpose of putting into service and maintenance a digital solution for the integration of information on the Carris services, with presentation layers for different environments. The goal of Carris was to develop new channels of information and communication with its users, and thus the Carris application appears as a way of informing the users of the "before and during" a trip. The application was developed using Xamarin, for Android and iOS, through Visual Studio and in the same language (C#) as the other Tecmic products, which eases the integration and reuse of shared library code (dynamic link libraries—DLLs). The applications have their own local DB, to cache resources (Favorites, estimated time to the next stop, etc.). This DB is implemented using SQLite. Sporadically, the official editors (Android Studio and XCode) are used as support tools. As an example, Android Studio is sometimes used to viewing logs (for comparative purposes between applications) and creating layouts. XCode is mostly used to manage certificates and to create application layouts.

4.1   Main Features

This chapter describes some of the main features of the Carris application. The goal was to make the application interface as simple as possible, thus facilitating the use by people from different contexts and ages. Given the diversity of devices within the Android and iOS systems, it was decided, based on the Android and iOS market distribution study, to support Android devices with a version equal to or higher than API 17 (Android Jelly Bean) and iOS with version 8.0 or higher. The application supports two languages, Portuguese and English, adapting according to the language defined in the device.

Home Screen

The home screen (Fig. 1) is the point of entry into the application, making it possible to perform most of the options from this screen. To promote the interpretation of the information by the user, the information was grouped into four tabs—Home, Stops, Routes and Institutional Information, according to their purpose. To organize the tabs in the Android application, the TabLayout component was used and the layout of the tabs was at the top of the screen. In the iOS application, the TabBarController component was applied and in contrast to the Android application, the tabs were placed at the bottom of the screen to respect Apple's standards and guidelines [9]. Also, this screen has a global map, through the implementation of Google Maps, where are presented the Carris network stops. By default, the application starts in the first tab (Tab Home).

Fig. 1. Home screen

Tab Home

This screen lets the user to access the travel planner with shortcuts to plan a trip to his favorite places—Home and Work. These bookmarks are also managed through these shortcuts. The user can also plan a trip by setting the source location and the destination of the user choice. Trips saved on the planner are also available on this screen for reference. In situations where the user subscribes notifications for a trip, they are also listed on this tab.

Tab Stops

This screen (Fig. 2) shows the most relevant stops for the user, according to the context, as well as the time estimations. Stops are displayed according to the following criteria:

Fig. 2. Tab stops screen

1. If the user's location is active, the nearest stops will be displayed first;
2. If the user has selected a stop on the map, stops near the selected stop appear;
3. Stops that the user selected as favorites appear next to the nearest stops.

Tab Routes

This tab lists the routes that the user has declared as favorites so that they can quickly see their route and their stop list. It is also possible to search for a line between the various routes of the Carris network.

Trip Planner

The user can navigate directly through the planner to its favorites in Tab Home, or he can choose the origin and destination point manually. For this purpose, the user can choose on the map or also search for a location. When planning the trip, the application assumes the current date as the desired time of departure. The user has the option to choose another date for the planning, for example, to schedule a trip at the end of the workday. After obtaining the travel plan, the application lists the various steps that the user must follow for this trip, including information on the type of route (by foot, bus or subway), price, travel time, etc. The route is also drawn to the map so that the user can follow it. Finally, the trip can be saved for later reference, even when the application is offline.

5 Conclusion and Future Work

The development of the mobile application Carris granted a ground new knowledge in developing Android and iOS applications. With the use of the Cross-Platform Xamarin tool, there was the advantage of having code reutilization, with shared libraries between the platforms. By this way, these libraries can be used in future mobile applications or in already existing projects. The objective of this stage was to develop a "Carris" application from scratch. It was implemented with the main aim of putting into service and maintenance, a digital solution for integrating information about Carris services. Additionally, it was a way for Carris to promote its image and use of public transportation in the Lisbon area. Several characteristics were faced on how to develop mobile applications to show the difficulties that may arise during the development, limiting the use of certain tools. The Carris application was officially launched on the March 20, 2018 and placed in their respective stores, Google Play and Apple Store. Heretofore, the Android application has 46,000 installations and the iOS application counts with 26,000 installations. As the applications are placed in the stores, it is necessary to make the analysis of possible errors or crashes. Along with this, a user observation should be made on the comments for possible enhancements of existing features. Likewise, it is necessary to keep applications up to date to be compatible with the latest operating system versions.

References

1. Zavitsas, K., Kaparias, I., Bell, M. G. H., & Tomassini, M. (2010). Transport problems in cities. *ISIS, 6,* 05.
2. Samsel, C., Beul-Leusmann, S., Wiederhold, M., Krempels, K. H., Ziefle, M., & Jakobs, E. M. (2014, April). Cascading Information for Public Transport Assistance. In *WEBIST* (Vol. 1, pp. 411–422).
3. Anes, J. D. A. (2014). Evolução Move-Me.
4. Eder, S., Conforto, E. C., Amaral, D. C., & da Silva, S. L. (2015). Diferenciando as abordagens tradicional e ágil de gerenciamento de projetos. *Production, 25*(3), 482–497.
5. Centers for Medicare & Medicaid Services. (2008). Selecting a development approach. *Centers for Medicare & Medicaid Services,* 1–10.
6. dos Santos Soares, M. (2004). Metodologias ágeis extreme programming e scrum para o desenvolvimento de software. *Revista Eletrônica de Sistemas de Informação, 3*(1).
7. Leal, A. F. S. (2015). *4Forces Smart Teams: inteligência nas forças de segurança* (Doctoral dissertation, Instituto Politécnico de Leiria).
8. Fernandes, J. R. (2013). Relatório de Estágio: Mestrado em Engenharia Informática-Computação Móvel: SAFER Response (Doctoral dissertation).
9. Guidelines and Resources—App Store—Apple Developer. Retrieved May 19, 2018, from https://developer.apple.com/app-store/guidelines/.

Hardware and Software for Learning IoT Technologies

Daniel J. White[(⊠)]

Valparaiso University, Valparaiso, IN 46383, USA
dan.white@valpo.edu,dan@whiteaudio.com

Abstract. This paper describes the design and implementation of a junior and senior undergraduate course studying technologies used in the Internet of Things. It uses Flipped Learning and Problem-Based Learning approaches to structure student learning with many opportunities for practice and self-directed study in a collaborative environment. The course was divided into two modules.The first was a series of small projects to gain experience with Linux-based embedded systems, shell scripting and command lines, and the Python programming language. The second was an all-course project to build a wireless sensor network from the ground up. Sub-groups of students were responsible for each major aspect of the network including the sensing nodes, communications protocols, wireless transmission, and data collection servers.

Keywords: Project-based learning · Flipped learning · IoT · Wireless · Raspberry Pi

1 Introduction

The skills of rapidly learning a new technology and situating that learning within a larger project is critical to many areas of work. These competencies by definition are impossible to grow without a challenging and new application. Using a large course-wide project about Internet of Things topics allow students to practice the skill of learning new technologies in a structured and collaborative learning environment.

There is a balance of curriculum content between fundamental concepts required for a certain discipline and giving students the opportunity to learn about and gain experience working with current technology trends. This balance does not imply, however, an either-or strategy for selecting content. Indeed, current technology trends require the use of fundamental concepts.

The paper's goal is to provide a specific, detailed example and demonstration of an Internet of Things technology course built from project-based learning and flipped learning principles. Lessons learned and improvements are included to help readers be aware of and anticipate issues in student perceptions and course logistics.

© Springer Nature Switzerland AG 2019
M. E. Auer and T. Tsiatsos (eds.), *Mobile Technologies and Applications for the Internet of Things*, Advances in Intelligent Systems and Computing 909,
https://doi.org/10.1007/978-3-030-11434-3_32

The Flipped Learning Network community uses the following definition for this approach [1]:

Flipped Learning is a pedagogical approach in which direct instruction moves from the group learning space to the individual learning space, and the resulting group's space are transformed into a dynamic, interactive learning environment where the educator guides students as they apply concepts and engage creatively in the subject matter.

"Rapid Electronic Development" was created as a pilot course to give students hands-on experience working with wireless and embedded system technologies related to the Internet of Things. The course was available to junior and senior level undergraduate students in electrical and computer engineering.

The pedagogical approach taken for the course was that of primarily flipped learning for the first half and project-based learning (PBL) for the second half. In place of the traditional paper-based textbook, the students purchased their own Raspberry Pi 3 (Raspi) board and collection of necessary accessories, and were provided with other sensors, microcontroller development boards, and wireless modules.

A Google Classroom was created for the course to allow students access to a central location for reference materials and collaboration tools the Google Apps for Education ecosystem. Students were progressively given autonomy and discretion as part of their execution of their part of the course projects. Google Docs provided an overview of the week or day's tasks and concepts and would have links to reference materials, code examples, and an outline of the main activities for class time.

The course was partitioned into two halves to first provide a structured, hands-on introduction to the hardware and software tools, then to practice their new skills on a large class project to build an Internet of Things wireless sensor network. The first half included introductions to operating a Linux-based system from the command line, shell scripting, and the Python programming language. Each topic included pointers to web-based tutorials and help documentation required to prepare for the in-class activities for the next class time. A project to build a complete wireless sensor network from the ground up provided a rich source of opportunities for applying skills from the first half of the course.

Frequent formative assessment was used during class times in the form of quick surveys and one-on-one help during individual work intervals. Deliveries of code and hardware designs provided summative methods of evaluating student progress. Functional specifications provided a means for students to self-evaluate their work.

2   Course Description

2.1   Learning Objectives

Rapidly building the hardware and software required to make a proof-of-concept electronic embedded system.

1. Design circuits to interface with embedded processors.
2. Compare hardware and software implementations of the same signal processing operation.
3. Assemble complex tasks from simple software tools.
4. Analyze collected data using python plotting tools.
5. Be aware of real-time versus non-realtime processing differences.
6. Utilize development boards and modules to rapidly assemble a system prototype and understand its limitations.
7. Physically modify pre-made modules to fit current needs.

2.2 Logistics

This course specifically made use of open licensed and web-accessible reference materials. Instead of purchasing a traditional textbook, students were required to purchase their own Raspberry Pi 3 (Raspi) [2] along with a recommended kit of accessories from SparkFun [3] for about 90 USD. The Raspi is a popular and inexpensive single-board computer with built-in WiFi, Bluetooth, and several programmable general-purpose input/output pins. The starter kit of parts included a case, power supply, breakout board and cable for the GPIO pins, a USB-to-serial converter, and a small assortment of electronic components. All other additional hardware for later course modules were provided at no cost to the students including Arduino modules [4].

A custom distribution of Raspbian, a Linux distribution for the board, was provided to students. It had the major software components and necessary system settings already pre-installed on top of the base distribution. This ensured every student's device started with a known and useful configuration for the course requirements.

The course met three times per week in during a standard 50-min class session.

3 Course Content and Activities

Activities and content were organized into four major modules, described in the following sections. Over time, the activities and assignments progressed in the level of given detail. Knowledge and skills from previous modules were required to be successful in later modules.

3.1 Tools and Techniques

The purpose of this first segment is to give students exposure to and practice using the Linux command line and scripting with Bash, the Python programming language, and the digital input/output pins available on the Raspi.

Table 1. Four NAND gate implementations

#	Platform	Language	Library
1	Raspberry Pi 3	Bash	WiringPi `gpio` utility
2	Raspberry Pi 3	Python	`import wiringpi`
3	Raspberry Pi 3	C	`#include <wiringPi.h>`
4	Arduino Nano	Arduino (C++)	(built-in)

Four Ways to Implement a NAND Gate: Tools only realize their utility when being applied to solving a specific problem. A series of four assignments implementing a 2-input NAND gate function using various platforms and languages, summarized in Table 1.

To emphasize the operation and usage of the available environments, the functional specifications were common to all four assignments:

- Two inputs, one output using I/O pins on the embedded system.
- Verify the function's truth table using lab equipment (oscilloscope, voltmeter, etc.)
- Measure the propagation delays of all transitions and collect statistics such as minimum, maximum, and average delay.

A typical class time during this module began with students booting and connecting to their Raspberry Pi. They would also open the Google Doc for the day on a PC which contained an outline of the day's activities, homework due next time, and links to references. Following the outline, all students would follow the instructor while they gave live examples of creating and running shell scripts at a Linux command line, for example.

Initial exploration of the relevant commands, such as `gpio`, was performed in an interactive Bash session. Laboratory test equipment or attached buttons and LEDs were used to generate input and show GPIO pin output states, which would change in response to the terminal commands. Listing 1.1 shows an example final script turned in after a class session and following homework for NAND gate implementation #1 from Table 1.

Selected Readings In addition to obtaining hands-on experience using common Linux-based tools and techniques, this module included several required readings.

The Art of UNIX Programming [5] describes the major principles behind the operation of POSIX-like operating systems and their associated tools.

The Art of Insight in Science and Engineering [6] covers estimation and extraction of technical insights from "back-of-the-envelope" type calculations, also called Fermi estimates.

Listing 1.1. Example NAND gate implementation in Bash

```
#!/bin/bash

PINA = 18
PINB = 19

while true; do
  INA = $(gpio -g read $PINA)
  INB = $(gpio -g read $PINB)

  if [ $INA -eq 1 -a $INB -eq 1 ]; then
    gpio -g write 20 0
  else
    gpio -g write 20 1
  fi
done
```

3.2   Project 1—Wireless Chat and Sensor Logging

The final 2 weeks, or six class sessions, before spring break was utilized for students to create a wireless keyboard-to-keyboard chat system without using any existing Wi-Fi or other infrastructure. Inexpensive 2.4 GHz transceivers, the Nordic Semiconductor nRF24L01+, provided the basis for the wireless functionality [7].

As this was the first project in the course, student activities were scaffolded into three phases of increasing system integration. After the initial familiarization interval in Phase 1, class meeting times were used for relevant tutorials and live, follow-along demonstrations of additional techniques and technologies. Example subjects included a taxonomy of common serial communication standards such as UART, I2C, and SPI, or using standard input and output from python scripts.

Later into the project, class times were used as group meeting times to work on the project where the instructor was available for real-time and one-on-one questions and troubleshooting.

A frequent comment during these individual help times was "Why didn't we learn YYY in [previous course(s)]". Discussed later in Sect. 5, these are ideal opportunities to highlight the different nature of the course's approach, with the short instructor's answer as "You just learned YYY now, right? Which way do you imagine will stick with you better?"

Phase 1—Bootstrap This phase occupied the first two class sessions. In-class time was utilized for students to become familiar with communicating with devices using SPI, the libraries available on the Raspi for this form of serial connection, and the features of the nRF24L01+ transceiver.

Activities began with running example code to ensure the wiring was correct and the modules were working properly. They then progressed to modifying the code to change various aspects such as the RF channel, the packet addressing headers, or the data payload. Working in pairs, one student would run example transmitting code while the other would run example receiving the code.

Phase 2—Keyboard-RF-Keyboard Chat After demonstrating successful hardware setup and software tools operation, the next phase was to demonstrate a keyboard-to-keyboard chat system, lasting two sessions and included a weekend. This required students to leverage their new experience about selecting RF channels and packet addressing modes to make system design decisions. The first demonstration at the beginning of the project's second week was for a one-way system where text typed into one terminal session of a Raspi would be transmitted to a second Raspi and output to a terminal screen by a receiving program.

Immediately after more than one group had functional one-way transmission, issues such as RF channel and packet addressing were made apparent. The first instructor suggestion was for groups to coordinate with each other to use unique addresses. With 14 transmitters, with different addresses but still utilizing the same RF channel, there was a drastic reduction in error-free throughput due to packet collisions. Such situations prepare students for hearing and learning about topics like media access protocols, listen-before-talk schemes, and frequency coordination. They can directly identify with the problems addressed by these techniques and immediately begin to implement them or independently seek out additional information.

Example code and a demonstration were provided at this point to show how a few techniques for sending and receiving with the radio and handling both input via `stdin` and output via `stdout` at the same time. Students then demonstrated bi-directional, keyboard-to-keyboard wireless chat functionality between two partners. True full-duplex communications was not a requirement for this project, only the ability to both send and receive text using the same program. The specific method of achieving this was left up to each group, their own research, and discussions among student groups.

Phase 3—Data Collection The final phase during the last two meeting times was to modify the chat system to timestamp and log received packets and to transmit packets of data derived from sensor readings.

The log file format was specified as the following:

```
<timestamp> <RF-channel> <TX-ID> <hex-data> <ASCII-data>
```

By specifying a reasonable format, students were able to experience designing systems to meet specific requirements. Part of demonstrating a functional system was to save and turn in these log files.

Because of the space-delimited format, except for possibly the last data field, the files provided an application for the introduction of pipes and commands-as-filters concept, specifically `awk` (gawk) in this case [8]. Further application of these tools were later homework questions to extract various information from their own log files.

By the end of the class time before spring break, students effectively demonstrated the data transmission aspect of a wireless sensor network. The next project was designed to build on this experientially-gained knowledge.

3.3   Project 2—Wireless Sensor Network Subsystems

The four weeks following spring break, which takes place at precisely the halfway point at our institution, were spent learning about specific sub-systems which make up an operational wireless sensor network. The high-level goals of this Project 2 were for students to deep-dive into a specific aspect of a wireless sensor network and then to be able to function as subject matter experts for the final Project 3. Students were split into four teams, with each team assigned a sub-system:

- **Power**: Power supplies using solar or batteries.
- **Networking**: Networking sensor nodes using the RF24Network libraries [9].
- **RFM69**: Exploring the RFM69HW 915 MHz transceiver modules [10].
- **Sensors**: Sensors for measurement of environmental or other conditions.

Project 2 represented a clear break from the traditional paradigm of uniformity of student work. This interval had four parallel tracks of effort, where an individual student worked only on their sub-system with their smaller group of five people. This emulates an engineer's work on a realistic project, where individuals work in sub-teams with various *areas of responsibility*, which are managed as part of a larger effort. The instructor then functioned as a project manager having responsibility over several sub-teams.

Each sub-group was provided additional information and assistance in setting their goals for weekly team deliverables and requirements for the end of the Project 2 interval. Class times were utilized for group meetings, individual help for specific issues, and short lectures/discussions about topics such as:

- Engineering project documentation and notebooks.
- Areas of Responsibility by project management software company Asana [11].
- *Interactive student feedback to select additional short lecture topics.*
- Temperature measurements and time constants.
- Battery technologies and system-level power estimation.
- How to extract relevant information from chip datasheets.

Each individual was required to write a status report post each week to the course's mailing list (a Google Group) containing:

- Work accomplished during the past week and how it related to their prior declared plans and the sub-group's goals.
- Experimental data and notes.
- Work plan for the next week.

Every 2 weeks, or mid-project and at the end, each sub-team was responsible for a 10 min demonstration of their work and current knowledge of their sub-system. Students were instructed to evaluate the demonstrations and ask questions as if they would later be responsible for using this information when constructing their own sensor node.

3.4   Project 3—ValpoSensorNet

The final four weeks of the course had each individual student responsible for building and operating their own node as part of the class-wide implementation of a wireless sensor network. Being successful in this task required students to become familiar with and utilize the knowledge gleaned by each of the four sub-teams.

ValpoSensorNet used a hybrid star and tree network topology [12] consisting of several node types:

- Leaf nodes utilizing a specified transceiver and sensor (11, 60% of the students)
- Communication relays for RF network range extension (4).
- Gateways to receive transmissions from one type of transceiver and RF channel and forward packets to an Internet-connected database (3).
- Bridge to route packets between the two different RF networks (1).
- Server to collect all packets and make the data available to others (1).

Individuals who had responsibility for writing example code then provided help to all others using their work. Students who learned about the RF24Networking addressing scheme gave tutorials and assistance to select individual routing parameters (each device was assigned a unique ID number, to be included or appended to all packets traversing that node). Local experts in battery power helped provide estimates of node lifetime and tips on software techniques to reduce energy consumption. Students could refer back to the archived progress reports for reference information or contact the person responsible for that bit of knowledge.

Weekly, individual status reports of the same format as Project 2 were still required and served to provide a public record of activity. This was especially helpful to monitor the progress of students with few questions to know when to "check in" with them at various points.

Further work scaffolding for consistent project progress was given in the form of three-goal dates:

- **Goal 3a**: Functional node which successfully reads sensor data and transmits properly formatted packets.
- **Goal 3b**: Leaf nodes battery-powered and demonstration (server logs containing the node ID) of at least 72 hours of continuous operation.
- **Goal 3c**: Complete ValpoSensorNetwork operation by last day of classes.

Student preparation for the final exam session was to use the tools learned throughout the course to analyze data collected from their own node's operation. The final exam session then consisted of two data-intensive slides about this analysis with an accompanying 5— min discussion.

4   Assessment

The primary means of determining student progress through the course material was through demonstrations of hardware and software. Nearly all students consistently met the given specifications by the deadline. Time to achieve the performance metrics and additional assistance required by each student to achieve functional status varied widely, however.

Individual status reports were evaluated for conformance to the prescribed content (description of work accomplished, next week plans, etc.) and not scored based on a measure of work or progress toward the goal. They served to give indications of students needing extra help or those progressing well.

Because of the approach of progressive implementation used throughout the projects section, students received continuous feedback on their work from peers and the instructor. This came in the form of refining and updating code in each phase, peer feedback when others utilized their code, and peer tutorials and help with hardware designs and troubleshooting. As seen later in the student comments about the course, several desired to have the more familiar framework of scored, individual, homework and did not recognize the operational feedback as an additional gauge of their progression.

4.1   Selected Student Perceptions

Student perceptions of the course and their learning varied widely. As a new course in a new format that was operated in a manner different than a traditional lecture, students had many comments for improvement. Criticism seemed to originate from the lack of readily identifiable coursework which resembled typical "homework" activities.

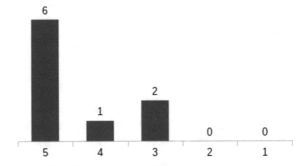

Fig. 1. Student survey answers to "Was the course challenging enough to advance your problem-solving ability?" Scale: 5—Reached a Higher Level, 1—Not Challenged (N = 9)

Figure 1 reflects that students felt challenged and increased their problem-solving ability. Most of the other questions on the standardized student survey focus on aspects of traditional classroom time and activities, which do not map well to this course's format. As a result, responses to those questions had a much wider spread than normal due to the mismatch.

Student Textual Responses to "What Aspects of This Course Were Most Beneficial to You?"

- Working on a real project and seeing what we are trying to do. It was great feeling the accomplishment of getting something done.
- Learning Linux and bash. I was able to land an internship because of those skills.
- Getting hands-on experience with various devices. Felt like one of the most real-world applicable classes I have taken, being able to use things like the Raspberry Pi, Arduino, Python, and other tools that are very widely used today.

Of the 19 students in the course, only 9 (47%) completed the course survey and, further, only 6 (32%) provided text responses of any kind. Since the surveys were anonymous, it is not possible to determine which comments were written by the same student across the text response questions.

Suggestions for improvement had much to do with group dynamics and the distribution of work among members. This is the area of greatest potential improvement for the course's operation and a for other courses which have a large focus on group or teamwork.

Student Textual Responses to "What Do You Suggest to Improve This Course?"

- There was too much of a reliant on other people to do their work adequately and to know more than they did. If there was someone not pulling their weight everyone was let down. Make it more individual based early on so people who are willing to go the extra mile can and others can be left behind with no true consequence to others.
- The project-based style was somewhat frustrating, especially as all of the students had widely varying backgrounds in the subject. For one of the projects, I would try to tell my classmates that I wanted to do something (because I wanted to contribute in a meaningful way to the project). When we would get together the next day, I would have the part that I had previously claimed done, but another groupmate would already have it implemented "because it was so easy for them to whip up" (even though it was extremely challenging to me). It was very frustrating to have my work reduced to nothing and not be able to add in a significant manner.

There were a few other comments relating to the wide range of students' prior knowledge of the various topics covered in the course that is not included here for brevity.

This response below is curious since nearly all of the programming work involved using the Raspberry Pi platform to communicate with wireless modules, or to develop software and program the smaller Arduino-based nodes. There was another comment that also mentioned a desire to have a larger number of smaller projects instead of the half-semester progression to build a complete wireless sensor network.

Student Textual Response to "Additional Comments?"

- Overall this course felt a little falsely advertised. The descriptions of the course made it sound like we would be designing prototypes of multiple electrical systems and making use of the raspberry pi to simulate and test these prototype circuits, yet we barely used the Pi and the only project we really worked on was the wireless sensor network which we worked on for over half of the semester.

To close this section, an unsolicited comment received recently from a student who took the course (16 months afterward):

Your class structure is pretty on par with my grad classes from what I'm seeing, and the familiarity is definitely helping a lot with the transition. Im not going to say that I absolutely adored your classes ... because they required a lot of work on your own to go out, find the questions you had, search for the answers, and validate the solution ... but I can say that I definitely did appreciate your technique more than ever right now as I begin my first semester of grad school.

5   Discussion

Students seemed generally satisfied with having a final project scoring based on meeting functional specifications for demonstrations, which was typically on a binary Meets/Doesn't Meet scale.

None of the textual comments recognized the (non-graded) work scaffolding provided as having a relationship to project goals. Post-course criticisms had a large focus on the absence of activities which mapped to the familiar activity of *homework*—short, graded work turned in by individuals. Though there were frequent in-class discussions of the different nature of the course's operation, in retrospect, it would have helped students describe in more direct and contrasting terms these differences.

Of the nineteen students in the course, fifteen earned an A or A− and only two ended the course with a C or below. The two lowest-scoring students consistently turned in work or reports late or not at all and had multiple demonstrations which failed to meet the given specifications.

The instructor was familiar with the majority of the students in the course by being their academic advisor, and having taught a required course with the same individuals in a previous semester. Particularly interesting was the group of about 6 students who would typically have earned a B grade in a traditional course setting, but ended up earning a A− or A. This was because of this group, particularly, seemed to engage with the different format and view demonstrations as a challenge to overcome. They also asked the most questions, requested the most help with troubleshooting, and seemed the most intent on meeting the requirements.

As the performance of the software or hardware was an externally-measured metric, it was observed that students did not seem to internalize the scores as a reflection of themselves. Instead of being a reason for negative self-talk, these performance metrics were treated more as factual statements that the system was not working *yet*. This was the motivation to continue working or seek additional help.

6 Conclusion

After completing the course, students had enough familiarity with the technologies make productive use of them in the following course project work, including their capstone design course. The use of single-board computer systems in capstone design projects increased after the course. They demonstrated an ability to use Python scripts to control embedded hardware and communications transactions and shell scripting to automate tasks.

The pilot course provided a platform to implement flipped learning and project-based learning concepts. For most students, this was their first exposure to a course not built on the traditional lecture-based class time model. Experience gained from the course included insights into the scaffolding students need to transition their expectations for inside and outside class time.

References

1. Flipped Learning Network. https://flippedlearning.org.
2. Raspberry Pi Foundation. https://www.raspberrypi.org/.
3. SparkFun. https://www.sparkfun.com.
4. Arduino. https://www.arduino.cc/.
5. Raymond, E. S. (2003). *The art of UNIX programming.* Addison-Wesley Professional. http://www.catb.org/esr/writings/taoup/.
6. Mahajan, S. (2014). *The art of insight in science and engineering.* MIT Press. https://mitpress.mit.edu/books/art-insight-science-and-engineering.
7. Semic, N. https://www.nordicsemi.com/eng/Products/2.4GHz-RF/nRF24L01P.
8. Foundation, F. S. Gawk: Effective AWK programming. https://www.gnu.org/software/gawk/manual/.
9. Optimized network layer for nRF24L01(+) radios on Arduino and Raspberry Pi. https://github.com/nRF24/RF24Network.
10. HOPE Microelectronics CO., L.: RFM69HW −120dBm 433/868/915Mhz RF Transceiver Module. http://www.hoperf.com/rf_transceiver/modules/RFM69HW.html.
11. Hickey, K. F. Rethinking the org chart: Areas of responsibility (AoRs). https://wavelength.asana.com/workstyle-aors/.
12. Yick, J., Mukherjee, B., & Ghosal, D. (2008). Wireless sensor network survey. *Computer Networks, 52*(12), 2292–2330. http://www.sciencedirect.com/science/article/pii/S1389128608001254.

A Comparison of Time Series Databases for Storing Water Quality Data

Muntazir Fadhel, Emil Sekerinski$^{(\boxtimes)}$, and Shucai Yao

McMaster University, Hamilton, ON, Canada
{fadhelm,emil,yaos4}@mcmaster.ca

Abstract. Water quality is an ongoing concern and wireless water quality sensing promises societal benefits. Our goal is to contribute to a low-cost water quality sensing system. The particular focus of this work is the selection of a database for storing water quality data. Recently, time series databases have gained popularity. This paper formulates criteria for comparison, measures selected databases and makes a recommendation for a specific database. A low-cost low-power server, such as a Raspberry Pi, can handle as many as 450 sensors' data at the same time by using the InfluxDB time series database.

1 Introduction

Smart sensor devices are about to change the way how we see and control our environment [12,16]. While smart sensor devices are gaining popularity in households, the focus here in on environmental sensing, in particular of water quality. Worldwide water quality is a concern for many communities. For example, in Canada an analysis by Indigenous communities revealed that in 2017, up to 72,000 people were affected by drinking water advisories, potentially affecting health.

This work is part of a multi-university project that addresses these issues by developing new inexpensive water quality sensor systems and by outreach to communities. The goal is to empower communities to keep track of water quality themselves through automated wireless sensors and to augment that with observations on mobile devices, in particular unusual natural phenomena, like level of streams, appearance of water surfaces, wildlife, vegetation, that may be indicative of environmental changes. Rather than having official agencies monitor the water quality, the intention is to let communities do this themselves, as a matter of cost, self-motivation, and trust. Thus, the focus is on inexpensive open hardware and open-source software. The ongoing work also includes outreach activities to high schools and to communities that teach how to build and maintain sensor networks, including associated databases, and how to analyze the sensor data for anomalies.

The plan is to deploy sensors (starting in Fall 2018) for dissolved oxygen, turbidity, temperature, conductivity and, if there is need, for ammonium, nitrate,

© Springer Nature Switzerland AG 2019
M. E. Auer and T. Tsiatsos (eds.), *Mobile Technologies and Applications
for the Internet of Things*, Advances in Intelligent Systems and Computing 909,
https://doi.org/10.1007/978-3-030-11434-3_33

and chloride close to households to monitor drinking water and in first-, second-, third-, and fourth-order streams to monitor water flow and sources of potential contamination. The sensor data together with other observations are stored in a community-run database. The current readings and basic statistical analysis are done on mobile devices, while more sophisticated analysis of time series data is done programmatically in Jupyter notebooks [13], through standard Python and R packages and through dedicated machine learning techniques that are also developed as part of this project. Mobile devices will also be used to capture further observation textually or as pictures. The outreach activity in form of high school activities and continuing education courses will teach communities on the hardware and software of sensors, on networks and wireless transmission, on servers with databases and web servers, so communities can themselves maintain and quickly deploy water quality sensors as the need arises. This paper is on the selection of an appropriate database for water quality data. Companion papers are on the selection of an appropriate web frontend for mobile and desktop applications, the selection of low-cost microcontroller platform, and the selection of inexpensive sensors. This series of papers, starting with the present one, documents the rationale of developing an open hardware–software solution that is both enough robust for critical assessment of water quality and inexpensive enough to be used for lab-based teaching.

Relational databases are common for storing and retrieving large amounts of data. Atomicity, consistency, isolation (for concurrent access) and durability (persistence) are the defining characteristics of database transactions (insertions, updates, searches). Recently, time series databases (TSDB) emerged as an alternative. While the assumption behind relational databases is that queries are frequent compared to insertions and updates and that queries can be composed, the assumption underlying time series databases is that insertions are frequent compared to queries and that queries do not target individual entries but summarize entries in a time interval. The motivation behind recent time series databases is logging of high-frequency events, such as requests to servers and trades in stock markets, which can be in the thousands or even millions per second.

Monitoring of water quality falls between the typical uses relational and time series databases: being a time series, time series databases would be the obvious choice. However, sensors are sampled at most once per minute and the cost limits the number of sensors, so relational databases would be as suitable for this frequency of insertions. Incoming data is to be analyzed for outliers, which involves ongoing queries of exact values over time ranges as insertions progress, so queries are more common than typical for time series databases and are more complex than simple summarization. Water quality data includes also textual descriptions of observations and of procedures used for cleaning and maintaining sensors, but may also include pictures and videos of events, e.g., floods or wildlife, which are normally not part of time series databases. This work addresses the question which kind of open-source database is most suitable for water quality data.

2   Related Work

Previous publications are motivated by releases of new TSDBs or the search for a database for processing or storing time series data for a specific scenario. Some of these publications compare performance. In 2016, Acreman published an overview over 19 existing TSDBs online [8], which is the latest and most complete overview that could be found on existing TSDBs.

Wlodarczyk compares four solutions for storing and processing time series data. The results are that OpenTSDB is the best solution if advanced analysis is needed and that TempoIQ can be a better choice if a hosted solution is desired [19]. No benchmark results are provided, only features are compared.

Deri et al. present tsdb as a compressed TSDB that handles large time series better than three existing solutions [10]. They discovered OpenTSDB as the only available open-source TSDB, but do not compare it to tsdb. The reason is the architecture of OpenTSDB, which is not suitable for their setup. MySQL, RRD-tool, and Redis are compared against tsdb with the result that the append/search performance of tsdb is best out of the four compared databases.

Pungila et al. search for a database for a set of households with smart meters [18]. They compare three relational databases (PostgreSQL, MySQL, SQLite3), one TSDB (IBM Informix—community edition with TimeSeries DataBlade module), and three NoSQL databases (Oracle BerkeleyDB, Hypertable, MonetDB) in two scenarios. Their conclusion is that if the data and therefore the queries are based on a key like client identifier, sensor identifier, or timestamp, then some databases result in increased performance. The performance is logarithmically decreased for these databases when using a combination of tags. Two of the compared databases are interesting for their scenario, depending on whether the focus lies on INS queries, READ queries, or both. If the focus lies on the highest rate of executed INS queries in combination with a worse SCAN performance, Hypertable is the best choice. BerkeleyDB, which has a lower rate of executed INS queries but executes more SCAN queries, is the second best choice.

Acreman [8] compares 19 TSDBs (DalmatinerDB, InfluxDB, Prometheus, Riak TS, OpenTSDB, KairosDB, Elasticsearch, Druid, Blueflood, Graphite (whisper), Atlas, Chronix Server, Hawkular, Warp 10 (distributed), Heroic, Akumuli, BtrDB, MetricTank, Tgres) including a performance comparison. For measuring performance, a two node setup generated about 2.4–3.7 million INS queries per second on average. As a result, DalmatinerDB can execute two to three times as much INS queries than other TSDBs in this comparison, followed by InfluxDB and Prometheus. Although Acreman is the co-founder of Dataloop, a company that is connected to DalmatinerDB, he tries to provide as much transparency as possible by releasing the benchmark. As a summary, it can be concluded that none of the existing publications focus on comparing TSDBs specifically for environmental data or for low-cost hardware.

Table 1. Sample of time series data

Time (UTC)	Reading
11/1/2017 2:20:00 PM	29.20630074
11/1/2017 2:21:00 PM	29.36000061
11/1/2017 2:22:00 PM	29.36050034
11/1/2017 2:23:00 PM	29.50049973
11/1/2017 2:24:00 PM	29.65399933

3   Time Series Databases

In relational databases, data with uniform characteristics are organized in tables linked by relationships [14]. Each table is logically divided into columns and rows, where a row is uniquely identified by means of a primary key. Moreover, data stored on the database can be modified and deleted. For typical time series workloads, however, the data is not characterized by many relationships and does not require modification after being inserted [9]. Additionally, time series data has a natural temporal ordering, and are generally written in a predefined order. At every measurement interval, tables are populated with fresh data that increases the cardinality of the tables; obviously, the space on disk increases continuously. Additionally, when table indices become large enough to prevent themselves to be cached, data retrieval slows down significantly [17].

In order to address these issues with relational databases, TSDBs [15] were created. They are optimized for time series data and are built specifically for handling metrics, events or measurements that are time-stamped. As a result, the only type of data supported is time series data, often represented in the form of time-measurement, key-value pairs. More specifically, the key consists of a time stamp represented as a nanosecond time stamp, while the value consists of either a 64-bit float, 64-bit integer, string or Boolean data type. Sample time series data is illustrated in Table 1. Clearly, data input types and format supported by TSDBs are much more limited in comparison to typical relational databases. However, the result of this imposed model gives TSDBs significant advantages over relational databases in the following ways:

Efficiency A TSDB co-locates chucks of data within the same time range on the same physical part of the database cluster and hence enables quick access for faster, more efficient analysis. As a result, range queries, even over many records can complete quickly.

Usability TSDBs typically include functions and operations that are common to time series data analysis. First, they utilize data retention policies, continuous queries, flexible time aggregations, and range queries. Additionally, they offer mathematical querying functionality that allows users to understand their data in ways common to the analysis of time series data. As a result, the overall user experience in dealing with time series data is signifi-

Table 2. Summary of open-source licenses, compatibility with Raspberry Pi, supported protocols, and default ports

Database	Version	License	Compatibility	Protocols	Default port
OpenTSDB	2.3.1	LGPL	Yes	HTTP	80
InfluxDB	0.10.0	MIT	Yes	HTTP	8086
Elasticsearch	6.4.0	Apache	No	HTTP	9200
Kdb+	3.5	Proprietary	Yes	HTTP	5001
RRDtool	1.5.5	GNU GPL	Yes	TCP	13900
Graphite	1.1.2	Apache	Yes	TCP/UDP	2003
Prometheus	2.3.2	Apache	Yes	HTTP	80
DalmatinerDB	0.3.0	MIT	No	TCP	2003
SQLite3	3.24.0	Public domain	Yes	–	–

Table 3. Growth of selected databases with number of data points. After 10 million inserts, OpenTSDB slowed down to two insertions per minute and CPU usage increased to 100%, InfluxDB stopped working with low CPU and memory usage.

Data points	OpenTDBS	InfluxDB (WAL/TSM)	Elasticsearch	SQLite3	RRDTool	Graphite
1	132 KB	12 KB/20 KB	6.9 KB	4.0 KB	4.0 KB	4.0 KB
1,000	132 KB	12 KB/20 KB	100.2 KB	28 KB	12 KB	12 KB
10,000	132 KB	12 KB/68 KB	866.7 KB	256 KB	80 KB	120 KB
100,000	132 KB	12 KB/564 KB	8.4 MB	2.7 MB	784 KB	1.2 MB
1,000,000	43 MB	12 KB/5.4 MB	80.2 MB	28 MB	7.7 MB	12 MB
10,000,000	–	–/–	802.1 MB	294 MB	77 MB	115 MB
100,000,000	–	–/–	8.3 GB	3.1 GB	763 MB	1.2 GB

cantly improved in comparison to database systems in which this functionality needs to be manually configured and developed.

4   Comparison

Our goal is to evaluate various time series databases in their ability to be incorporated into a low-cost water quality monitoring system. In particular, we aim at Arduino-based sensors (which is not of further relevance here) and Raspberry Pi as a low-cost, low-power server. We consider a total of eight time series databases: OpenTSDB [5], InfluxDB [3], ElasticSearch [11], Kdb+ [4], RRDtool [7], Graphite [2], Prometheus [6], and DalmatinerDB [1]. Table 2 summarizes the main characteristics and Table 3 shows the growth of the database store with the number of data points.

OpenTSDB

OpenTSDB is a scalable, distributed TSDB written in Java and built on top of Apache HBase. OpenTSDB is not a standalone TSDB. Instead, it relies upon

HBase as its data storage layer, so the OpenTSDB Time Series Daemons (TSDs) effectively provide the functionality of a query engine with no shared state between instances. This can require a significant amount of additional operational cost and overhead to manage in a production deployment. As OpenTSDB uses HBase, it needs JVM, HBase, and Zookeeper as well. RAM becomes a bottleneck on the Raspberry Pi: these consume up to 650 MB RAM, making it impractical to run OpenTSDB on the current models of the Raspberry Pi.

In OpenTSDB's data model, time series are identified by a set of arbitrary key-value pairs, and each value belongs to exactly one measurement; each value may have tags associated with it. Tags allow for the separation of similar data points from different sources or related entities, enabling users to easily graph such points individually or in groups. One common use case for tags consists in annotating data points with the name of the machine that produced it as well as name of the cluster or pool the machine belongs to. This allows one to easily make dashboards that show the state of a service on a per-server basis as well as dashboards that show an aggregated state across logical pools of servers. For example, in the storage of sensor readings, pairs may be labeled with the id of the sensor the reading came from. This would allow for the aggregation of results by specific sensors and help understand when sensors have gone off scale and require replacement. Additionally, all data for any given metric is stored together, limiting the cardinality of metrics. OpenTSDB does not have a full query language but allows simple aggregation and math via its API. OpenTSDB supports up to millisecond resolution. This becomes increasingly important as sub-millisecond operations become more common, and additionally allows the freedom to accurately store timestamps for events that may occur in close temporal proximity to one another. One caveat about OpenTSDB is that it is primarily designed for generating dashboard graphs, not for satisfying arbitrary queries nor for storing data exactly.

InfluxDB

InfluxDB is an open-source TSDB developed by InfluxData Inc. (with commercial support). It is written in Go and has no external dependencies, meaning one can put a pre-compiled binary file on a server and run it without having to install or configure anything else. InfluxDB also provides an SQL-like language query language for querying measurements (tables), series, and points. Similar to OpenTSDB, the InfluxDB data model stores time series data consisting key-value pairs called the field set and a time stamp. When grouped together by a set of key-value pairs called the tag set, these define a series. Finally, series are grouped together by a string identifier to form a measurement. Values can be 64-bit integers, 64-bit floating points, strings, and Boolean values. If no timestamp is provided, InfluxDB uses the local server's timestamp. Retention policies are defined on a measurement and control how data is downsampled and deleted. Continuous queries run periodically, storing results in a target measurement.

Elasticsearch

Elasticsearch is a full-text search and analysis engine based on Apache Lucene [11]. Since its initial release in 2010, Elasticsearch has gained popularity as a fast and scalable document indexing and search engine with millions of users worldwide. Elasticsearch stores data in indices, similar to relational databases, where data is logically separated. A single index can contain data for users (personal information, hobbies, etc.), companies (for example, name, addresses, phone numbers), or other entities. The ability to split indices into one or many shards is a crucial feature that allows Elasticsearch to outperform InfluxDB in the horizontal scaling, distribution, and parallelization of data. Elasticsearch is designed as a distributed system in which it is easy to add more instances to the cluster—it will move shards and replicate to new instances automatically in order to maximize the cluster's availability. Internally, Elasticsearch relies on Lucene's implementation of inverted indices, which can be viewed as a map of terms and the documents in which these terms can be found. This is useful in that it can return a subset of documents containing terms specified in a search query. Unlike the other TSDBs evaluated in this report, Elasticsearch is a document search engine and was not built specifically for working with time series data. Therefore, in order to store time series data using Elasticsearch, time series data points are stored as individual documents. Unfortunately, Elasticsearch cannot be installed on current models of the Raspberry Pi: Elasticsearch is based on JVM, which requires around 805 MB in memory.

Kdb+

Kdb+ is a column-based relational time series database with in-memory features. It aims to serve for high-frequency trading in financial markets. It can process billions of records and analyze the data in real time. It also supports a query language similar to SQL, called Q, that can process streaming, real-time, and historical data. Because Kdb+ uses a proprietary licence, it is not analyzed further.

RRDtool

RRDtool is a circular buffer-based time series database. It is designed to manage time series data such as temperature or CPU load. The data is stored in a fixed-size circular buffer. The size is specified when the database is created and cannot be changed afterwards. That is, RRDtool stores time-variable data in intervals of a certain length and it can efficiently implement summarization functions, such as average, minimum, maximum. However, this fixed-size buffer is not suitable for environmental data because the historical data is of significance. RRDtool is therefore not analyzed further.

Graphite

Graphite is an open-source monitoring system with a time series database. Graphite contains three components: Graphite-Web, Carbon, and Whisper. Graphite-Web is a Django-based web application which provides analysis results to the users. Carbon is a daemon processing the data. Whisper is also a fixed-size database. Same as RRDtool, it also uses a round-robin structure, which provides fast, reliable storage for time series data. Whisper features an adjustable resolution of data. It allows higher resolution for recent data than long-term retention of historical data. However, this fixed-size buffer is not suitable for water sensor project because the historical data is of significance. Graphite is therefore not analyzed further.

Prometheus

Prometheus is a free, open-source event monitoring system with a time series database. It collects data from registered nodes at given intervals, evaluates rule expressions, displays the results. Besides, it can trigger alerts if some condition is true. It supports a pull model via HTTP for time series data and provides an intermediary gateway which accepts the time series data pushed from the external source. However, Prometheus accepts only one timestamp (local server time) per sample via the push gateway. Any pushes with timestamps (sensor time) will be rejected. Prometheus is therefore not analyzed further.

DalmatinerDB

DalmatinerDB is an open-source high-performance time series database written in Erlang. It provides a front end which consists of an HTTP endpoint, the query engine as well as a simple website for results. It supports sequential metrics reads and writes. DalmatinerDB is designed to run on the ZFS file system, for which the Raspberry Pi is not well suited; it is short on memory and I/Os.

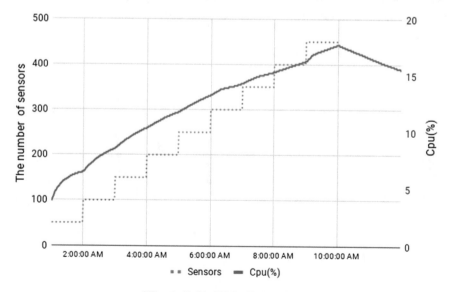

Fig. 1. InfluxDB's CPU usage

SQLite3

SQLite is an open-source, single file, lightweight relational database. Unlike the previously discussed databases, SQLite is implemented as a C library. There is no SQLite process for communicating with other applications. SQLite stores the data as a single file and provides databases functionality through functions calls. In this case, writing data to SQLite amounts to writing a file in one process, rather than IPC. The synchronization of SQLite among multiple processes relies on file system locks. Because of its simplicity, SQLite is widely used by embedded systems. SQLite is a possible choice for the water sensor data project. The disadvantage for SQLite is that it does not provide specific support time series data, so more effort is needed by programmers.

5 Result

Based on the attributes discussed previously, InfluxDB emerges as the first choice among the selected TSDBs. It is easy to install, provides multiple interface formats, and is the easiest to maintain. Additionally, we feel that it is better suited for the intended water quality monitoring system over Elasticsearch due to its incorporation of an HTTP GUI: Such a feature is valuable for individuals coming from a nontechnical background who wish to analyze their data in a visual and intuitive manner. Both OpenTSDB and DalmatinerDB were difficult to install on the Raspberry Pi as they relied on multiple external packages. Additionally, they lack some advanced features such as continuous calculations.

In order to test the performance of InfluxDB on Raspberry Pi, we created several virtual sensors to send data at the speed of one sample per second. The database receives the HTTP requests from sensors.

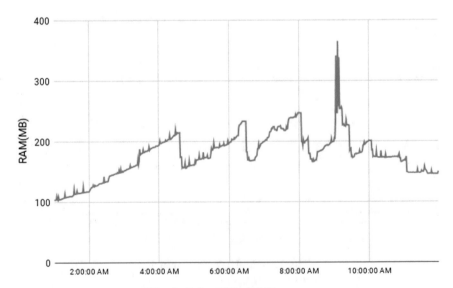

Fig. 2. InfluxDB's RAM usage

All the tests start with 50 sensors and increase 50 sensors by each hour until 450 sensors. We monitor the user CPU usage, RAM usage, and database size in real time. Figure 1 shows the total CPU usage: even with 450 sensors, the user time is less than 20%, meaning that a Raspberry Pi can comfortably handle that load.

Figure 2 shows the user RAM usage with increasing number of sensors (the times have to be correlated to Fig. 1): the RAM usage does not increase significantly with the number of sensors. With 1 GB of RAM, the Raspberry Pi can comfortably handle this load.

InfluxDB maintains two primary structures: Time Structured Merge Tree (TSM) and Write-Ahead Logging (WAL). WAL is suitable for fast writing while TSM provides an efficient approach to querying. In InfluxDB, updates are treated as normal writes. Writes, deletes, and updates occur by writing an entry to the current WAL segment. All data points currently stored in the WAL segment have an in-memory cache. When a segment reaches 100 MB, it triggers a flush to permanent storage. The TSM files are a collection of read-only files. A query searches the files that contain a time range matching the query and InfluxDB does a binary search against each TSM index to find the location. Figure 3 confirms that the size of the TSM continuously increases while the size of the WAS is capped at 100 MB.

6 Conclusion

While carrying out this study, the first attempts of using the Raspberry Pi failed due to the overhead of the web server, to our surprise: initially, the architecture was that the sensors would send HTTP requests to the web server, which would

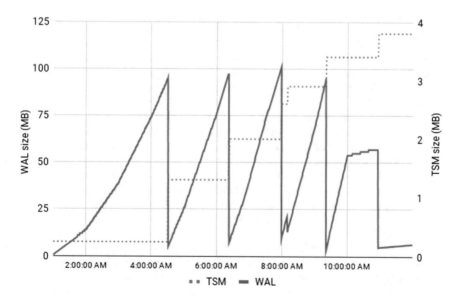

Fig. 3. Size of InfluxDB's TSM and WAL

forward those to the database. This way, the specifics of the database requests would be hidden from the sensors. With both flask and Django, that limited the number of sensors to about 80. The current architecture uses a web server only for visually querying the database and makes the sensors send the requests directly to the port of the database. Thus, this work revealed that with an adequate architecture and a carefully selected TSDB, even a low-cost low-power server like the Raspberry Pi can handle water quality monitoring installations of substantial size.

References

1. DalmatinerDB. https://dalmatiner.io/. Accessed May 30, 2018.
2. Graphite. https://graphiteapp.org/. Accessed August 26, 2018.
3. InfluxDB. https://www.influxdata.com/. Accessed: May 30, 2018.
4. Kdb+. https://kx.com/. Accessed August 26, 2018.
5. OpenTSDB. http://opentsdb.net/. Accessed May 30, 2018.
6. Prometheus. https://prometheus.io/. Accessed August 26, 2018.
7. RRDtool. https://oss.oetiker.ch/rrdtool/. Accessed August 26, 2018.
8. Acreman, S. (2018). *Top10 time series databases*. https://blog.outlyer.com/top10-open-source-time-series-databases. Accessed August 26, 2018.
9. Bader, A. (2016). *Comparison of time series databases*. Diploma thesis, Institute of Parallel and Distributed Systems, University of Stuttgart.
10. Deri, L., Mainardi, S., & Fusco, F. (2012). tsdb: A compressed database for time series. In A. Pescapè, L. Salgarelli, & X. Dimitropoulos (Eds.), *Traffic monitoring and analysis* (pp. 143–156). Berlin, Heidelberg: Springer.
11. Gormley, C., & Tong, Z. (2015). *Elasticsearch: The definitive guide: a distributed real-time search and analytics engine*. O'Reilly Media, Inc.
12. Hsu, L., & Selvaganapathy, P. R. (2014). Stable and reusable electrochemical sensor for continuous monitoring of phosphate in water. In: *SENSORS* (pp. 1423–1426). 2014 IEEE. IEEE.
13. Kluyver, T., Ragan-Kelley, B., Pérez, F., Granger, B. E., Bussonnier, M., Frederic, J., et al. (2016). Jupyter notebooks-a publishing format for reproducible computational workflows. In: F. Loizides & B. Schmidt (Eds.), *Positioning and power in academic publishing: Players, agents and agendas* (pp. 87–90). IOS Press.
14. Maier, D. (1983). *The theory of relational databases* (Vol. 11). Rockville: Computer Science Press.
15. Namiot, D. (2015). Time series databases. In *Selected Papers of the XVII International Conference on Data Analytics and Management in Data Intensive Domains (DAMDID/RCDL 2015)* (pp. 132–137), Obninsk, Russia, October 13–16, 2015, http://ceur-ws.org/Vol-1536/paper20.pdf.
16. Nokovic, B., & Sekerinski, E. (2015). Automatically quantitative analysis and code generator for sensor systems: The example of great lakes water quality monitoring. In: *International internet of things summit* (pp. 313–319). Springer.
17. Pescapè, A., Salgarelli, L., & Dimitropoulos, X. (2012). *Traffic Monitoring and Analysis: 4th International Workshop, TMA 2012* (Vol. 7189), Vienna, Austria, March 12, 2012, Proceedings. Springer Science & Business Media.
18. Pungilă, C., Fortiş, T. F., & Aritoni, O. (2009). Benchmarking database systems for the requirements of sensor readings. *IETE Technical Review, 26*(5), 342–349. Taylor & Francis. https://www.tandfonline.com/doi/abs/10.4103/0256-4602.55279.

19. Wlodarczyk, T. W. (2012). Overview of time series storage and processing in a cloud environment. In *4th IEEE International Conference on Cloud Computing Technology and Science Proceedings* (pp. 625–628).

Assessment and Quality in Mobile Learning

Poster: Evaluation of the Skills for M-learning Environments in Higher Education: Case Study Students of Information Technology of the Autonomous University of Baja California Sur (UABCS)

Jesús Andrés Sandoval-Bringas[1]([envelope]), Francisco Javier Rodríguez-Álvarez[2], and Mónica Adriana Carreño-León[1]

[1] Universidad Autónoma de Baja California Sur, B.C.S., La Paz, Mexico
{sandoval,mcarreno}@uabcs.mx
[2] Universidad Autónoma de Aguascalientes, Aguascalientes, Mexico
fjalvar@correo.uaa.mx

Abstract. This research paper presents the results of a study that applied with a sample probability and random sample to students of technologies of the information of the Academic Department of Computational Systems (DASC, by its initials in Spanish) of the Autonomous University of Baja California Sur (UABCS, by its initials in Spanish) in order to analyze what is the level of skills for the use of m-learning environments, i.e., the set of knowledge that has managed to acquire the student without any difficulty for the use of m-learning environments. The study considered a methodological approach of descriptive quantitative, where were collected data on different aspects of the use of mobile devices. The development of research was a group of students at university level of Educational Programs (PE) of the DASC: Engineering in Development of Software (IDS, by its initials in Spanish), Engineering in Computational Technology (ITC, by its initials in Spanish), and Bachelor of computing (LC, by its initials in Spanish). Of the total number of students enrolled in the DASC (496) randomly selected a representative sample of 115 students. Factors related to the evaluation of mobile learning in higher education were classified into three main categories: basic operations, management of information and communication and collaboration.

Keywords: M-learning · Mobile devices · Digital skills

1 Introduction

Nowadays, there is a fairly widespread tendency to investigate the use of mobile devices in education, since this type of device makes learning more flexible given the possibility of use anywhere and anytime. The United Nations Educational, Scientific and Cultural Organization (UNESCO) consider that mobile technologies can expand and enrich educational opportunities in different contexts [1].

© Springer Nature Switzerland AG 2019
M. E. Auer and T. Tsiatsos (eds.), *Mobile Technologies and Applications for the Internet of Things*, Advances in Intelligent Systems and Computing 909,
https://doi.org/10.1007/978-3-030-11434-3_35

Mobile learning is the process that links the use of mobile devices to teaching-learning practices in classroom or distance learning [1]. Mobile learning (m-learning), "is an innovative form of teaching-learning that shows the applicability of mobile learning through a broad spectrum of activity" [2], since it allows the design of activities in a real context and a greater interaction between "student-teacher".

Mobile devices have had a growth both in the development of their own technologies (software and hardware), and the acquisition of these by a large segment of the population. In the field of education, educational tools of great help in this sector are being developed, because mobile devices allow both teachers and students to connect to the network from any place and at any time.

Considering that many of the students currently have a mobile device, the use of m-learning as a form of teaching and learning is made possible in an important way. This technology allows strengthening the interaction and support to the teaching and learning processes, particularly in the different communication processes in the educational model that is being implemented.

There are numerous projects that refer to the use of m-learning environments and their incorporation into educational processes. However, the inadequate use, as well as the lack of skills for its management, causes the disinterest on the part of students and teachers, not reaching the expected results according to the trends of higher education. On the other hand, the lack of updating of the teacher in technological aspects does not contribute to the expected success, wasting the opportunity to exploit the new technological advances [1]. Only some projects have been dedicated to know the necessary skills to learn using m-learning environment. What can be considered as a problem and at the same time an opportunity for study.

Therefore, the correct use of m-learning environments could help to improve education, awakening the interest of students and motivate them with the use of new technological resources, managing to improve their academic performance during the teaching-learning process.

In [1], reference is made to ICTs in the educational context and states that they can help students acquire certain skills to become competent in them, such as search engines, analyzers, and evaluators of that information; have the ability to solve problems; be creative, communicators, collaborators, publishers, producers, and capable of contributing to society.

In recent years, the use of mobile technology for educational purposes, known as m-learning, has had a great development in higher education, since there are universities in Europe and America that have mobile education systems [3]. Through mobile learning, digital convergence of mobile devices is exploited, focusing on: the capacity of applications that allow recording information from real environments; retrieve information available on the web, and relate people to collaborative work [4].

The use of mobile devices requires and demands, like all human activity, a set of minimum or basic skills and abilities on the part of the student, that is, knowing how to do in a context, for which a certain theoretical and practical knowledge is required and of a digital competence that allows the student to develop his/her capacity of analysis and critical sense having as the main base the technology [5]. In [6], it is mentioned that the first necessary competence for mobile learning knows how to use the device.

Perrenoud [7] defines competence as the ability to mobilize a set of cognitive resources (knowledge, skills, information, among others) to face a family of situations with relevance and effectiveness. The European Higher Education Area (EEES, for its acronym in Spanish) has considered the importance of defining a set of key competences in the information age. The tuning project has proposed a list of the skills that a student must acquire during their university studies, divided into two large blocks: generic, common to all careers, and specific, depending on the career. Some of the generic skills are linked to the use of ICT [8].

On the other hand, in [9], it is mentioned that there are some skills that have greater relevance, necessary for virtual training and that should constitute prerequisites to virtual training, among them: search skills, assessment, quality and selection of the information in the network, skills of analysis, treatment, representation, and information of digital information, skills for the elaboration and structuring of the own production in digital format, skills of work in the team within virtual environments.

In [10], it is stated that students do not reach higher education with an optimum level of digital competence and therefore it is essential to design and develop training and accreditation processes, which allow demonstrating the level of this skill. On the other hand, it can also be observed that they do not know how to use mobile devices adequately for their academic training, since they have not yet developed digital skills. One of these competences is the interactive use of mobile devices and the Internet, as well as applications, which consists of having the skills to search, obtain process and communicate information, and transform it into knowledge [11].

Some researchers propose a new profile of students who enter universities, called "digital natives", who present certain competences developed by the "natural use" of ICT [12–14]. However, it can be affirmed in this framework that students who arrive at higher education, assuming they are digital natives, in practice manifest difficulties in manipulating and using the information they find, demonstrating that they do not have all the competencies to use it. TIC. On the other hand, not all students come to a mobile learning course with the same level of digital domain; Some of them immediately recognize the need to have the technological training that puts them at the required level for the subject, but some others do not recognize this need, presenting difficulties in carrying out the digital tasks that are entrusted to them [15].

Based on the aforementioned considerations, it has been determined to analyze in this investigation different aspects related to the level of practical appropriation of knowledge for the use of m-learning environments of students.

2 Methodology

This research considers a descriptive-quantitative methodological approach, where data or components on different aspects of the use of mobile devices were collected to show the main findings about the level of practical appropriation of knowledge for the use of m-learning environments students of the Academic Department of Computational Systems (DASC), of the Autonomous University of Baja California Sur (UABCS, for its acronym in Spanish).

The study was conducted during the months of January–June 2017, the survey being the instrument used to collect the assessments and perceptions of the students analyzed. The survey consists of a set of questions regarding one or more variables to be measured [16].

The research group involved a group of university-level students of the Educational Programs (PE) of the DASC: Software Development Engineering (IDS), Computational Technology Engineering (ITC) and Computer Science (LC). From the total DASC enrollment (496), a representative sample of size (n) was randomly selected based on the algorithm described in [17], in a study conducted by the Autonomous University of Baja California (UABC) to know the appropriation and educational uses of the cell phone by university students and teachers.

We used (1) to perform the sample size calculation, because it considers a finite population, and the study population is known (496 students).

$$n = \frac{N z_{\alpha/2}^2 P(1 - P)}{(N - 1)e^2 + z_{\alpha/2}^2 P(1 - P)} \tag{1}$$

where

n   Estimated sample size
N   Size of the population
Z   The value of z corresponding to the chosen level of confidence
P   Frequency/probability of the factor to be studied
e   maximum error estimate

The estimated sample size was 115 based on the following values: [N = 496; Z = 1.96; P = 0.5; e = 0.08]. Said sample size represents 23.18% of the total enrollment of the Academic Department of Computational Systems (DASC). The members of this population have as main characteristic to have a professional profile focused on the use of technology, that is, Engineers in Software Development and Computational Technology, and Graduates in Computing.

Based on the above, we used a random selection criterion of the participants, considering complying with the estimated minimum sampling percentage in each of the DASC careers.

An experiment was designed in order to analyze what is the practical appropriation of knowledge for the use of m-learning environments of students. Figure 1 shows the phases of the experiment carried out.

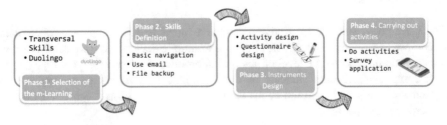

Fig. 1. Phases of the experiment performed. *Source* Authors

Phase 1. An interactive system was selected in an m-learning environment, whose competence was transversal, that is, a competence that covers most of the study disciplines. Due to the above, the Duolingo™ application (http://www.duolingo.com), a language learning platform created by Professor Luis von Ahn of Carnegie Mellon University (CMU), was chosen as a sophisticated solution to the need to translate Internet content while training its users in languages.

Phase 2. The skills that would be evaluated through the use of m-learning environment were defined: ability to send files by email, store files on the device, store files in the cloud, modify files on the device, modify files in the cloud, basic navigation in the device, basic device configurations, installing applications in the device, search for files in the device, search for files in the cloud, participate and interact in social networks, use browsers in the device. From the classification of dimensions in [18], skills were classified into three dimensions, basic operation, communication and collaboration, and information management.

Phase 3. The instruments used in the experiment were designed: The activity that was presented to the students was designed, which was carried out in groups of 10 students, on the heuristic evaluation and to explore usability in mobile devices.

For the activity, mobile devices owned by the institution were used; the personal devices of the participants were not used.

A questionnaire was designed to evaluate the student's perception in relation to the activity developed, which consists of a set of questions regarding one or more dimensions to be measured [17]. The questionnaire was elaborated using reagents that were evaluated with a Likert scale. It had 10 statements in which students were asked to express their opinion by choosing one of the four points or categories of the scale (almost never, 4, rarely, 3, sometimes, 2, and almost always, 1). For the design of the questionnaire reagents, three dimensions were considered: basic operations (reactive 1, 6, and 7), communication and collaboration (reagents 5, 9, and 10), information management (reactive 2, 3, 4, and 8) [19].

Phase 4. The designed activity was executed. In each session, the students were asked to perform the assigned tasks in relation to each of the dimensions: Basic Operations, Communication and Collaboration, and Information Management.

The assigned tasks consisted of: 1. Configure the device to access the Internet, 2. Download the DUOLINGO app, and install it on the mobile device, 3. Configure the device for the use of the downloaded application, 4. Run the application, 5. Select the English language, as well as the daily goal, 6. Perform the recognition exam, 7. Create a profile for the use of the tool, 8. Send by email the screenshot of the completed activity. Additionally, they were told that they should generate evidence for each of the activities developed, and store them on the device. The estimated time for the development of the activity was 40 min.

At the end of each session, the questionnaire was applied collectively to the participating students, the data was codified and analyzed through descriptive statistical analysis.

3 Results

The questionnaire was designed and applied, which measured the skills of each item on a scale of one to four. The products of the three particular skills or dimensions considered were analyzed: basic operations, communication, and collaboration, information management, resulting in:

Descriptive analysis. With this analysis it is exposed that experts in the management of interactive systems in m-learning environments were the students, yielding the following information: the global average in the scores of the 115 students reflected a high intermediate competence, this shows the obtained result of 2.92 (Table 1), taking into account that the highest qualification that can be obtained is 4.

Table 1. Skills for the use of m-learning environment.

Particular skill	Average	Standard deviation
Basic operation	3.675362	0.449086
Communication and collaboration	2.275362	0.582829
Information management	2.817391	0.662532
Average	2.922705	0.564816

Source Authors

At the same time, it was observed that the particular skills that most students dominate are the basic navigation ability, with an average of 3.68 (Table 1). While the population of the research was highly heterogeneous in terms of knowledge for the management of m-learning environments, this was warned by the global index of 0.56 (Table 1), which shows that the further away from zero is the result, more diverse is the population studied. Figure 2 shows the average level in the domain of particular competencies.

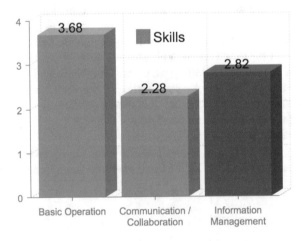

Fig. 2. Skills in the management of m-learning environment. *Source* Authors

Figure 3 shows the number of students for each of the levels of skills by dimension analyzed. In this graph, it is evident that the majority of students do not have problems with the basic navigation of the device, however, not all students dominate communication and collaboration skills, and information management.

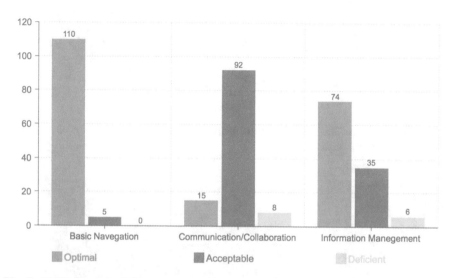

Fig. 3. Skill's levels in the management of interactive systems in m-learning environments by dimension. *Source* Authors

In the reliability statistics, the measurement instrument obtained a Cronbach alpha coefficient of $\alpha = 0.784$, which indicates an acceptable level of reliability.

For the semester of the participants, 10% (n = 12) were first, 24% (n = 28) were third, 31% (n = 36) were fifth, 21% (n = 24) were seventh, 13% (n = 15) were ninth.

On the other hand, in the study carried out it was found that 50% of the students had an optimum level of competencies (58 students), 46% an acceptable level (53 students), and compared to only 4% of the level of poor competence (4 students). This can be seen in Fig. 4, where the frequencies of the averages obtained by the students are shown. For the optimum level, an overall score greater than or equal to 30 was considered, for the acceptable level an overall score greater than or equal to 20 and less than 30, and for the deficient level an overall score less than 20.

In Fig. 5, an image of the participation of students in the developed experiment is shown, where the use of the mobile device can be appreciated.

4   Results

The quantitative results of the research carried out reflect a heterogeneous population in terms of knowledge for the management of m-learning environment. That is, there are variations in the level of competence for the same student for each of the dimensions.

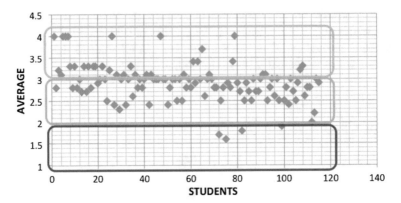

Fig. 4. Distribution of the average level of knowledge per student in the use of m-learning environments. *Source* Authors

Fig. 5. Students of the DASC conducting the experiment. *Source* Authors

While 95% of students dominate the competences of the basic navigation dimension; 64% dominate the competence of information management, only 13% dominate the competence of communication and collaboration. On the other hand, it was also found that there is no direct relationship with the level of skills with the semester that is currently studying.

In [20], it is suggested that current university students are "digital natives", that is, young people born after the eighties who have had contact with new technologies throughout their lives communication what has had an effect on their habits: (1) they are people who seek to be connected all the time by means of new technologies; the "hyper connection". (2) They prefer visual communication over texts. (3) Privileging immediate satisfaction over activities that mean a long-term effort, tend to lose concentration more easily. (4) There is a trend difference in technological capabilities and access to information between students and their teachers.

According to the results, it is assumed that a significant percentage of students, despite being "digital natives", reach their university studies without the necessary level to potentiate and consequently obtain the greatest benefit in the use of the devices mobile and m-learning environments; This can be seen as only 50% of students have an optimal level of competencies for the three dimensions analyzed; therefore, it is essential to design and develop training programs to provide the necessary competencies for the use of m-learning environment.

References

1. UNESCO. (2012). *Turning on mobile learning in Latin America* (pp. 1–69). UNESCO.
2. Naismith, L., Lonsdale, P., Vavoula, G., & Sharples, M. (2004). *Literature review in mobile technologies and learning.* http://www.futurelab.org.uk/sites.
3. Zamarripa, R. (2015). M-learning: El aprendizaje a través de la tecnología móvil, desde la perspectiva de los alumnos de educación superior, 1–15.
4. Aguilar, G., Chirino, V., Neri, L. J., Noguez, J. J., & Robledo, V. F. (2010). Impacto de los Recursos Móviles en el Aprendizaje. 9ª Conferencia Iberoamericana en Sistemas, Cibernética e Informática. Orlando, Florida.
5. Herrera, J., Lozano, F., & Ramirez, M. (2008). Competencias aplicadas por los alumnos para el uso de dispositivos m-learning. Memorias del XVII Encuentro Internacional de Educación a Distancia. Guadalajara, Jalisco.
6. Sandoval, E., García, R., & Ramírez, M. (2012). Competencias tecnológicas y de contenido necesarias para capacitar en la producción de recursos de aprendizaje móvil. EDUTEC Revista Electrónica de Tecnología Educativa, 1–16. https://doi.org/10.21556/edutec.2012.39.379.
7. Perrenoud, P. (2008). La evaluación de los alumnos. Editorial Colihue.
8. Guitert, M., Romeu, T., & Pérez-Mateo, M. (2007). Competencias TIC y trabajo en equipo en entornos virtuales. Revista de Universidad y Sociedad del Conocimiento.
9. Zapata, M. (2010). Evaluación de competencias en entornos virtuales de aprendizaje y docencia universitaria. RED Revista de Educación a Distancia, 1–34.
10. Gisbert, M., Espuny, C., & González, J. (2011). INCOTIC. Una herramiente para la autoevaluación diagnóstica de la competencia digital en la universidad. Profesorado, 75–90.
11. Barragán, A., Martín, A., & Peralta, A. (2016). Análisis del Smartphone como herramienta de apoyo en la formación académica de alumnos universitarios. *Pistas Educativas, 38,* 135–155.
12. García, F., Portillo, J., Romo, J., & Benito, M. (2007). Nativos digitales y modelos de aprendizaje. *SPDECE,* 1–11.
13. Prensky, M. (2011). *Enseñar a nativos digitales.* Madrid: SM.
14. Matilla, M., Sayavedra, C., & Alfonso, V. (2014). Competencias TIC en alumnos universitarios: Dimensiones y Categorías para su análisis. Congreso Iberoamericano de Ciencia, Tecnología, Innovación y Educación, Buenos Aires, Argentina.
15. Vargas, L., Gómez, M., & Gómez, R. (2013). Desarrollo de habilidades cognitivas y tecnológicas con aprendizaje móvil. Revista de investigación educativa de la escuela de graduado en educación, 30–39.
16. Hernández, Fernández, & Baptista. (2014). *Metodología de la Investigación.* Mc. Graw Hill.

17. Organista-Sandoval, J., Serrano-Santoyo, A., McAnally, L., & Lavigne, G. (2013). Apropiación y usos educativos del celular por estudiantes y docentes universitarios. Revista Electrónica de Investigación Educativa, 139–156.
18. Restrepo, S. (2015). Desarrollo de la Competencia Digital en Educación Superior. XVIII Congreso Internacional EDUTEC Educación y Tecnología desde una visión transformadora, 1–12.
19. Montoro, J., Morales, G., & Valenzuela, J. (2014). Competencias para el uso de tecnologías de la información y la comunicación en docentes de una escuela normal privada. Virtualis, 9, 20–34.
20. Salado, L., Velazquez, M., & Ochoa, R. (2016). La apropiación de las TIC en los estudiantes universitarios: Una aproximación desde sus habitus y representaciones sociales. Estudios Lambda. Teoría y práctica de la didáctica en lengua y literatura, 214–234.

An Active Learning Strategy for Programming Courses

Seshasai Srinivasan[✉] and Dan Centea

School of Engineering Practice & Technology, Hamilton, Canada
{ssriniv, centeadn}@mcmaster.ca

Abstract. The objective of this research is to employ a simple active learning strategy in conjunction with the principles of cognitive psychology to enhance student learning in an undergraduate programming course. The investigation was done for a first-year course in C++. The course was taught over a period of 13 weeks during which the instructor meets with the students for 4 h every week. Specifically, quantitative investigations were made in which the course was taught using the *intervention*- and *reinforcement*-based active learning method to a group of approximately 120 students. The two key components of these strategies include: (a) hands-on programming while teaching the concepts and (b) group debugging exercises. In the former, the students were taught concepts via collective coding exercises in the classroom as a part of active learning strategy. Additionally, to periodically recall and reinforce concepts, the following were done: (1) each new class begins with a group debugging exercise where the students collectively participate in debugging or constructing a program, requiring them to recall earlier concepts. (2) Weekly programming labs and assignments are assigned that the students must complete for a certain grade. The student learning is measured via assessments through two term tests and a comprehensive final exam. From these assessments, it has been found that the active learning strategy has benefitted the students in terms of their learning. Further, this is attributed to the *effective* reinforcement of the concepts as well as the intervention strategy employed in the classroom. The former was achieved via (a) practice, teaching/learning/communications with their peers inside the classroom and the latter via compulsory labs and assignments that the students undertook outside the classroom.

Keywords: Active learning · Programming · Cognitive psychology

1 Introduction

Across the globe, there is an active struggle to provide good quality and affordable education to children. Depending upon the setting of the learning environment, the researchers have adopted a variety of teaching and learning techniques [1, 2]. In establishing the best practices, numerous experimental investigations have been conducted inside the classroom. This has resulted in a variety of teaching and learning methods, each pertinent for a particular group of students in a specific learning environment.

© Springer Nature Switzerland AG 2019
M. E. Auer and T. Tsiatsos (eds.), *Mobile Technologies and Applications for the Internet of Things*, Advances in Intelligent Systems and Computing 909,
https://doi.org/10.1007/978-3-030-11434-3_36

Of the several methods that have emerged, some of the most commonly employed techniques in the classroom include cooperative and small group learning [2–7], problem-based learning [8–14], inquiry-based learning [8, 15], challenge-based learning [16], and undergraduate research-based learning [17, 18].

Among these, the teaching methodology that is of interest in this work is cooperative active learning. In cooperative learning, Springer et al. [3] concluded that small group learning promotes academic achievements, and improve the persistence levels of the students and their attitudes towards learning. Further, there is ample evidence in the literature to show that student performance is further enhanced if there is an active component in the classroom, wherein, there is an interactive education process [5, 6]. Some investigations have advocated the use of computer-based active instruction to enhance student learning [19, 20].

On a different front, studies have used the principles of cognitive science to deliver lectures and have found success in improving student learning and retention [21–27]. In implementing this inside the classroom, three key iterative processes that are incorporated into the learning environment include *Reinforcement, Spacing,* and *Instant feedback.* The motivation behind the integration of these principles in the intervention strategy is their positive effect on learning and long-term retention of the material, which are as follows:

(a) *Reinforcement*: By repeatedly recalling the concepts from the memory, the information is more permanently stored in the memory.
(b) *Spacing*: To aid the retention of the material for a longer duration of time, the material must be practiced over a longer span of time.
(c) *Instant Feedback*: An immediate corrective feedback can help in better understanding of the material more

This work presents the use of active learning in teaching a first-year undergraduate engineering course in programming using C++. The positive results of active learning as observed in the literature and the fact that in the traditional lecturing approach only 20% of the students master the material [4], prompted the authors to make this shift in the teaching and learning environment.

The rest of the paper is organized as follows: Sect. 2 presents the materials covered in the course. Section 3 describes the experiments conducted and the data collection methods inside the classroom. The results and discussion of the experiment are presented in Sect. 4 followed by summary and conclusions in Sect. 5.

2   Materials

The course in which the active learning strategy is introduced is a first-year undergraduate engineering course in C++. The course is offered over a period of 13 weeks. The students are divided into groups of 40 based on their schedules. Each class has an enrollment of approximately 40 students and each student has access to a dedicated computer equipped with Microsoft Visual Studio environment for coding the C++ programs. The course covers the following topics:

1. Data types, math operators, logic operators, libraries
2. Decision structures
3. Loop structures
4. File I/O
5. Functions
6. Pointers
7. Arrays
8. Strings
9. Data structures

3   Methods

The topics listed above are covered over a period of 13 weeks. The class met two times a week for a duration of 2 h in each meeting. All lectures were delivered in a classroom equipped with computers for each student and a projector to share the instructor's screen as well as presentation slides with the students. The performance of the students was assessed via seven lab homeworks, three assignments, two term tests, and a comprehensive final exam. The lectures, programming demonstration, classroom activities, and interactions, as well as assessments were computer based.

3.1   Active Learning Component

Active learning was an extremely important aspect of learning in this course. This was introduced in the course via group debugging exercises. In this, each week, the lecture would begin with a coding exercise in which the entire class participates. More precisely, the instructor defines a problem, which would require the students to construct a code that produces certain desired outcomes. The students were then asked to dictate the code that should be entered by the instructor to make it functional. As the instructor codes the program, it is visible on the projector screen for the entire class to view. Every student had to participate in this exercise, and the participation was enforced by asking each student based on their seating sequence, to help contribute to the code building. This could include dictating some small aspect of the code, or editing a possible wrong contribution by his/her peer with proper justification or present an alternative coding strategy. Eventually, after much deliberations over the formulation of the logic, applicable syntax, etc., a functional code is constructed collectively by the class. Put differently, during the process, the students could suggest new parts of the code, edit existing parts of the code, propose alternative approaches and also debate with the other students in the class. This exercise was conducted once every week for about 20 min of a 2-h class.

3.2   Cognitive Psychology Component

Each week, students were required to develop a program for one or two programming questions. These questions involve concepts taught in previous cases as well as new concepts that were to be taught in the present week. Thus, while solving programming problems in class, the students were required to recall the concepts taught in the previous class. This approach that involves repeated recalling and implementation of the concepts enables the *reinforcement* of the concepts. The fact that this was done once every week introduced a *spacing* that is important for the synthesis of information. Further, while solving the problem in class, the solutions developed by the students were immediately evaluated and the students could rectify their mistakes and correct their codes. This *instant feedback* enabled weed out the chances of students cementing wrong concepts that will be difficult to rectify at a later stage.

3.3   Data Collection

Student performance and learning was measured using a variety of data. These include weekly lab homework and assignments that the students submit for a grade. Specifically, the students submit a total of 6 labs and 3 major assignments through the term. The labs are generally simpler than the assignments and require the students to program and implement codes that primarily focus on a set of 3–4 concepts taught in the recent week. On the other hand, assignments require a major development of codes to solve programming problems and require recalling and implementation of numerous concepts taught in the recent few weeks. An assignment is usually assigned once every three weeks.

Apart from the labs and assignments, two terms tests were administered. The tests were computer based, have a duration of 1.5 h, and are held approximately on weeks 5 and 10, respectively. The topics on which the students were tested on the two term tests are the following:

Test 1: primitive data types, string object, variables (naming conventions), constant, typecast, arithmetic operators, unary operators, compound operators (+=, −=, *=, / =, %=), integer division, output to console, input from keyboard, comments if, if/else, if/else if/else, switch, nested control structures, relational operators, Boolean logic, random number generator, C++ library functions, formatting output reading input from a file, (no output to a file), while, do/while, for loop, counters, sentinel, data validation (pre-test and post-test), flags, nested control structures.

Test 2: Defining and calling a function, prototype, pass by value, return using return statement, local and global variables, scope address of a variable, initializing pointers, passing pointers as parameters in a function

Each term test typically required them to develop a code for two problems. In solving the questions on the test, the students developed codes for the problems directly on the computers and submit the developed code as a solution to the question.

Final exam was a comprehensive computer-based exam that is for about 2.5 h. In the final exam, the students were required to solve two or three programming problems. Collectively, they covered almost every topic taught in the course through the 13 weeks. As part of their solution, the students submitted their programs that they developed for the problems.

The labs and assignments were graded by a qualified teaching assistant who grades them for execution as well as correctness. The labs were collectively worth 10% of the total course grade. Similarly, each assignment was also graded by the teaching assistant. The three assignments were collectively worth 10%.

The term tests and final exams were graded by the instructor. While Test 1 was weighted at 15%, the second test was weighted at 25%. The final exam was worth 35%. These were graded for their execution as well as the correctness of implementation.

4   Results and Discussion

The average performances of the students in the various components of the course are shown in Fig. 1. As seen in this figure, student performance is quite satisfactory in labs, averaging around 81%. This indicates that the students are able to grasp and implement the latest concepts taught in the class in the programs. The dip in the average lab scores around weeks 5–7 is because the students are taking multiple courses during the term and constantly try to optimize their time by dividing it between different courses to maximize their performance. With high initial scores in the labs, they divert their attention to the other subjects and this results in a dip in the lab scores. Closer to the end of the term, when the course becomes more challenging, the students start focusing their attention again and this reflects on higher grades in the last 3 labs. Nevertheless, this is not as high as the first few labs. This is also expected and is due to the fact that the topics covered in the last few labs are quite challenging for the students.

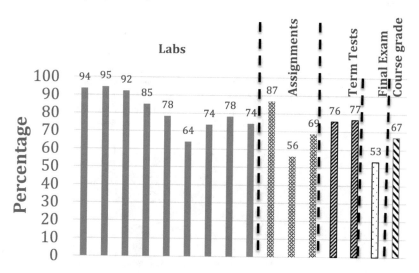

Fig. 1. Average performance of the class in labs, assignments, term tests, final exam, and the overall course grade

Similarly, the average assignment score of the students ranges from 56 to 87%. The dip in the second assignment is because the assignment is during the 6th week of the course when the students have midterm tests from several other courses. Given that the weight of the individual assignment is about 3.3%, the students do not put a lot of time and efforts in this assignment, channeling their time and efforts to the midterms from the different courses instead.

The justification presented for the dip in the student scores in the labs and assignments is supported by the performance of the students in the two term tests. Specifically, in Test 1, which is administered during week 5 of the course, the average score of the students is 76%. Among others, this test also includes topics from the labs that have a low score. Similarly, the average performance of the students in Test 2 is nearly the same at 77%. Once again, among others, this test contains topics of the second assignment and the labs where the student performance was not good.

It must be added that unlike the labs, the term tests are more challenging requiring students to develop a detailed program for each question. A typical term test question is as follows:

Question [Marks 60%]: We want to develop a C++ application that will study the projectile problem in the adjoining figure. The key variables indicated in the figure are as follows: V_0 is the initial velocity in m/s, θ is the angle in degrees, d is the distance in meters, h is the max height in meters, and $g = 9.8$ m/s^2. The user has to study the angle in which the player should shoot so that the ball lands in the basket. This problem is studied as follows: The main program will get the values of V_0 and d_{target} (the distance of the ball from the basket) from the user. These variables must satisfy the following conditions $4 \leq v_0 \leq 6$ m/s and $1.7 \leq d_{target} \leq 2.2$ meters. The main program will prompt the user to enter the values until these constraints are satisfied. A function **releaseAngleRange** will use these variables and calculate values of d for all values θ in the interval [20, 30] degrees in steps of 0.1 degree using the formula

$$d = \frac{v_0^2 sin2\theta}{g}. \tag{1}$$

For each angle, if d is within 0.05 m of d_{target} then it is assumed that the ball falls into the basket and that shooting angle is acceptable. In sifting through the range of [20, 30] degrees, the function calculates the minimum shooting angle (θ_{min}) and the maximum shooting angle (θ_{max}) for the ball to fall into the basket. The main program prints the values of all the parameters provided by the user to 2 decimal places. The minimum and maximum shooting angles are printed to 1 decimal place.

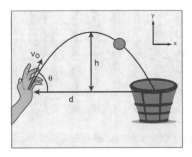

Solving this question, for instance, to create the application requires a good understanding of looping structures, decision structures and functions.

The final exam is also computer based, comprehensive and covers several topics in each question. Further, the final exam questions are typically assessing the ability of the students to design and create applications. Since the assessments pertain to the highest levels of Bloom's taxonomy, the magnitude of difficulty would be much more requiring design and creation. There are typically two such questions and the students are required to solve them within 2.5 h.

As seen in Fig. 1, the student performance on the final exam is significantly lower at 53%. This is expected and can be attributed to several reasons:

1. The final exam period requires students to work simultaneously on several other subjects. This means that the students are unable to allocate the amount of time and effort needed to prepare for the final exam.
2. The programming final exam is always the last exam since it is a computer-based exam and after a grueling exam season, the students are mentally exhausted. This has been our observation over several terms. Unfortunately, moving C++ exams earlier would mean some other subject will be facing this outcome.
3. The final exam focuses on the highest levels of Bloom's taxonomy, requiring the students to design and create difficult programs in a stressful examination environment. As a result, students find the final exam very challenging.
4. Students do not get a chance to adequately practice some of the topics covered in the past 2–3 weeks of the course since they are busy optimizing their time with other courses with the impending final exam period.

Finally, we also present the overall course grade of the students in this subject. Among the first-year students in this program, a course average of 67% is considered a very good performance It must be highlighted that a majority of the students taking this course in the first year of their engineering program have no prior programming experience. Further, C++ itself is a very challenging programming language that adds to the level of difficulty of the course. In conclusion, given that our objective is to create a learning environment that is conducive to student learning, as measured by their performance in various assessments, the active learning environment coupled with the principles of cognitive psychology is a reasonably good approach.

5   Summary and Conclusions

In this work, an active learning strategy in conjunction with principles of cognitive psychology has been implemented in a first-year undergraduate engineering course in C++, and student learning is measured using a variety of assessment techniques.

The study was performed with 120 students split into three sections based on their schedule. Each week, the class met for 4 h in a computer lab setting. An active learning strategy was employed in all three sections. Specifically, there was a combination of live programming during the lecture, group debugging exercises as well as computer-based labs and assignments.

An important highlight of the lectures was the active learning strategy, i.e., group debugging exercise. In this exercise, a simple problem was proposed to the class and the students had to point out or construct the solution. Every student had to participate in this exercise, and the participation was enforced by asking each student based on their seating sequence to help contribute to the code building. During the process, the students could suggest new parts of the code, edit existing parts of the code and also debate with the other students in the class. In some instances, students would also propose alternative approaches.

Apart from the weekly debugging/constructing exercise, the students had to submit ten labs and four assignments through the term. The labs involved solving simple questions that were primarily designed for students to be able to implement concepts. The assignments included more challenging questions, were multiple concepts had to be implemented to solve the problems.

In the debugging exercises in the classroom, by repeatedly recalling the concepts from the memory, the information is more permanently stored in the memory. Further, to aid the retention of the material for a longer duration of time, the material is practiced over a longer span of time by assigning questions in labs and assignments that require the students to recall the concepts.

The active learning strategy (group debugging exercise) in conjunction with the principles of cognitive psychology enable the students to retain the concepts and perform well in the assessments. Specifically, the students performed extremely well in the initial part of the course involving simpler concepts. Their performance was a little lower in more advanced concepts of programming. Overall, the average course grade was around 66% which is a very satisfactory result for a first-year undergraduate course in C++, a very challenging course taken by students with little or no programming background.

References

1. OECD. (2012). *Education at a glance 2012. OECD indicators*. Paris: OECD.
2. UNESCO Institute for Statistics (UIS). (2011). *Global education digest 2011*. Montreal: UIS.
3. Springer, L., Stanne, M. E., & Donovan, S. S. (1999). Effects of small-group learning on undergraduates in science, mathematics, engineering, and technology: A meta-analysis. *Review of Educational Research, 69*(1), 21–51.
4. Wage, K. E., Buck, J. R., Wright, C. H. G., & Welch, T. B. (2005). The signals and systems concept inventory. *IEEE Transactions on Education, 48*(3), 448–461.
5. Hake, R. R. (1998). Interactive-engagement vs. traditional methods: A six-thousand-student survey of mechanics test data for introductory physics courses. *American Journal of Physics, 66*(1), 64–74.
6. Buck, J. R., & Wage, K. E. (2005). Active and cooperative learning in signal processing courses. *IEEE Signal Processing Magazine, 22*(2), 76–81.
7. Terenzini, P. T., Cabrera, A. F., Colbeck, C. L., Parente, J. M., & Bjorklund, S. A. (2001). Collaborative learning vs. lecture/discussion: Students' reported learning gains. *Journal of Engineering Education, 90*(1), 123–130.
8. Prince, M. J., & Felder, R. M. (2006). Inductive teaching and learning methods: Definitions, comparisons, and research bases. *Journal of Engineering Education, 95*(2), 123–138.
9. Prince, M., & Felder, R. (2007). The many faces of inductive teaching and learning. *Journal of College Science Teaching, 36*(5), 14–20.

10. Kolb, D. A. (2015). *Experiential learning: Experience as the source of learning and development* (2nd ed.). Pearson Education, Inc.
11. Freeman, S., O'Connor, E., Parks, J. W., Cunningham, M., Hurley, D., Haak, D., et al. (2007). Prescribed active learning increases performance in introductory biology. *Cell Biology Education, 6,* 132–139.
12. Sidhu, G., Srinivasan, S., & Centea, D. (2017, July 3–5). Implementation of a problem based learning environment for first year engineering mathematics. In Guerra, A., Rodriguez, F. J., Kolmos, A., & Reyes, I. P. (red.) (Eds.), *PBL, social progress and sustainability.* Aalborg: Aalborg Universitetsforlag, *6th International Research Symposium on PBL (IRSPBL 2017),* Bogota, Colombia (pp. 201–208).
13. Centea, D., & Srivavasan, S. (2015, July 6–9). Problem based learning in the conceptual design of hybrid electric vehicles—Development of a global network for PBL and engineering education. In *Proceeding of the International Joint Conference on the Learner in Engineering Education (IJCLEE 2015),* Donostia-San Sebastian, Spain (pp. 149–154) (full paper).
14. Centea, D., & Srinivasan, D. (2016). A comprehensive assessment strategy for a PBL environment. *International Journal of Innovation and Research in Educational Sciences, 3* (6), 364–372.
15. Farrell, J. J., Moog, R. S., & Spencer, J. N. (1999). A guided inquiry general chemistry course. *Journal of Chemical Education, 74*(4), 570–574.
16. Roselli, R. J., & Brophy, S. P. (2006). Effectiveness of challenge-based instruction in biomechanics. *Journal of Engineering Education, 95*(4), 311–324.
17. Russell, S. H., Hancock, M. P., & McCullough, J. (2007). Benefits of undergraduate research experiences. *Science, 316,* 548–549.
18. Seymour, E., Hunter, A.-B., Laursen, S. L., & Diatonic, T. (2004). Establishing the benefits of research experiences for undergraduates in the sciences: First findings from a three-year study. *Science Education, 88,* 493–534.
19. Redish, E. F., Paul, J. M., & Steingberg, R. N. (1997). On the effectiveness of active-engagement microcomputer-based laboratories. *American Journal of Physics, 65,* 45–54.
20. Murray, T. (1999). Authoring intelligent tutoring systems: An analysis of the state of the art. *International Journal of Artificial Intelligence in Education, 10*(98–129), 1999.
21. Deslauriers, L., Schelew, E., & Wieman, C. (2011). Improving learning in a large-enrollment physics class. *Science, 332,* 862–864.
22. Butler, A. C., Marsh, E. J., Slavinsky, J. P., & Baraniuk, R. G. (2014). Integrating cognitive science and technology improves learning in a STEM classroom. *Educational Psychology Review.* https://doi.org/10.1007/s10648-014-9256-4.
23. Centea, D., & Srinivasan, S. (2016). A comprehensive assessment strategy for a PBL environment. *International Journal of Innovation and Research in Educational Sciences (IJIRES), 3*(6). 2349-5219.
24. Sidhu, G., & Srinivasan, S. (2018). An intervention-based active-learning strategy to enhance student performance in mathematics. *International Journal of Pedagogy and Teacher Education* (accepted).
25. Srinivasan, S., & Centea, D. (2015, July 6–9). Applicability of principles of cognitive science in active learning pedagogies. In *Active Teachers—Active Students—Proceedings of the 13th Active Learning in Engineering Education Workshop (ALE), International Joint Conference on the Learner in Engineering Education (IJCLEE 2015),* Donostia-San Sebastian, Spain (pp. 99–104).

26. Sidhu, G., & Srinivasan, S. (2015). An intervention-based active learning strategy employing principles of cognitive psychology. In *CEEA15, Proceedings of Canadian Engineering Education Associate (CCEA15) Conference*, Hamilton, Canada.
27. Srinivasan, S., & Sidhu, G. (2014). Technology and intervention-based instruction for improved student learning. In *ICEER2014 International Conference on Engineering Education and Research*, Hamilton, Canada.

Assessment in Problem-Based Learning Using Mobile Technologies

Dan Centea[(⊠)] and Seshasai Srinivasan

McMaster University, Hamilton, Canada
{centeadn, ssriniv}@mcmaster.ca

Abstract. In this paper, the authors present the assessment methodology on a problem-based learning (PBL) course offered at McMaster University. In the PBL model implemented in this course, student groups work on solving weekly problems. They make various decisions but are aware that, while trying to solve an open-ended problem, there are multiple solutions. At the end of each week, all groups present their solution to the problem. Each presentation is followed by peer evaluations using paper-based rubrics provided by the instructor. The assessment approach proposed in this paper is to develop a mobile application that replaces the paper-based peer evaluation and captures students' interest in providing an accurate assessment of their peers.

Keywords: Classroom response system · Clickers · Problem-based learning · PBL · Assessment in PBL · Mobile-based assessment · Peer evaluation

1 Introduction

University educational practices have evolved over several generations through the introduction of teaching practices that improve student learning and increase retention. This evolution has resulted in several active-learning strategies to be applied in the classroom that include, and are not limited to, experiential learning [1–5], co-operative and small group learning [6–8], problem-based learning [9–13], project-based learning [14], active learning [15–19], inquiry-based learning [20], challenge-based learning [21], peer-led team learning [22], and undergraduate research-based learning [23]. Although the number of courses taught in traditional ways in which professors talk and students listen outnumbers by far the number of courses taught with nontraditional approaches [24], an increasing number of faculty members deliver courses using active-learning strategies as they encourage deep learning and provide superior learning outcomes.

PBL is a pedagogical approach in which students master a subject by applying fundamental concepts to solve problems. This teaching and learning strategy facilitate the development of self-directed learning skills through a structured problem-solving strategy. The course instructor, whose role is to mentor and advise students, designs good ill-structured problems that tend to have multiple solutions. After understanding the problem and identifying the current concepts that are applicable to solve the problem, students work in groups to evolve their knowledge and solve the problem in hand. The ill-structured problems stimulate students to identify the topics of interest to

© Springer Nature Switzerland AG 2019
M. E. Auer and T. Tsiatsos (eds.), *Mobile Technologies and Applications for the Internet of Things*, Advances in Intelligent Systems and Computing 909,
https://doi.org/10.1007/978-3-030-11434-3_37

them and outline the topics that they need to study. This approach allows students to take control of their education by defining their learning needs, assessing their own progress, and assessing the progress of their peers.

The course instructor also needs to design good evaluation techniques to assess the level in which students have met the learning outcomes of the course. In PBL, the assessment of students' work consists of evaluating the active learning process instead of the ability of the student to reproduce content from memory [25]. Having this in mind, the assessment methods need to be carefully designed using various types of assessment strategies.

The course described in this paper is *Conceptual Design of Electric and Hybrid Electric Vehicles,* and is delivered to full-time students enrolled in the last semester of an Automotive and Vehicle Technology program. By employing the principles of PBL in conducting this course, students use previously taught knowledge, investigate various forms of published information, use problem-solving activities, identify relevant design approaches, distinguish between acceptable and non-acceptable solutions, and assess the applicability of innovative ideas. Students' performance to problem development and problem solving are assessed with several strategies. The paper briefly describes the assessment strategies used in this PBL course and focuses on peer evaluations. Considering the extensive use of mobile devices by today's students, this paper describes the use of clickers and mobile technologies to capture students' interest in providing an accurate assessment of their peers.

The assessment methodology is presented in Sect. 2. Section 3 describes in detail the paper-based peer evaluations that are currently used and identifies a problem: the limitation of the paper-based approach. The approaches proposed by this paper to address the problem by using clickers or mobile technologies and the expected outcomes of these approaches are discussed in Sects. 4 and 5, respectively. The conclusions of the proposed clicker-based and smartphone-enabled solutions are presented in Sect. 6.

2 Assessment Methodology

The evaluation of the student work in the PBL course that is the subject of this paper include three major assessment methods, namely, individual assessments, group assessments, and peer evaluations. Each assessment method has its own merits for a certain type of course deliverable. The choice of the proper assessment for a certain deliverable is important to ensure that students' mastery of the learning outcomes is properly measured. Group assessments and individual assessments are the most commonly used evaluation method by course instructors. Group assessments reduce the amount of work done by the course instructor when compared with the individual assessment. However, individual assessments provide a better evaluation of the knowledge of individual students. Weaker students have an opportunity to hide their level of knowledge through group assessments, and therefore they prefer group assessments. On the other hand, students with high academic performances prefer individual assessments; they claim that group assessment downgrades their performance. Studies indicate that, although students generally prefer group assessments instead of individual assessments [26], students adopting a deep approach to learning prefer assessment procedures that allow them to independently demonstrate their understanding [27].

Each of the three assessment methods used in the course is more suitable than others to measure different learning objectives. The assessments are done by instructors (weekly report, presentation skills, assessment skills, and final reports), by peers (content and presentation skills of the weekly presentations) and by a panel of engineering managers for the automotive related industry (final presentations and final report). A full description of the assessments used in the course described in this paper is presented in reference [28].

The PBL course whose assessment is described in this paper includes eight problems related to the conceptual design of electrical and hybrid electrical vehicles. The students are divided into groups of four, each group solving the same problem. One problem is solved each week and includes a literature review, brainstorming activities, identification of the theoretical elements that are known and/or have to be studied in detail, face-to-face and online meetings between the group members, discussions and decisions on the approaches to solve the problem, writing a group report, and delivering an individual presentation.

All students enrolled in the class work on solving the weekly problems. They make various decisions, but they are aware that, while trying to solve an open-ended problem, there are multiple solutions. They have no way of knowing if their solution is the best one. Their goal is to provide a solution that addresses the given problem to the best of their knowledge. Simultaneously, students are encouraged to think of innovative solutions. To satisfy these expectations, students sometimes propose design solutions that are either too futuristic and impossible to be technically obtained within the suggested 3–5 years, or have a depth and breadth of knowledge below expectations. These issues are kept in check through the feedback and assessment provided by the course facilitator and by peers. Finding out about the solutions proposed by other groups and agreeing or disagreeing on the arguments that support them is an important learning process that encourages students to think about their decisions. The questions and answers that follow each presentation allow students to reflect on their solution and decide, if needed, to slightly change their solution to provide a better match between the requirements of the problem and the outcomes of their approach and solution.

Peer evaluation has been introduced in the course in order to increase students' engagement and to allow them to critique their colleagues. All students are asked to assess the technical content of the weekly presentations and the presentations skills of the group manager who delivers the presentation. To ensure that these peer evaluations are performed in a fair and consistent manner, students have been made aware that (a) a student is not allowed to assess his/her group; (b) the peer assessments of the presentations will not be included in the final grade of the presenter; and (c) the peer assessments will be evaluated by the course facilitator and the corresponding marks will be included in the peer assessor's final mark.

The peer evaluations and rankings of each student group are posted in order to enable students to judge their relative performance in comparison to the other groups, encouraging the weaker group to put in additional efforts for the upcoming projects, while encouraging the groups performing well to keep the momentum.

3 Paper-Based Peer Evaluations

Each presentation of a groups' solution to the problem is followed by peer evaluations. They allow students to take responsibility for evaluating other projects. Each student is required to separately assess the technical content of the presentation (assessing the group's work) and the presentation skills (assessing the presenter). At the beginning of a presentation period, all students receive an evaluation sheet specific for the weekly problem. They use such a sheet to assess four elements of the technical content and four elements of the presentation skills by using a numeric grading method of their choice (ex. 1–10).

In the sample peer evaluation sheet shown in Fig. 1, five groups (A to E) deliver their presentations within the 50 min of the class. Students are expected to assess only four groups as they are not allowed to evaluate their own group. After evaluating all the presentations, students calculate the total score of each group as they have recorded, and use this to rank the groups from 1 to 5. Their rankings of the groups for the technical content and the presentation are recorded on a peer evaluation summary sheet, as shown in Fig. 2. The rankings are posted on the Learning Management System after four presentations so that nobody can identify who gave low or high ranking to their group.

Technical Content	A	B	C	D	E	Presentation	A	B	C	D	E
Drivetrain						PowerPoints					
Power sources						Presentation skills					
Vehicle control						Persuasive speech					
Power requirements						Reasons for solutions					
Total						**Total**					
Rank						**Rank**					

Fig. 1. Sample peer evaluation sheet

Your name			Ranking Presentations		
			Group	Content	Presentation
			A		
Your group:			B		
Problem #4			C		
Powertrain			D		
Date			E		

Fig. 2. Sample peer evaluation summary sheet

The grades and ranking of the peer evaluations are not included in the student's evaluation of the course. The purpose of the ranking is to let students know how their peers assess their work. The goal is to encourage the groups with lower rankings to put more effort into preparing their reports and presentations, and to show the groups with high rankings that their level of work is recognized by their peers.

After several deliveries of the course, it has been discovered that there is a problem with using paper-based rubrics. The first issue is that the students do not have the time to add all the grades and to provide a correct ranking of the contents and presenters. Some students, although they provide individual grades, end up ranking the group based on their general feeling of the presentations. This issue can be solved in two ways, either by providing more time for the peer assessment or by automating the ranking process. The second issue is that the peer evaluation is only numeric and students cannot provide feedback to the presenters. This can be addressed by providing space on the marking rubrics to provide written comments, or by replacing the paper-based rubrics with electronic rubrics. Although the first issue of the problem can be addressed through the use of clickers, both issues can be solved by using smartphone-enabled mobile technologies to conduct peer evaluations.

4 Peer Assessments Using Clickers

One way to address the length of time needed by students to write the evaluation marks on the paper-based rubrics and to use these marks to rank the presentations is to use an interactive classroom response system (CRS), also called student response system or classroom performance system, systems commonly referred "clickers". CRS allows each student to assess the technical content of presentations and the presentation skills of the presenter by pushing a button on hardware remote or by selecting an icon on a smartphone- or tablet-enabled polling software. After polling students' responses and recording them, the CRS software can record and automatically rank student inputs.

Hardware clickers are CRS based on remote devices with buttons that use radio-frequency signals to connect with a local receiver or a Wi-Fi connection to connect to the local receiver or to an internet-based software. The remotes allow students to select one out of five answers (A to E). Each remote is identified through a code provided by the manufacturer, and a numeric label that allows students to quickly identify the remotes assigned to them. The approach of providing a unique remote assigned to each student allows the system to record his/her inputs, but can also be used to keep track of participation and attendance. Various academic institutions use different hardware clicker systems such as eInstruction Classroom Performance System, Qwizdom, TurningPoint, Interwrite PRS, iClicker, and H-ITT [29, 30].

The peer assessments process in the course described in this paper will replace the paper rubrics with a set of hardware iClickers that have been purchased by the School of Engineering Practice and Technology at McMaster University. In order to provide numeric assessments, the five buttons of a clicker will represent a point on a 5 to 1 scale (ex. the button A represents 5 points and the button E represents 1 point). After each presentation, all students are expected to provide four inputs for the technical content and four inputs for the presentation skills, each input corresponding to a page shown on screen and being collected individually. The approach is expected to be more efficient than the former paper-based assessment and is expected to take less time.

Although the hardware clickers with several keys do not allow students to write any text, they can be used for a simplified form of feedback by selecting sentences suggested by the course instructor. The feedback suggestions are grouped on several pages, each page being related to one of the topics that the instructor, through his/her experience, knows that often generate class discussions after presentations.

One of the limitations of the hardware clickers systems is the cost. For example, each iClicker2 hardware remote costs about $60 CAD. Students could be asked to buy remotes and use it for their entire studies if there are enough courses that use them. Alternatively, for academic programs that include classes with low numbers of students, an academic department can purchase these remotes and provide them to the students for the duration of a lecture. This option incurs only one initial cost, but no subsequent costs for students or the department.

Although the hardware clickers address some of the issues identified in the course, they are also limited to pressing a hardware key that represents one answer from a multiple choice list, and consequently, they cannot be used to provide proper feedback with suggestions written by students. However, the peer evaluation process using hardware clickers can be improved if their limitations related to a limited number of keys, impossibility to enter a text or numeric values and the price are somehow addressed. A possible solution to address these limitations is the use of specialized programs or apps installed on mobile devices. The wide use of smartphones and tablets by the student population makes this approach very attractive.

5 Peer Assessments Using Mobile Technology

Mobile-enabled CRS that use Wi-Fi or data-enabled devices such as smartphones, tablets or laptops provide a good alternative to hardware clickers. The two possible approaches for using mobile technologies for peer evaluations are commercial mobile-enabled CRS, and self-designed mobile-enabled CRS. The advantage of mobile-enabled CRS is the possibility to address the limitations of the hardware clickers by allowing students to provide various types of feedback. Mobile-enabled CRS enhances the peer evaluation process by allowing students to answer several other question types such as multiple-choice with more possible answers, numerical, sequence/ordering, short answer, fill in the blanks, and matching questions.

Feasibility studies to assess the use of smartphone- or tablet-enabled CRS have been carried out for various commercial CRS software packages that include but are not limited to Socrative [31–35]; Top Hat Monocle [36, 37]; ResponseWare [38, 39]; Poll Everywhere [40, 41]; and iClicker Cloud [42, 43]. These studies show that clickers positively impact the student's learning during the course; mobile-enabled clickers provide students with another mechanism to reflect upon their own learning.

The disadvantages of using commercial mobile-enabled CRS are the cost and the development of the question database. While the hardware clickers, paid by students or by an academic department have only one initial cost, the use of commercial mobile-enabled CRS incurs costs for each student and for each course. The cost of the academic department to work together with the companies that provide mobile-enabled CRS to create questions databases and the willingness of the commercial companies to create question databases also need to be considered. A company would develop course-specific questions with complex drawings and simulations and a corresponding answering interface for their soft CRS only if they would be able to sell their product to multiple academic clients. A business model would involve the development of a complex user interface for a commercial soft CRS that could be used for only one course and one client would not be successful. This is an important limitation for very specialized courses like the one described in this paper.

Having these limitations in mind, a second approach proposed by this paper is to develop a mobile application that replaces the paper-based peer evaluation. The first phase of the mobile application, which is web-based for mobile devices, is currently in development. The interface uses the same rubrics as the paper-based peer assessment form but also allows students who want to provide feedback to enter a text. The final costs incurred for these developments are expected to be less than the equivalent costs for developing a question database with the companies offering commercial CRS systems. Meanwhile, the second development of the self-designed CRS is expected to use iPhone and Android platforms for the peer assessment.

The mobile web-based interface or App is expected to provide a series of advantages that address the issues identified in the paper-based peer evaluation. The expected advantages include obliging students who start the peer evaluation to provide a grade for each rubric; performing automated ranking of the presentation for each student evaluator; providing room for written peer feedback; creating an environment that is familiar with students' everyday life; keeping track of the attendance and participation of the students in the course.

Although most students own and use mobile devices, the use of mobile technology for peer assessment will be complemented with the classical approach of using the traditional paper-based rubrics for students who, for some reason, cannot use the mobile-technology approach for peer evaluation.

The self-designed CRS is expected to be further developed for other courses, including upper-year highly specific courses for which companies selling commercial soft CRS might be reluctant to develop. The development effort for a certain course mainly includes the development of the initial course-specific question database. New questions can be added every year, while the existing questions can be frequently refined with student scholars.

Realistic solutions need to be found for students who change their phone during the course and need to register another device, for students who forget to bring their smart device to class, for students whose smart devices are not charged enough, and for students who do not have smartphones.

Another possible approach for performing peer evaluations with self-designed mobile-based CRS is to use existing commercial CRS and work with its developers to allow the use of a self-designed user interface. The authors of this paper intend to also look at this approach as it might reduce the time when the soft CRS could be deployed in a real classroom environment.

6 Summary

The paper presents an intervention to a course delivered with the PBL approach. The intervention addresses two issues that make one of the course assessments, namely a peer assessment, possibly inaccurate, incomplete, or time consuming. The proposed approach is using mobile technologies that are expected to encourage students to provide fair and fast feedback to their peers in an environment that is comfortable for most of them.

References

1. Kolb, D. A. (2015). *Experiential learning: Experience as the source of learning and development* (2nd ed.). Pearson Education, Inc.
2. Kolb, D. A. (1984). *Experiential learning: Experience as the source of learning and development*. Upper Saddle River, NJ: Prentice Hall.
3. Kolb, D. A., Boyatzis, R. E., & Mainemelis, C. (2011). Experiential learning theory: Previous research and new directions. In *Perspectives on thinking, learning, and cognitive styles*. Routledge, Taylor and Francis Group.
4. Centea, D, Singh, I., & Yuen, T. K. M. (2015, May 31–June 3). A framework of the bachelor of technology concept and its significant experiential learning component. In *Proceeding of the Canadian Engineering Education Association Conference (CEEA 2015)*. Hamilton.
5. Balan, L, Centea, D., Yuen, T. K. M., & Singh, I. (2015, May 31–June 3). Capstone projects with limited budget as an effective method for experiential learning. In *Proceedings of the Canadian Engineering Education Association Conference (CEEA 2015)*. Hamilton. Paper #150.
6. Springer, L., Stanne, M. E., & Donovan, S. S. (1999). Effects of small-group learning on undergraduates in science, mathematics, engineering, and technology: A meta-analysis. *Review of Educational Research, 69*(1), 21–51.
7. Buck, J. R., & Wage, K. E. (2005). Active and cooperative learning in signal processing courses. *IEEE Signal Processing Magazine, 22*(2), 76–81.
8. Prince, M. (2004). Does active learning work? A review of the research. *Journal of Engineering Education, 93*(3), 223–231.
9. Barrows, H. S., & Tamblyn, R. M. (1980). *Problem-based learning: An approach to medical education*. Springer Series on Medical Education.
10. Capon, N., & Kuhn, D. (2004). What's so good about problem-based learning? *Cognition and Instruction, 22*(1), 61–79.
11. Dochy, F., Segers, M., Van den Bossche, P., & Gijbels, D. (2003). Effects of problem-based learning: A meta-analysis. *Learning and Instruction, 13*, 533–568.
12. Gijbels, D., Dochy, F., Van den Bossche, P., & Segers, M. (2005). Effects of problem-based learning: A meta-analysis from the angle of assessment. *Review of Educational Research, 75*(1), 27–61.
13. Centea, D., & Srinivasan, S. (2015, July 6–9). Problem based learning in the conceptual design of hybrid electric vehicles. In *Development of a Global Network for PBL and Engineering Education—Proceeding of the Conference on the Learner in Engineering Education (IJCLEE 2015)* (pp. 149–154). Donostia/San Sebastian, Spain.

14. Lima, R. M., Mesquita, D., Fernandes, S., Marihno-Araujo, C., & Tabelo, M. L. (2015, July 6–9). Modelling the assessment of transversal competencies in project based learning. In *Development of a Global Network for PBL and Engineering Education* (pp. 12–23). Donostia/San Sebastian.

15. Beichner, R. J., Saul, J. M., Abbott, D. S., Morse, J. J., Deardorff, D. L., Allain, R. J., et al. (2007). *The student-centered activities for large enrollment undergraduate programs (SCALE-UP) project.* Retrieved August 15, 2018, from https://www.researchgate.net/publication/228640855.

16. Burrowes, P. A. (2003). A student-centered approach to teaching general biology that really works: Lord's constructivist model put to a test. *The American Biology Teacher, 65*(7), 491–502.

17. Freeman, S., O'Connor, E., Parks, J. W., Cunningham, M., Hurley, D., Haak, D., et al. (2007). Prescribed active learning increases performance in introductory biology. *Cell Biology Education, 6,* 132–139.

18. Srinivasan, S., & Centea, D. (2015, July 6–9). Applicability of principles of cognitive science in active learning pedagogies, active teachers—active students. In *Proceedings of the 13th Active Learning in Engineering Education Workshop (ALE), International Joint Conference on the Learner in Engineering Education (IJCLEE 2015)* (pp. 99–104). Donostia/San Sebastian, Spain.

19. Centea, D., Yuen, T., & Mehrtash, M. (2016, July 6–8). Implementing a vehicle dynamics curriculum with significant active learning components. In *Sustainability in Engineering Education—Proceeding of the 8th International Symposium on Project Approaches in Engineering Education and 14th Active Learning in Engineering Education Workshop (PAEE/ALE'2016)* (pp. 189–198). GuimarÃ£es, Portugal.

20. Lewis, S. E., & Lewis, J. E. (2005). Departing from lectures: An evaluation of a peer-led guided inquiry alternative. *Journal of Chemical Education, 82*(1), 135–139.

21. Roselli, R. J., & Brophy, S. P. (2006). Effectiveness of challenge-based instruction in biomechanics. *Journal of Engineering Education, 95*(4), 311–324.

22. McCreary, C. L., Golde, M. F., & Koeske, R. (2006). Peer instruction in the general chemistry laboratory: Assessment of student learning. *Journal of Chemical Education, 83* (5), 804–810.

23. Hunter, A.-B., Laursen, S. L., & Seymour, E. (2007). Becoming a scientist: The role of undergraduate research in students' cognitive, personal, and professional development. *Science Education, 91,* 36–74.

24. Bonwell, C. C., & James, A. E. (1991). *Active learning: Creating excitement in the classroom.* ASHE-ERIC Higher Education Report No. 1. Washington, DC: The George Washington University. Retrieved August 16, 2018, from https://files.eric.ed.gov/fulltext/ED336049.pdf.

25. Reynolds, F. (1997). Studying psychology at degree level: Would problem-based learning enhance students' experiences. *Studies in Higher Education, 22*(3), 263–275.

26. Kolmos, A., & Hoolgard, J. E. (2007, June 22–24). Alignment of PBL and assessment. In *1st International Conference on Research in Engineering Education* (pp. 1–9). Honolulu.

27. Gijbels, D., & Dochy, F. (2006). Students' assessment preferences and approaches to learning: Can formative assessment make a difference? *Educational Studies, 32*(4), 399–409.

28. Centea, D., & Srinivasan, S. (2016). A Comprehensive assessment strategy for a PBL environment. *International Journal of Innovation and Research in Educational Sciences (IJIRES), 3*(6), 2349–5219.

29. Bojinova, E., & Oigara, J. (2011). Teaching and learning with clickers: Are clickers good for students? *Interdisciplinary Journal of E-Learning and Learning Objects, 7,* 169–184.

30. Bojinova, E., & OIgara, J. (2013). Teaching and learning with clickers in higher education. *International Journal of Teaching and Learning in Higher Education, 25*(2), 154–165.
31. Coca, D., & SliÅ¡ ko, J. (2013). Software Socrative and smartphones as tools for implementation of basic processes of active physics learning in classroom: An initial feasibility study with prospective teachers. *European Journal of Physics Education, 4*(2), 17–24.
32. Dervan, P. (2014). Increasing in-class student engagement using Socrative (an online Student Response System). *The All Ireland Journal of Teaching and Learning in Higher Education (AISHE-J), 6*(3), 1801–1813.
33. Awedh, M., Mueen, A., Zafar, B., & Manzoor, U. (2015). Using Socrative and smartphones for the support of collaborative learning. *International Journal of Integrating Technology in Education (IJITE), 3*(4), 17–24.
34. Wash, P. (2014). Taking advantage of mobile devices: Using Socrative in the classroom. *Journal of Teaching and Learning with Technology, 3*(1), 99–101.
35. Dakka, S. M. (2015). Using Socrative to enhance in-class student engagement and collaboration. *International Journal on Integrating Technology in Education (IJITE), 4*(3), 13–19.
36. Alemohammad, H., & Shahini, M. (2013, November 15–21). Use of mobile devices as an interactive method in a mechatronics engineering course: A case study. In *Proceedings of the ASME International Mechanical Engineering Congress and Exposition IMECE2013, Education and Globalization* (Vol. 5). Can Diego, California, USA.
37. Lucke, T., Keyssner, U., & Dunn, P. (2013). The use of a classroom response system to more effectively flip the classroom. In *2013 IEEE Frontiers in Education Conference* (pp. 491–495).
38. De Vos, M. (2018). *Using electronic voting systems with ResponseWare to improve student learning and enhance the student learning experience—Final report.* Retrieved August 14, 2018, from https://www.researchgate.net/publication/267250288.
39. Gong, Z., & Wallace, J. D. (2012). A comparative analysis of iPad and other m-learning technologies: Exploring students' view of adoption, potentials, and challenges. *Journal of Literacy and Technology, 13*(1), 2–29.
40. Popescu, O., Chezan, L. C., Jovanovic, V. M., & Ayala, O. M. (2015, June 14–17). The use of polleverywhere in engineering technology classes to student stimulate critical thinking and motivation. In *122nd ASEE Annual Conference and Exposition, Making Value for Society.* Seattle, WA, USA.
41. Tregonning, A. M., Doherty, D. A., Hornbuckle, J., & Dickinson, J. (2012). The audience response system and knowledge gain: A prospective study. *Medical Teacher, 34*(4), 269–274.
42. Wu, X., & Gao, Y. (2011). Applying the extended technology acceptance model to the use of clickers in student learning: Some evidence from macroeconomics classes. *American Journal of Business Education, 4*(7), 43–50.
43. Whitehead, C., & Ray, L. (2018). *Using the iClicker classroom response system to enhance student involvement and learning.* Retrieved August 15, 2018, from https://www.researchgate.net/publication/265192585.

Full Lecture Recording Watching Behavior, or Why Students Watch 90-Min Lectures in 5 Min

Matthias Bauer[✉], Martin Malchow, and Christoph Meinel

Hasso Plattner Institute (HPI), University of Potsdam, Potsdam, Germany
{matthias.bauer,martin.malchow,christoph.meinel}@hpi.de

Abstract. Many universities record the lectures being held in their facilities to preserve knowledge and to make it available to their students and, at least for some universities and classes, to the broad public. The way with the least effort is to record the whole lecture, which in our case usually is 90 min long. This saves the labor and time of cutting and rearranging lectures scenes to provide short learning videos as known from Massive Open Online Courses (MOOCs), etc. Many lecturers fear that recording their lectures and providing them via an online platform might lead to less participation in the actual lecture. Also, many teachers fear that the lecture recordings are not used with the same focus and dedication as lectures in a lecture hall. In this work, we show that in our experience, full lectures have an average watching duration of just a few minutes and explain the reasons for that and why, in most cases, teachers do not have to worry about that.

Keywords: e-Learning · Lecture video recording · e-lecture · Attention span · Learning behavior

1 Introduction

In University teaching, it is common to record lectures with a video recording solution. These recorded lectures can be used by students who were absent from the lecture, or by students who actually attended the lecture, but would like to re-watch some passages for obtaining a better understanding of the taught topic. For those students, it is very helpful to have the lecture recordings. However, different from video units in Massive Open Online Courses, lecture recordings are often not cut in smaller units and hence usually have a full lecture length of 90 min. In many cases, this would cause too much work to be done for every single lecture. Especially, when taking into account that the lecture recordings should be available as quickly as possible. It would not be acceptable if the students would have to wait a couple of days to (re-)watch a lecture online. In this work,

© Springer Nature Switzerland AG 2019
M. E. Auer and T. Tsiatsos (eds.), *Mobile Technologies and Applications for the Internet of Things*, Advances in Intelligent Systems and Computing 909, https://doi.org/10.1007/978-3-030-11434-3_38

we concentrate on full length recordings and do not look at cut and rearranged lecture recording excerpts.

The availability of an increasing number of lecture recordings raises the question of how students use those recordings. Do they watch full lectures, or do they watch only certain chapters? Do they navigate to those chapters? By skipping through the video, by seeking to a certain time code, by clicking a link to the specific scene or by using a search function based on optical character recognition (OCR) and automated speech recognition (ASR)? We have run a lecture video platform with recordings of ca. 7000 lectures which were held at our university and recorded with the recording system we have developed and produced. Those videos usually consist of two video files being played in sync—one showing the lecturer recorded by a camera and one showing the presentation (e.g., slides with text and diagrams or other video content) recorded by grabbing the video output of the presenter's computer.

Recording lectures and making them available to students is one approach of fostering video learning. However, providing those videos as they are is only one of the things to do when intending to support learners with video content and technological tools.

We are interested in discovering out how the videos and the platform they are hosted on are used. We would like to learn more about how our students learn with lecture recordings. The goal is to obtain insights about the video watching behavior based on measuring certain metrics such as clicks, time spent on the platform, videos watched, local time, etc.

Moreover, there is an attempt to find certain categories for different learning styles or behaviors, e.g., "watchers" versus "clickers". There are students who watch a smaller number of videos, but for longer time periods and there are others who watch a larger number of videos, but only short chunks and they navigate more frequently between different videos.

The users on our lecture video platform can use several ways of finding an answer to their question inside a video. The simplest, but slowest way is to watch a whole lecture or even more than one. This can be sped up by skipping through the video. We provide lecture slide preview thumbnails (generated by our lecture slide transition recognition) which can be seen by hovering over the video player timeline or (when using a touch device) by pressing the finger on the respective preview block on the timeline. In this way, the learners can oversee the (slide) presentation video visually in a quick manner. Another way of finding a topic, when knowing a specific search term to look for, is to use the search function that does not only search through static metadata such as lecture topic, description, tags, presenter's name, etc. It also examines the terms gathered via optical character recognition (OCR) and automated speech recognition (ASR). The search results are linked to the exact time codes where the terms were found. So the learners can jump to the exact position in the result video with one click on the search result. Thus, the question arises which of these functions are used most and whether that influences the watching behavior and the time spend with lecture recordings.

2   Overview of Lecture Recording and Related Work

In the broad research area of lecture recording, a lot of research is done with a variety of different aspects being investigated by many esteemed researchers which we will discuss in this section. The focus of the research in this area is very diverse. Some researchers concentrate on the question if lecture recording makes sense and how it can be measured and improved. Others investigate the differences between attending live lectures in the lecture hall or watching the lecture recording online. There are differences in the student's perception and in their attainment. But different studies show different and often opposing results. This is—even though the studies are mostly well designed and try to eliminate uncontrollable variables—often the case because there are still pre-conditions and variables that are either ineliminable or not being thought of as many researchers (including the authors) often tend to approach a problem within the limitations of their own universe defined by their personal experience, watching their own system, contents, and learners. But, there are big differences between learning platforms, delivery formats, subjects, learners, etc.

Also, it is important to consider the fact that technology has evolved resulting in major improvements in quality, accessibility, and availability. Hence, the results of studies older than a few years should be looked at critically being aware that the findings might not be valid to the same extent today as they were at the time of gathering the data.

Obviously, for both recorded lectures and live lectures, advantages and disadvantages can be found. Inside a lecture hall when attending the live lecture, it is easier to concentrate and set a reserved time slot for the lecture [24], there is (depending on the lecturer) also the opportunity to ask questions during or after the lecture and the possibility to talk to classmates. In addition, there is social interaction with the other classmates before, during and after lectures, which—in this intensity—is not possible when watching lecture recordings at home. But on the other side, they also extol the advantages of recorded lectures such as independence from time and place and the possibility to make breaks while watching being able to continue right where the learners left off. Both live and recorded lectures can be combined whereby the advantages can be maximized [24]. On the one hand, they describe the positive effect and perception that students can gain a better understanding if they attend the actual lecture and later re-watch it at home. On the other hand, other researchers describe in their work that this combination of attending the live lecture and watching the lecture recording afterwards does not have an influence on the learning gains [22]. They at least recommend lecture recordings as a feedback for teachers and for creating recordings for flipped classroom purposes.

2.1   Attendance

Some studies find that student's attendance at the actual lectures drop if lecture recordings are available [8]. However, others found that this is almost not the case [14]. Many lecturers fear that the availability of lecture recordings makes the attendance in the lecture hall drop. Some students use lecture recordings as a replacement for attending the actual lecture [6]. In contrast, also others find that the attendance does not significantly drop when recordings are available [15]. Or that it is (only) the case for lectures where the presenter does not answer questions [24], which would add value to lecture attendance.

2.2   Video Length and Attention Span

Unarguably, there are valid reasons for producing shorter learning videos for university students or MOOC learners—the human attention span. Pi and Hong [19] conducted a study observing students when watching lecture videos with the help of eye-tracking and they looked at the blink duration as a measure for student's fatigue. They found out that when watching a lecture, students fatigue after ca. 10 min, which they can still overcome. The next fatigue comes after approximately 22 min and this cannot be overcome easily. They conclude that learning videos should not be longer than this duration, which makes sense, but is not applicable to our case.

This paper concentrates on working with lectures recorded in a lecture hall. This approach does not cause the speaker more work or interfere too much with his or her teaching style and habit—except for the fact that they should have a wireless microphone attached to their clothes [11] or wear a headset or at least have a microphone nearby.

But this is a field of work and research that is still in movement. As of 2011, less than 10% of the universities applied lecture capture [10]. Today, the percentage is significantly higher meaning that there are numerous universities performing lecture recording which do not have many years of experience with this topic and are still in a process of finding the ideal way for themselves.

One other thing being in movement is the legal parameters. As the Internet evolves and grows, also data protection and privacy are being improved gradually. As of 2018, the new General Data Protection Regulation (GDPR) has been introduced for protecting everyone's personal data, which also makes it more difficult and time-consuming for researchers to gather online data linked to a learner and tracking him or her for longer than a single action or session.

2.3   Attainment

There is research investigating the differences in attainment of students attending the actual lecture, students who watch the lecture recording and students who (at least for some parts) do both. Some found that attainment was worse in their cohort if lectures were available as recordings [8]. But they also assume that attendance and attainment are linked to each other, and that less attendance leads to worse attainment, which means if lecture recording leads to decreasing attendance, it also leads to worse attainment. Others find that this is not the case [6,17,22] or that attainment is even better with lecture recordings if students can select themselves if they have the choice between attending the lecture or watching the recording [7]. This shows that granting students self-determination regarding their learning preferences can lead to better performance [9].

2.4 View Duration

Regarding the duration learners spend watching a video recording, different studies have found different results. The calculated view duration we have found ranges between 95 s [2] when also considering the users who do not watch a video for more than a few seconds (for example, because they just download the podcast video for offline use and leave the page) and 7 min [3,25]. These numbers are understandable considering the human attention span and the platform's search and navigation features.

When comparing research results and considering how studies and conclusions differ, it should also be taken into account that there are various understandings of video lectures, podcasts, and lecture recordings. The authors have the understanding that a lecture recording or lecture capture, respectively, is a recording of a full lecture length, which in our university is about 90 min long and does not have a break in between. In contrast, some works refer to shorter videos [16,19] taken from lectures or even recorded and produced with the purpose of showing it to the students. This is much more time-consuming and requires more work than simply recording lectures that are being held anyway.

3 Approach and Experiment

In this section, we will describe how lectures are recorded, post-processed, and distributed at the Digital Engineering Faculty at the University of Potsdam, Germany including the necessary tools that had to be developed and built for these purposes.

3.1 Lecture Recording Tools

For recording lectures, a particular approach was chosen. We have developed a recording system built in a portable case. This box contains a computer with a touchscreen, video grabbers, a wireless microphone receiver, etc., neatly built in with customized frames in the bottom and top part of the case. The idea was to have a portable recording solution that can indeed stay in the lecture halls, but also can be taken to smaller seminar rooms and also to off-site events such as conferences and workshops where traveling is necessary.

Our recording software runs on Windows. We have developed the software by ourselves. The project dates back to the year 2002 (2003 ACM Einstein Award). Since then, several new generations of the recording hardware and software have been implemented—always making use of latest technology and research findings. At the moment, we are considering making the recording software platform-independent, so it can be ported to other operating systems easily in the future. It is possible and advisable to record two videos at the same time as well as an

Fig. 1. Portable recording system and lecture video web portal

audio stream that can be mixed from different audio sources (wireless microphone, HDMI output, etc.). In our faculty, we record one video of the presentation and one video of the presenter. The two videos are recorded at the same time and saved as separate files on the recording systems' solid-state drive (SSD). These two videos can later be played back synchronously [4].

According to [18], 84% of the students find the availability of presentation slides most important or very important. As of August 2018, in our own collection of lecture recordings, approximately 95% are dual-stream videos (presenter and presentation). The rest are video-only recordings—if no slides were presented such as in Mathematics—and a vanishingly small amount (ten) of slide videos that include the speaker's voice (without depicting the presenter). Momentarily, there are 8733 presentation videos (e.g., slides) and 9211 presenter's videos.

With our recording solution, it is possible to record everything presented on the speaker's computer, it not bound to a PowerPoint plug-in as opposed to other solutions [17]. It is not necessary to present slides at all. Teachers can also show videos, or present certain software systems or demos, or do live coding exercises in computer science lectures, etc. This gives the teachers a bigger variety of things they can show during a recorded or live-streamed lecture.

3.2 Lecture Distribution

The recorded lectures are uploaded to an in-house server via SFTP and then published on our lecture video web portal. The web portal is basically a big library of all recordings of our faculty. It has been developed by ourselves and has recently been updated aiming for higher compatibility with different viewer's devices and an improved user interface as well as (perceived and actual) speed improvement as described in earlier work [2]. A screenshot of the web portal can be seen in Fig. 1. Lectures are organized in collections (here called a lecture "series"). Each series is a course taught in a specific semester meaning that it will be recorded every year if the lecture is repeated annually. Recording lectures every time they are given has advantages such as updated slides, or the possibility to tell the students to watch last year's recording in case the teacher is unavailable.

3.3   Lecture Watching Analysis

The lecture videos considered and being examined here are hosted on a server at the faculty. They can be played back using the web portal. The majority of videos are available publicly to everyone. A small amount of approximately 10% is available to enrolled students and staff only. In the web portal, the learners can navigate through semesters, lecture collections, lectures, chapters, or use the multisource search function. In order to analyze the watching numbers and details, we basically apply techniques known from web analytics adding the awareness that we have videos with a certain length and varying degree of interestingness (even inside a single lecture). There are several ways videos are streamed to the watchers. The video player has access to four versions of each video (if the transcoding process has already run, which might not be the case for videos recorded earlier on the same day): an HD version, an SD version, an HLS version, and a podcast version for download. Thus, there are several files that have to be monitored: the MP4 video files of the HD and SD versions and the TS files (.ts) referenced in the lecture's M3U8 file for the HLS version. More information about HTTP live streaming (HLS) using M3U8 playlist files can be retrieved from RFC 82161. The two MP4 files have the length of the full lecture whereas the TS files are short chunks of the full lecture with a length of (in our case) 6 s. That means that a full lecture of 90 min consists of 900 TS files that have to be monitored. This requires a different approach than back in the time all the lecture videos had been streamed out by an Adobe Media Server which provided extensive log files with events such as play, pause, seek (including the seek position) that could be facilitated to monitor students' watching behavior and even to detect difficult spots inside a lecture [5].

In the works of [13, 16], and others, learners have been classified into four main categories ("Not watching", "Looking over", "Watching by zapping", "Watching completely"). We think it is possible to find more categories that can characterize the learners and their behavior more precisely. Also, the facts and resources used to group the learners and assign them to a category should be verified and revised, if appropriate.

An example is the categorization "Watchers" versus "Clickers" as shown in Fig. 2, which can be introduced by examining the time per action on the video platform and correlating the number of actions performed with the time spent on the platform. At the moment, this has been done for groups of users distinguished by their operating system (OS) as announced by their browsers. This should also be investigated for single users.

Even this rough segmentation of users shows that most users spent an average of approximately 60 s per action, whereas there are outliers: GNU/Linux (Linux on a Desktop Computer) users seem to watch longer while clicking less and users of mobile devices such as iPhones and iPads (iOS) and Android Smartphones (Android 6.0) spend the least time on the video platform performing more actions in less time. Here, it seems more probable that users do brief visits and leave the platform quicker than users with desktop operating systems. This is partly caused by users visiting the respective page to download the podcast video (a one-file video-in-video variant of the lecture we provide by transcoding it) manually, or automatically if they use an iOS device.

Seconds per Action

OS	actions	time in s	s per action
GNU/Linux	3	240	80
Windows 10	4	253	63
Windows 7	4,7	289	61
Windows 8.1	3,8	229	60
Android 7.0	4,7	280	60
Ubuntu	4,6	274	60
Mac 10.13	3,9	191	49
Android 6.0	3,9	183	47
iOS 11.3	2,4	94	39

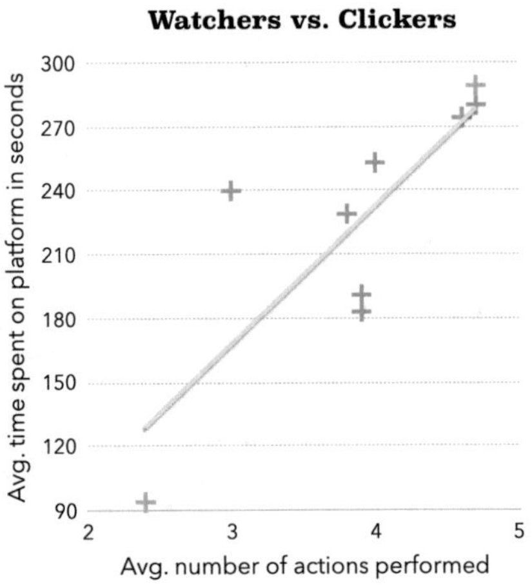

Watchers vs. Clickers

Fig. 2. Average seconds per action

Another thing we have examined was the time and day the learners used the lecture recordings on our platform. In Fig. 3, it can be seen that basically, the normal work or university week is being visible. Friday is the day with the least lectures and seminars in our faculty. Students often use it to work in their part-time job such as programmer in a company, in their own start-up or as a student assistant. Some whose parents live further away also use it for traveling home for the weekend. Saturday, as the first day of the weekend after the workweek, is the day with the least video watchers. On Sunday, the number increases again as it is time to prepare for the next study week or use the videos to solve the homework tasks.

The distribution of video views over the day as depicted in Fig. 4 shows that our students use the videos during the normal university hours—here, the course offer begins at 09:00 and ends at 16:45 or sometimes at 18:30. It is obvious that the usage of lecture recordings is highest during this time. But, it can also be seen that after the university courses, a large number of students continues learning with the offered lecture recordings. From 18:00 until 22:00, there is an about steady number of views. When night approaches, the number decreases but remains higher in relation to the morning hours before lecture start showing that many students are evening learners.

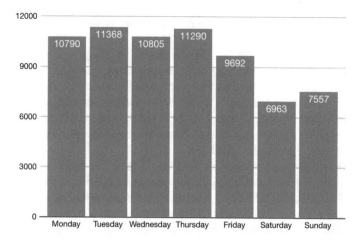

Fig. 3. Number of views by day of week August 16, 2017–August 15, 2018

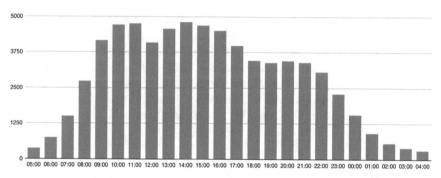

Fig. 4. Number of visits by hour of day August 16, 2017–August 15, 2018

3.4   Advantages of Lecture Recordings

Many of the disadvantages students find with attending actual lectures in the lecture hall are advantages of lecture recordings at the same time. Petrović and Pale [18] found that lecture recordings have more potential supporting the students' personal learning styles and habits such as self-paced learning, which is not possible when attending a live lecture. Especially not, if the lecture (like in our university) does not have a break after 45 min as described by Petrović and Pale [18].

Pausing the video playback to rethink gives the students a better chance to understand the topic. They also studied the reasons for not attending a lecture. Some of the students are learning for other courses, some skip a lecture out of a lack of sleep, and others mentioned that the course can be mastered without attending the lectures (62% of the study group) [18].

In contrast to the attendance of an actual lecture in a lecture hall, lecture recordings can be enriched with manifold technological tools to support learners and teachers: It is possible to analyze the presented slides automatically and extract thumbnail pictures for every single slide, run optical character recognition (OCR) and automated speech recognition (ASR) on the presentation video and presenter's video, respectively, as demonstrated by Yang et al. [23]. This helps the students navigate through long lecture recordings and even find the distinct spot when a search term was first shown on a slide or said by the lecturer. Also, the thumbnail creation can speed up students' online research significantly.

There are advantages for teachers as they can benefit from an automated quality analysis by detecting accumulations of certain events such as jump-back [5], and changes in playback speed [20].

Also, additional features can be integrated in the video player such as self-tests [12] as known from MOOCs. It is possible to offer links, frequently asked questions (FAQs) and additional reading material to the videos [17]. Even results of a semantic analysis of the presented slides [1] can automatically be offered within the video (at the distinct time where detected) such as links for detected URLs, RFCs, ISBNs, etc.

Lecture recordings offered online also allow collaborative learning if applied as described for lecture video platforms [3] and also for MOOCs [21].

We think that the possibility to subscribe to a lecture collection via RSS feed is also important to make it easier for the students to follow the course over the semester. That is why we have implemented RSS feeds and a subscription form inside the web portal for every course and offer them on our own web portal as well as on iTunes/Apple Podcasts. Otherwise, it would cause the students more time and effort accessing the latest lecture recording after it has been published online [22].

4   Conclusion

In this work, we have given an overview of lecture recording in universities, their advantages, and challenges and compared it to attending the actual lectures in person. We showed that video learners can be categorized in certain ways and that there is room for finding more behavioral categories. Also, we pointed out that most of the users do not access a video lecture platform to watch the whole lecture and that platform providers do not need to worry about users spending only a few minutes on their platform. The time spent with a lecture recording should not be taken as a measure of the importance of providing learning material and knowledge in form of videos online. Offered tools have become much better and students' learning behavior has changed dramatically with the digital revolution of the last two centuries.

We showed and discussed that many of the problems students have with attending live lectures can be solved with lecture recordings such as missing knowledge prior to the begin of the actual lecture, the need to repeat a part of the lecture, pause, and have some extra time to rethink. Many of those can be solved with self-paced learning with recordings and with the enrichment through our Learn Together approach where learners can watch a video together online and synchronize their interactions over the Internet, semantic search, automatically generated links for further research, integrated self-tests in the lecture video player, and more.

References

1. Bauer, M., Malchow, M., & Meinel, C. (2015). Enhance teleteaching videos with semantic technologies. In: V. L. Uskov, R. J. Howlett, L. C. Jain (Eds.), *Smart education and smart e-Learning* (Vol. 41, pp. 105–115), Chap. Smart Educ. Springer. https://doi.org/10.1007/978-3-319-19875-0_10.
2. Bauer, M., Malchow, M., & Meinel, C. (2018). Improving access to online lecture videos. In *IEEE global engineering education conference*, EDUCON (Vol. 2018-April, pp. 1161–1168).
3. Bauer, M., Malchow, M., Staubitz, T., & Meinel, C. (2016). Improving collaborative learning with video lectures. In *INTED2016 Proceedings. 10th International Technology, Education and Development Conference*, IATED (pp. 5511–5517). https://doi.org/10.21125/inted.2016.0322.
4. Bauer, M., & Meinel, C. (2013). Using the tele-TASK lecture recording system to improve e-Learning. In *The 7th International Conference on Software, Knowledge, Information Management and Applications (SKIMA 2013)* (pp. 1–10).
5. Bauer, M., & Meinel, C. (2014). A concept to analyze user navigation behavior inside a recorded lecture (to identify difficult spots). In *13th International Conference on Information Technology Based Higher Education and Training (ITHET)*.
6. Bos, N., Groeneveld, C., van Bruggen, J., & Brand-Gruwel, S. (2016). The use of recorded lectures in education and the impact on lecture attendance and exam performance. *British Journal of Educational Technology, 47*(5), 906–917.
7. Bosshardt, W., & Chiang, E. P. (2016). Lecture capture learning: Do students perform better compared to face-to-face classes? *Southern Economic Journal, 82*(3), 1021–1038.
8. Edwards, M. R., & Clinton, M. E. (2018). *A study exploring the impact of lecture capture availability and lecture capture usage on student attendance and attainment* (pp. 1–19). Higher Education.
9. Gilardi, M., Newbury, P., Gilardi, M., Holroyd, P., Newbury, P., & Watten, P. (2015). The effects of video lecture delivery formats on student engagement (September).
10. Green, K. C. (2011). Campus computing 2011: The national survey of e-Learning and information technology in American higher education. https://www.campuscomputing.net/s/Green-CampusComputing2011_5-gawh.pdf/. Accessed 18 Aug 2018.
11. Grünewald, F., Yang, H., Mazandarani, E., Bauer, M., & Meinel, C. (2013). Next generation tele-teaching: Latest recording technology, user engagement and automatic metadata retrieval. Human Factors in ... 7946 LNCS (pp. 391–408). https://doi.org/10.1007/978-3-642-39062-3_25.
12. Malchow, M., Bauer, M., & Meinel, C. (2015). Self-test integration in lecture video archives. In *ICERI2015 Proceedings. 8th International Conference of Education, Research and Innovation, IATED* (pp. 7631–7638).
13. Mongy, S. (2007). A study on video viewing behavior: Application to movie trailer miner. *International Journal of Parallel, Emergent and Distributed Systems, 22*(3), 163–172.

14. Newton, G., Tucker, T., Dawson, J., & Currie, E. (2014). Use of lecture capture in higher education–lessons from the Trenches. *Trenches: Linking Research and Practice to Improve Learning, 58*(2), 32–45. https://doi.org/10.1007/s11528-014-0735-8.

15. Owston, R., Lupshenyuk, D., & Wideman, H. (2011). Lecture capture in large undergraduate classes: Student perceptions and academic performance. *Internet and Higher Education, 14*(4), 262–268.

16. Ozan, O., & Ozarslan, Y. (2016). Video lecture watching behaviors of learners in online courses. *Educational Media International, 53*(1), 27–41. https://doi.org/10.1080/09523987.2016.1189255.

17. Pale, P., Petrović, J., & Jeren, B. (2014). Assessing the learning potential and students' perception of rich lecture captures. *Journal of Computer Assisted Learning, 30*(2), 187–195.

18. Petrović, J., & Pale, P. (2015). Students' perception of live lectures' inherent disadvantages. *Teaching in Higher Education, 20*(2), 143–157. https://doi.org/10.1080/13562517.2014.962505.

19. Pi, Z., & Hong, J. (2016). Learning process and learning outcomes of video podcasts including the instructor and PPT slides: A Chinese case. *Innovations in Education and Teaching International, 53*(2), 135–144. https://doi.org/10.1080/14703297.2015.1060133.

20. Renz, J., Bauer, M., Malchow, M., Staubitz, T., & Meinel, C. (2015). Optimizing the Video Experience in Moocs. In *EDULEARN15 Proceedings. 7th International Conference on Education and New Learning Technologies*, IATED (pp. 5150–5158, No. July).

21. Staubitz, T., Pfeiffer, T., Renz, J., & Willems, C. (2015). Collaborative learning in a MOOC environment. In *Proceedings of the 8th International Conference of Education, Research and Innovation* (pp. 8237–8246).

22. Williams, A. E., Aguilar-Roca, N. M., & O'Dowd, D. K. (2016). Lecture capture podcasts: Differential student use and performance in a large introductory course. *Educational Technology Research and Development, 64*(1), 1–12.

23. Yang, H., Grünewald, F., Bauer, M., & Meinel, C. (2013). Lecture video browsing using multimodal information resources. In *Lecture Notes in Computer Science (including subseries Lecture Notes in Artificial Intelligence and Lecture Notes in Bioinformatics)*, LNCS (vol. 8167, pp. 204–213). https://doi.org/10.1007/978-3-642-41175-5_21.

24. Yoon, C., Oates, G., & Sneddon, J. (2014). Undergraduate mathematics students' reasons for attending live lectures when recordings are available. *International Journal of Mathematical Education in Science and Technology, 45*(2), 227–240. https://doi.org/10.1080/0020739X.2013.822578.

25. Zupancic, B., & Horz, H. (2002). Lecture recording and its use in a traditional university course. In *Proceedings of the 7th Annual Conference on Innovation and Technology in Computer Science Education—ITiCSE'02* (p. 24). http://portal.acm.org/citation.cfm?doid=544414.544424.

Active Learning Strategy Using Mobile Technologies

Dan Centea$^{(\boxtimes)}$ and Moein Mehrtash

McMaster University, Hamilton, Canada
{centeadn, mehrtam}@mcmaster.ca

Abstract. Active learning strategies represent an important approach to increase student learning. One of these strategies is the use of classroom response systems, also known as clickers, for testing the knowledge that students learned in previous lectures, for checking students' understanding during the development of solutions to technical problems, for checking their understanding of the new concepts presented in class, and for personal or anonymous student surveys. This paper describes the use of hardware clickers in two technical courses, presents their advantages, identifies their limitations, and proposes a solution to reduce or eliminate restrictions through the use of soft clickers.

Keywords: Classroom response system · Clickers · Mobile-based clicker · Active learning · Mobile technology · Student learning · Participation

1 Introduction

Preparing university graduates for the workplace by providing them the knowledge and skills needed by for today's employment expectations is a vital role in educational systems. The old employment model of asking the fresh graduates to do simple tasks or modifying existing solutions has been often replaced with problem-solving activities that engage them in brainstorming discussions, peer decisions, project developments, solution implementations, and presentations. In order to provide these skills to the graduates, the traditional university practices based on lecture-intensive approaches have evolved over several generations by introducing alternative teaching practices that enable deep learning and retention, and engage students in various experiential learning activities. This evolution has resulted in several teaching techniques that can be applied inside the classroom [1–8]. Researchers have described various classroom practices that include but are not limited to cooperative and small group learning [1], problem-based learning [2], project-based learning [3], active learning [4], inquiry-based learning [5], challenge-based learning [6], peer-led team learning [7], and undergraduate research-based learning [8].

The development of such a variety of classroom practices is attributed to the fact that many instructors consider that the traditional lecturing approach is a suboptimal method of delivering the content. In fact, in a traditional lecturing approach, only 20% of the material is mastered successfully by the students [9]. Given that each student is

© Springer Nature Switzerland AG 2019
M. E. Auer and T. Tsiatsos (eds.), *Mobile Technologies and Applications for the Internet of Things*, Advances in Intelligent Systems and Computing 909,
https://doi.org/10.1007/978-3-030-11434-3_39

different in his/her learning ability, there is no unique way of conducting the classroom sessions that is equally applicable for all courses and all students.

There is significant published literature related to active learning strategies. Various authors show the improvements in student learning produced by different active learning approaches. This paper focuses on an active learning strategy based on interactive classroom response systems (CRS), also called student response system or classroom performance system, commonly referred to as clickers. Section 2 presents the way in which CRS are used in the classroom. Section 3 presents the use of a CRS with hardware clickers in three technical courses offered in the automotive engineering technology program at McMaster University and discusses their advantages and limitations. A possible solution to addresses these limitations using commercial smartphone-enabled CRS, also referred to soft clickers, are presented in Sect. 4. A proposed self-designed software CRS that also addresses these limitations is presented in Sect. 5. Conclusions are presented in Sect. 6.

2 Using Clickers in the Classroom

Although active learning strategies are an important mean to increase student learning, some university students do not engage in providing responses to questions asked in class for different concerns that include passive personalities, cultural norms, lack of knowledge, or fear of humiliation. Clicker technology encourages participation in class while addressing some of these concerns. In the absence of a classroom response system, the vocal students that usually dominate the discussions are the ones answering most questions, while some students might have a fear of being ridiculed by their colleagues if their answer is wrong. The existence of a personal answering system forces every student to answer.

Individual student responses acquired by the CRS are not identified on the screen. The anonymous response approach guarantees a high level of participation. If assessment points are allocated for answers, the system provides total participation. The participation varies from a personal answer to an answer based on a peer discussion that involves deep thinking.

CRS actively engage students during the entire class period. They can be used to assess student preparation and provide prompt feedback [10]. Immediate feedback is extremely important to students. Most of the existing CRS provide bar graphs or pie charts of the aggregated responses. Students can see how their responses are compared with the responses of their colleagues and with the correct answer. Time for discussions is usually allocated by the instructor. These discussions are expected to be longer if the percentage of the students who provided the correct answer is low. This approach is a very important benefit of the CRS as it allows the instructor to concentrate on the topics that are either not clear or not perfectly understood by students. Testing via clicker questions helps instructors determine which material would be most useful to be covered more elaborately during lecture time in order to maximize learning benefits for the majority of students [11].

The efficiency of using clickers in the classroom has been assessed by several authors. The use of clickers has been shown to increase student engagement and achievement compared to traditional lecture format instruction [12], improve test scores, provide a positive and significant impact on student final grades, and reduce attrition [13, 14]. They improve the classroom experience and classroom environment [15, 16]. When combined with collaborative peer-aided learning, clickers have shown promising results for high-order thinking [17]. Students perceive high levels of the constructs when using the clickers and especially high levels of learning performance; they perceive that using clickers in the class facilitates the understanding of the concepts and class materials and significantly improves their learning process [16]. Research shows that students are generally satisfied with the use of clickers [18, 19].

3 Hardware Clickers and Their Limitations

CRS allow each student to respond to questions asked by the course instructor by pushing a button on a hardware remote or by selecting an icon on a smartphone- or tablet-enabled polling software. Students' responses are processed by the clicker software and can be displayed in the form of percentages or pie charts.

Hardware clickers are remote devices with buttons that use radio-frequency signals to connect with a receiver or a Wi-Fi connection to connect with an Internet-based software. Various academic institutions use different hardware clicker systems such as eInstruction Classroom Performance System, Qwizdom, TurningPoint, Interwrite PRS, iClicker, and H-ITT [20, 21]. This section focuses on the iClicker system [22] that uses iClicker2 hardware remote for each student and a receiver connected to the instructor's computer. The remotes allow students to select one out of five answers (A–E). Each remote is identified through a unique code provided by the manufacturer and a simple numeric label that allows students to quickly identify the remotes assigned to them. The approach of providing a unique remote assigned to each student allows the system to record the answers for various uses such as assessments, participation, or attendance.

One of the limitations of the system is the cost. Each iClicker2 remote costs about $80 CAD. Students could be asked to buy remotes and use it for their entire studies if there are enough courses that use CRS. Alternatively, for programs that include classes with low numbers of students, an academic department can purchase these remotes and provide them to the students for the duration of a lecture.

The iClicker system described in this paper is used in three automotive engineering technology courses offered at McMaster University. The initial iClicker system with 60 remotes that was initially purchased by the School of Engineering Practice and Technology has been followed by the purchase of two more similar systems. With an increasing number of students, the initial purchase price and maintenance cost became a limitation that needed to be addressed.

The CRS is used for different active learning strategies such as quizzes, application questions, and critical thinking questions. The initial active learning approach involved the assessment of students' knowledge through weekly quizzes. These quizzes are used to determine if students read the previous lectures and to assess if they understood the concepts, remember important points, and memorized key facts. Each quiz includes 10 questions: four questions to assess students' understanding of important concepts; four questions to assess how they memorized important points and facts; and two more difficult conceptual understanding questions. Students are asked to provide individual answers to simple questions, but are encouraged to talk with their peers before providing answers to the more challenging conceptual understanding questions. The time allocated for each quiz varies between 15 and 20 min and includes polling the answers from students, showing the distribution of the answers, and providing feedback through class discussions. Although the system is effective, it is limited to selecting one multiple-choice answer. The limitations include the impossibility to provide numeric values; to match two columns such as characteristics and concepts; or to provide a text.

The CRS is also used during lectures. In order to check if the students are actively thinking about the material, they use their clickers either to reinforce a new content or to connect a theoretical concept with a real-world application. Providing students a list of possible applications and asking them to select the one that has the strongest connection with the theoretical concept allows students to better understand and reinforce the new concepts. However, asking them to think about a possible application instead of asking them to select one from a predefined list would encourage them to think creatively. Furthermore, asking students to analyze the relationships between multiple concepts and make evaluations based on certain criteria is expected to develop their critical thinking skills. For both these applications of CRS, hardware clickers with five buttons help them grasp the concepts, but a system that would allow them to provide a text would definitely be better than selecting one multiple-choice answer.

The technical courses included in the current analysis contain a significant amount of problem-solving exercises. The approach of posting solved problems on the Learning Management Systems used by the course is not effective for many students as they should follow the step-by-step development of the mathematical solution. On the other hand, asking students to copy the detailed solution developed by the course instructor on the board does not encourage students to think. However, if students are encouraged to suggest the next steps during the development of a technical problem, they would have a better understanding of the approach and will have better abilities to develop a similar solution by themselves. Clickers can be used to ask them to suggest these steps, to suggest units, to accept, or not to accept the ranges of the numerical values of the results. If students could provide clicker answers that include numbers or a short text, the scope of the questions could be highly expanded.

Another successful experience attained by the authors of this paper uses clickers to break down long engineering problems into step-by-step solutions. The solution of many engineering problems includes extensive mathematical calculation, finding parameters from tables, and critical thinking. The clickers help students to solve each of these steps and allow them to figure out their mistakes in every step. Students can also learn from other's mistakes, while instructors explain the solutions.

The approach of using hardware clickers purchased by the academic department is a good approach to deliver the outcomes that are expected when using a CRS and incurs no cost to students. Although hardware clickers are a good active learning approach, the outcomes of using a CRS can be improved if their limitations are reduced. Many of the limitations described above could be addressed by replacing the hardware clickers that provide one of five possible answers with soft clickers that use programs or apps installed on mobile devices. The wide use of smartphones and tablets by the student population makes the soft clicker approach attractive.

4   Commercial Smartphone-Enabled Clickers Systems

Soft clickers using Wi-Fi or phone-data-enabled devices such as smartphones, tablets, or laptops provide a good alternative to hardware clickers. Feasibility studies to assess the level in which students understood the concepts taught in class and facilitate the exchange of opinions by polling students' responses using smartphone- or tablet-enabled CRS have been carried out for various specialized software packages that include but are not limited to Socrative [23–27], Top Hat Monocle [28, 29], ResponseWare [30, 31], Poll Everywhere [32, 33], and iClicker Cloud [34, 35]. These studies show that the approach is very feasible in today's classroom; students enjoy using their own smartphone devices; students stated that the interactive simulations and quizzes helped the concepts "stick in their memory"; the use of clickers positively impacted the students' learning during the course; smartphone-enabled clickers make lectures more interactive and provide students with another mechanism to reflect upon their own learning.

The main advantage of the soft clickers is the possibility to address the limitations of the hardware clickers by allowing students to provide various types of answers. Beyond true/false and multiple choice questions with only one possible answer, soft clickers can be used for other question types such as multiple choice with more possible answers, sequence/ordering, numerical, short answer, fill in the blanks, and matching questions. Furthermore, the questions can include 3D models that can be viewed by a student from different positions, animations in which the student can actively participate at different phases, and various simulations. These complex questions with student-controlled graphical interfaces can provide significant superior outcomes compared with the classic hardware clickers as they provide the possibility to involve critical thinking.

A disadvantage of using commercial soft clickers is the cost. While the hardware clickers, paid by students or by an academic department, have only one initial cost, the use of soft clickers includes a cost for each course. Furthermore, the cost of the academic department to work together with the companies that provide soft clickers to create elaborate questions that include complex 3D models or simulation is significant.

An important aspect that needs to be addressed before thinking about developing a commercial soft CRS is the generality of courses. A company would develop course-specific questions with complex drawings and simulations and a corresponding answering interface for their soft CRS only if they would be able to sell their product to multiple academic clients. A business model would involve the development of a complex user interface for a commercial soft CRS that could be used for only one course and one client would not be successful. This is an important limitation for upper year courses that are very specific for a certain academic specialization.

5   Self-designed Soft Clicker Systems

Commercial soft CRS presented in the previous section address most of the limitations of hardware CRS. Although they can provide an excellent experience, soft CRS have some limitations related to the costs and the generality of the course. A possible way to address these limitations is self-designed soft CRS personalized for certain courses. This soft CRS has three components: a software app that runs on a smartphone or tablet, polling software that runs on a classroom computer, and a user interface.

One of the approaches proposed by this paper is a soft CRS designed in-house that can be used for any course, combined with self-designed user interfaces for each specific course. Such a system is under development in the School of Engineering Practice and Technology at McMaster University. The first version of the soft CRS uses two software packages, namely, a smartphone app for students designed in Java for an Android platform and instructor software for Microsoft Windows. The future versions are expected to use iOS and macOS platforms. The software packages are developed, tested, and implemented by students employed over the summer, and refined during the year by student scholars. Meanwhile, course instructors are code-signing course-specific user interfaces that use various types of questions such as multiple choice questions with one or more possible answers, ordering, numerical, short answer, fill in the blanks, and matching questions. The final costs incurred for these developments are expected to be less than the equivalent costs for developing complex questions with the companies offering commercial CRS systems.

The self-designed soft CRS is expected to be used for many courses, including upper year highly specific courses for which companies selling commercial soft CRS might be reluctant to develop. The development effort for a certain course includes mostly the development of the initial course-specific question database. New questions can be added every year, while the existing questions can be refined often with student scholars.

The proposed self-designed soft CRS has a series of limitations that include the challenge of developing complex user interfaces, the software platform for smartphone or tablet and instructor's computer, and the software development time. Although the existence of complex questions can be ignored in the first phase of the software development, the self-designed CRS cannot be deployed to a live class until it runs on Apple-based software platforms. Furthermore, realistic solutions need to be found for students who change their phone during the course and need to register another device, for students who forget to bring their smart device to class, for students whose smart devices are not charged enough, and for student who do not have smartphones.

Realistic software development times needs also to be taken into consideration. Although software and questions database developments for one course have already been started, the development is expected to take about 3 years. This is a limitation that could be addressed if extra financial sources would become available and more developers could be hired.

Another possible approach for self-designed soft clickers is the use of an existing commercial CRS and working with its developers to allow the use of a self-designed user interface that would include various question types. The authors of this paper intend to look also at this approach as it might reduce the time when the soft CRS could be deployed in a real classroom environment.

6 Conclusions

This paper focuses on the use of classroom response systems, also known as clickers, in two technical courses. It describes the advantages of using this type of active learning approach in a university environment, presents its advantages, describes the existing approach that uses hardware clickers in three technical courses, lists the associated limitations, and proposes solutions to these limitations that use mobile technologies.

References

1. Slavin, R. E. (2010). Co-operative learning: What makes group-work work. In *The nature of learning: Using research to inspire practice* (pp. 161–178).
2. Savery, J. R., & Duffy, T. M. (1995). Problem based learning: An instructional model and its constructivist framework. *Educational Technology, 35*(5), 31–38.
3. Mills, J. E., & Treagust, D. F. (2003). Engineering education—Is problem-based or project-based learning the answer. *Australasian Journal of Engineering Education, 3*(2), 2–16.
4. Freeman, S., Eddy, S. L., McDonough, M., Smith, M. K., Okoroafor, N., Jordt, H., et al. (2014). Active learning increases student performance in science, engineering, and mathematics. *Proceedings of the National Academy of Sciences, 111*(23), 8410–8415.
5. Kirschner, P. A., Sweller, J., & Clark, R. E. (2006). Why minimal guidance during instruction does not work: An analysis of the failure of constructivist, discovery, problem-based, experiential, and inquiry-based teaching. *Educational Psychologist, 41*(2), 75–86.
6. O'Mahony, T. K., Vye, N. J., Bransford, J. D., Sanders, E. A., Stevens, R., Stephens, R. D., ... Soleiman, M. K. (2012). A comparison of lecture-based and challenge-based learning in a workplace setting: Course designs, patterns of interactivity, and learning outcomes. *Journal of the Learning Sciences, 21*(1), 182–206.
7. Tien, L. T., Roth, V., & Kampmeier, J. A. (2002). Implementation of a peer-led team learning instructional approach in an undergraduate organic chemistry course. *Journal of Research in Science Teaching: The Official Journal of the National Association for Research in Science Teaching, 39*(7), 606–632.
8. Healey, M. (2005). Linking research and teaching exploring disciplinary spaces and the role of inquiry-based learning. In *Reshaping the university: New relationships between research, scholarship and teaching* (pp. 67–78).
9. Wage, K. E., Buck, J. R., Wright, C. H. G., & Welch, T. B. (2005). The signals and systems concept inventory. *IEEE Transactions on Education, 48*(3), 448–461.
10. Johnson, J. T. (2005). Creating learner centered classroom: Use of an audience response system in paediatric dentistry education. *Journal of Dental Education, 69*(3), 378–381.
11. Anderson, L. S., Healy, A. F., Kole, J. A., & Bourne, L. E. (2013). The clicker technique: Cultivating efficient teaching and successful learning. *Applied Cognitive Psychology, 27,* 222–234.
12. Bojinova, E., & Oigara, J. (2011). Teaching and learning with clickers: Are clickers good for students? *Interdisciplinary Journal of E-Learning and Learning Objects, 7,* 169–184.
13. Bojinova, E., & Oigara, J. (2013). Teaching and learning with clickers in higher education. *International Journal of Teaching and Learning in Higher Education, 25*(2), 154–165.
14. Kennedy, G. E., & Cutts, Q. I. (2005). The association between students' use of an electronic voting system and their learning outcomes. *Journal of Computer Assisted Learning, 21,* 260–268.

15. Caldwell, J. (2007). Clicker in the large classroom: Current research and best-practice tips. *Life Sciences Education, 6*(1), 9–20.
16. Blasco-Arcas, L., Buil, I., Hernandez-Ortega, B., & Sesse, F. J. (2013). Using clickers in class. The role of interactivity, active collaborative learning and engagement in learning performance. *Computers & Education, 62,* 102–110.
17. Liu, C., Chen, S., Chi, C., Chien, K.-P., Liu, Y., & Chou, T.-L. (2017). The effects of clickers with different teaching strategies. *Journal of Educational Computing Research, 55* (5), 603–628.
18. Beckert, T., Fauth, E., & Olsen, K. (2009). Clicker satisfaction for students in human development: Differences for class type, prior exposure, and student talkativity. *North American Journal of Psychology, 11*(3), 599–611.
19. Barber, M., & Njus, D. (2007). Clicker evolution: Seeking intelligent design. *CBE—Life Sciences Education, 6,* 1–20.
20. Katz, L., Hallam, M. C., Duvall, M. M., & Polsky, Z. (2017). Considerations for using personal Wi-Fi enabled devices as "clickers" in a large university class. *Active Learning in Higher Education, 18*(1), 25–35.
21. Johnson, D., & McLeod, S. (2005). Get answers: Using student response system to see students' thinking. *Learning and Leading with Technology, 32*(4), 18–23.
22. https://www.iclicker.com/.
23. Coca, D., & Sliško, J. (2013). Software Socrative and smartphones as tools for implementation of basic processes of active physics learning in classroom: An initial feasibility study with prospective teachers. *European Journal of Physics Education, 4*(2), 17–24.
24. Dervan, P. (2014). Increasing in-class student engagement using Socrative (an online Student Response System). *The All Ireland Journal of Teaching and Learning in Higher Education (AISHE-J), 6*(3), 1801–1813.
25. Awedh, M., Mueen, A., Zafar, B., & Manzoor, U. (2015). Using Socrative and smartphones for the support of collaborative learning. *International Journal of Integrating Technology in Education (IJITE), 3*(4), 17–24.
26. Wash, P. (2014). Taking advantage of mobile devices: Using Socrative in the classroom. *Journal of Teaching and Learning with Technology, 3*(1), 99–101.
27. Dakka, S. M. (2015). Using Socrative to enhance in-class student engagement and collaboration. *International Journal on Integrating Technology in Education (IJITE), 4*(3), 13–19.
28. Alemohammad, H., & Shahini, M (2013). Use of mobile devices as an interactive method in a mechatronics engineering course: A case study. In *Proceedings of the ASME International Mechanical Engineering Congress and Exposition IMECE2013.* Education and Globalization (Vol. 5), November 15–21, Can Diego, California, USA.
29. Lucke, T., Keyssner, U., & Dunn, P. (2013). The use of a classroom response system to more effectively flip the classroom. In *2013 IEEE Frontiers in Education Conference* (pp. 491–495).
30. De Vos, M. (2018). *Using electronic voting systems with ResponseWare to improve student learning and enhance the student learning experience—Final report.* https://www.researchgate.net/publication/267250288. Accessed 14 Aug 2018.
31. Gong, Z., & Wallace, J. D. (2012). A comparative analysis of iPad and other M-learning technologies: Exploring students' view of adoption, potentials, and challenges. *Journal of Literacy and Technology, 13*(1), 2–29.
32. Popescu, O., Chezan, L. C., Jovanovic, V. M., & Ayala, O. M. (2015). The use of poll everywhere in engineering technology classes to student stimulates critical thinking and

motivation. In *122nd ASEE Annual Conference and Exposition, Making Value for Society,* June 14–17, Seattle, WA, USA.

33. Tregonning, A. M., Doherty, D. A., Hornbuckle, J., & Dickinson, J. (2012). The audience response system and knowledge gain: Λ prospective study. *Medical Teacher, 34*(4), 269–274.

34. Wu, X., & Gao, Y. (2011). Applying the extended technology acceptance model to the use of clickers in student learning: Some evidence from macroeconomics classes. *American Journal of Business Education, 4*(7), 43–50.

35. Whitehead, C., & Ray, L. (2018). *Using the iClicker classroom response system to enhance student involvement and learning.* https://www.researchgate.net/publication/265192585. Retrieved 15 Aug 2018.

"If you take away my phone, you take away my life..." Community Narratives about the Social Implications of Mobile phone Usage for Livelihood Security

Dianah Nampijja[✉]

University of Agder, Kristiansand, Norway
dianah.nampijja@uia.no; nampijjadianah@gmail.com

Abstract. Smallholder farmers in developing regions like sub-Saharan Africa still grapple with development challenges like poverty, illiteracy, food insecurity, poor infrastructures like roads and limited access to learning opportunities. Moreover, amidst all these challenges, these possess mobile phones as they continuously engage in diverse livelihood activities. Echoed in previous research, mobile phones are the most diffused technologies available in many rural communities. Yet, most mobile technological interventions are urban-based projects neglecting rural locales in harnessing mobile phone usage. More so, mobile for development research uses quantitative methods which limits proper interrogation of real perceived social implications of mobile phones for livelihoods. Thus, this study, with a more qualitative approach, seeks to understand community narratives about the social implications of mobile phone usage for livelihood security. Findings suggest the increased penetration of small end phones, with few smartphones. Noticeable social benefits include improved communication, financial inclusion, employment opportunities, increased business and market opportunities, increased access to information sharing and improved literacy practices. Conclusively, access to learning on mobile phones was an outstanding social benefit, amidst the presence of negative and dissenting voices about mobile phone use, fully embedded in cultural and religious beliefs within societies.

Keywords: Mobile phones · Social implications · Livelihood security · Smallholder farmers

1 Context

Smallholder farmers in developing regions like sub-Saharan Africa still grapple with development challenges like poverty, high population growth rates, illiteracy, food insecurity, poor health systems, poor infrastructures like roads, lack of access to electricity and most importantly, limited access to learning opportunities [1–3]. Moreover, amidst all these challenges, most smallholders especially those in rural communities have prioritized the possession of mobile phones compared to other available technologies as they continuously engage in several activities to secure and

© Springer Nature Switzerland AG 2019
M. E. Auer and T. Tsiatsos (eds.), *Mobile Technologies and Applications*
for the Internet of Things, Advances in Intelligent Systems and Computing 909,
https://doi.org/10.1007/978-3-030-11434-3_40

sustain better livelihoods. As noted by Furuholt and Matotay [4], the most widely spread Information and Communication Technology (ICT) across the world today, including developing regions, is a mobile phone. Highly documented as the fastest technological diffusion in communication history [5], mobile phones are easily accessible and can support African communities in accessing actionable information that can support adaptation for better livelihoods [6].

Significant and previous studies about the impact of mobile phones [7–11] undoubtedly show how mobile technologies are among means to leapfrog most sub-Saharan Africa states to the next-level development. However, there exists limited literature on how rural smallholder farmers' possession of mobile phones supports their livelihoods. In their study about how mobile technologies impact economic development in sub-Saharan Africa, Crossan and others found how most mobile technologies projects were mainly urban-based [10]. This broadens the digital divide; yet, ICTs like mobile technologies are among means to bridge this digital divide [12] and support those in rural locales. Thus, the potential of modern technologies to avail information access to such communities to act for secure livelihoods is an option worth understanding.

Today, we live in a 'hugely mobilized world as estimates put mobile subscriptions at more than 6 billion globally, with at least 75% of these being in developing countries' [13, 14]. In 2017, 5 billion people were connected to mobile services, where growth in the sector was driven by developing countries in sub-Saharan Africa [15]. The emergence of the connected mobile society with numerous information sources available at work, home, community and schools has arose considerable interests from educators and technical providers to exploit the capabilities these mobile technologies can offer for the new and engaging learning environments [16]. In developing countries like Uganda, there is an increasing trend in mobile phone ownership at 70.9%, with a significant increase at 65.7% in rural Uganda. This hype in mobile phone ownership (depicted in Fig. 1) offers a better justification for this study to explore and analyse how mobile phones impact on smallholder livelihoods in rural areas.

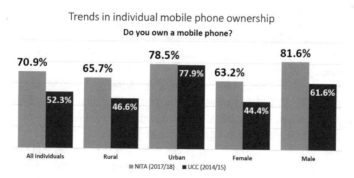

Fig. 1. Trends in mobile phone ownership in Uganda 2017. *Source* NITA [17]

Therefore, in this paper, community narratives of the social implications of mobile phones as contribution to their livelihood security are discussed. Current research on 'mobiles for development' focuses on mobile phones' impacts on farming, animal husbandry, availing market information and sharing weather updates; neglecting other social attributes, mobile phones extend or afford to rural livelihoods. Moreover, most of these studies exploring mobile phones' impacts employ quantitative survey methods which limit further interrogation of the real perceived usefulness from those using the technologies. More so, there is less exploration about the dissenting views and negative implications of mobile phone usage among rural livelihoods. Thus, this study seeks to understand the nature of mobile phones available in rural communities, and analyses the social implications of mobile phones for livelihood security in a developing country's context of Uganda. The rest of the paper discusses livelihood security, the concept of smallholder farmers, and engages with the study findings about the social implications (both positive and negative) of mobile phone usage among smallholder communities. A methodological section that discusses the mobiles for development projects and the nature of mobile phones possessed by the smallholder farmers then follows. The paper ends with a conclusion that situates the need for communication, access to learning on mobile phones, and mobile money financial transactions as outstanding social benefits, amidst the presence of negative and dissenting voices about mobile phone usage.

2   Livelihood Security

Development literature mainly situated in studies of poverty and rural development has had a different meaning of the term livelihoods. To begin with the English thesaurus dictionary, livelihoods denote the 'means to earn a living'. This implies a way of survival or a way of managing to exist and achieve the basic necessities of life like food, shelter and health.

Being considered 'a mobile and flexible term, "livelihoods" can be attached to all sorts of other words to construct whole fields of development enquiry and practice. These relate to locales (rural or urban livelihoods), occupations (farming, pastoral or fishing livelihoods), social difference (gendered, age-defined livelihoods), directions (livelihood pathways, trajectories), dynamic patterns (sustainable or resilient livelihoods), and many more' [18].

This shows how the term has been used in different development literature to capture different categorizations of resources, people and also occupations. Chambers and Conway [19] define livelihoods as capabilities, assets (in form of stores, resources, claims and access) and activities required to earn a living.

In this paper, by livelihoods, I imply 'the activities, the assets, and the access that together determine the living gained by an individual or household' [20]. More literally, the diverse activities people engage in to sustain a living and secure a socially and culturally acceptable way of living. Hence, the notion of livelihood viability is not only limited to survival and earning basic human needs but rather looks at the totality of how households survive both in the short term and long term. This ability to survive is livelihood security. In relation to mobile technologies, mobile phones are tools used to help people attain livelihood viability. The essence lies in understanding the role of technologies in responding to peoples' livelihoods options to enhance livelihood security.

Livelihoods in rural communities imply diversification, that is, 'rural families tend to adopt survival strategies composed of a diverse portfolio of activities that cut across orthodox economic sectors and transcend to rural urban divide' [20]. Their livelihoods tend to take the form of 'an asset-access' of activities where rural communities tend to cope with different circumstances as presented by the different shocks. Most smallholder farmer communities are vulnerable to many shocks, and this explains why most tend to engage in several livelihood activities. The potency rests on their ability not only to cope but rather stay resilient.

2.1 Smallholder Farmers

Smallholder farmers' livelihoods, especially those in rural communities of many developing states, are dependent on agriculture as a predominant activity. Agriculture (mainly crop and livestock) is the most important livelihood strategy and the primary source of income [21]. In the bigger global picture, smallholder farmers in sub-Saharan Africa and Asia occupy 70–80% of the total global farmland, producing 80% of the food that is consumed in developing countries [22, 23]. Smallholders 'play a crucial role in supplying food to the continent's population and bringing about economic transformation in rural areas' [24]. Women smallholders account for 50% of the agricultural labour force of developing countries, and thus, they offer productive resources as men to the different farmlands [23]. This therefore makes it opportune for the agriculture development strategy to invest in smallholder agriculture [25] worldwide. Most smallholder farmers engage in crop farming, livestock and fisheries, and some in export markets depending on the activities, ecological zones and availability of strong bridges and networks within their communities.

In Uganda, most communities are dependent on agriculture both subsistence and commercial. Agriculture contributes to 45% of GDP and supports 75% of the labour force [25]. In the country, over 75% of the agricultural sector is dominated by the smallholder farmers who largely grow at subsistence level as their composition is two-thirds of those engaged in agriculture for livelihoods. The agricultural sector has a development contribution to poverty reduction than non-agricultural growth given its strong linkages to the rural economy and the fact that it supports 80% of many livelihoods in developing regions [26]. In spite of the above observation, smallholder farmers, everywhere, are struggling for their survival as their livelihoods are greatly hampered by poor government policies and practices that are unresponsive to their needs [24]. Disenabling are the diverse weather conditions like droughts, floods, hailstorms and heavy rains given that smallholder agriculture is heavily dependent on rainfed cultivation [27]. Despite smallholder farmers being diverse groups, for this study, rural smallholders were a focus. This means interactions were with smallholder farmers in rural areas, mainly in western Uganda in the districts of Bushenyi, Ibanda and Kabale, majority in crops farming, apiary, with few livestock activities. Most crops were subsistence in nature like banana, cassava, potatoes, rice, millet, vegetables and maize, with some engaged in cash crops growing like coffee and cotton. Other farmer groups also engaged in export activities. For example, in the USAID CC projects, some farmer groups through Ibanda District Apiary Association exported local honey outside Uganda. The next section offers a further description of the multiple case sites of the 'mobiles for development' projects in Uganda.

3   Methodology

This study was conducted in western Uganda in three districts of Bushenyi, Ibanda and Kabale between July and December 2016. In these districts, parishes' characteristics of rural locales were purposefully selected, in an attempt to explore how mobile phones impacted on rural livelihoods. A qualitative study through an interpretivist and social constructivist perspective from multiple case studies of three mobiles for development organizations, that is, USAID Community Connector (CC) project, Grameen Foundation Community Knowledge Worker (CKW) project, and Lifelong learning for farmer (L3F) project, was adopted (as shown in Table 1). Six (6) Focused Group Discussions (FGD) with smallholder farmers, both men and women, were used to understand the general community narratives about the social implications of mobile phones' use. More so, given the fact that rural locales are busy, to generate personalized narratives, the study sought to employ forty (40) interviews including key informants, farmers in the case organizations and farmers outside organizational activities. The latter was to analyse how other farmers outside projects' operations were using mobile phones. Participants' observations were used in identifying the actual possession and use of mobile phones by the study participants. In total, 70 participants were a representative sample in a purely qualitative study given emphasis on thick data with socially constructed analyses in a multiple case study approach. NVivo tool aided the analysis of the field data through code classification themes like nature of the mobile phones, benefits of the mobile phones, problems with mobile phones to society and the social experiences about mobile phones in rural areas. For reliability and validity, several follow-up discussions with study participants were conducted. Also, given that some farmers had smartphones, continuous interactions on WhatsApp supported better understanding of mobile phone use for livelihood security.

Table 1. Mobiles for development organizations

	Grameen Community Knowledge worker	USAID Community Connector	L3F (Lifelong learning for farmers)
Macro Actor (Funder)	Grameen Foundation	USAID Feed the future (FTF), and Fhi 360	Common Wealth of Learning (CoL)
Project Goal	Help the world's poorest people reach their full potential, through connecting their determination and skills with the resources they need	Improve nutrition and hygiene, increase access to more diverse and quality foods, increase household assets and incomes.	Responds to a critical need that enables farmers to use ICT—particularly mobile technology—to access information
Approach	District project officer CKWs Super CKWs	Community Connectors, CKWs Service providers Grants to groups	Innovative platform leaders Self-help groups Small loans to groups
Nature of engagement	One-on-one meetings Group meetings @ 50 farmers Call centre	One-on-One meetings Family health schools Nutritional sites Saving with a purpose CC10 s	Personalized messages (text and audio messages) Trainings Horizontal learning Call centre
Mobile phones	Smartphones	Smartphones	Small end phones
Livelihood activities	Farming mainly in Coffee, Banana Data collection SIM registration	Farming (nutritional crops), savings, Apiary, poultry	Farming (Irish, Banana, Grapes) Marketing, value addition, credit activities
Case study sites	Bushenyi District (Katerera sub-county)	Ibanda District (Kicuzi sub-county)	Kabale (Bubaale sub-county)
Partner agencies	National Agriculture Research Organisation (NARO) MTN Uganda	Self Help Africa BRAC Mbarara University	Agriculture Innovations Systems Brokerage Association (AGNSBA)

4 Findings and Discussion

Findings from this study point to an increase in the penetration of mobile phones in most rural areas, citing the availability of network infrastructure and the fact that it is a social issue for one to have a mobile phone. Relatedly, the most mobile phones used by many smallholder farmers are small end phones (traditional non-smartphones), with limited penetration of smartphones. The available smartphones among the rural farmers are mainly those given to different change agents from various organizations using mobile phones for development-related activities. Also, youth farmers (below 25 years) possessed smartphones and occasionally visited online platforms mainly Facebook and WhatsApp for socialization and search for weather and market updates.

4.1 Social Implications of Mobile Phones

Many rural livelihoods are built on their ability to engage in a multitude of activities. They cannot just take part in one activity and survive. Thus, mobiles for development initiatives ought to appreciate the diverse portfolio of activities smallholder farmers engage in, as a way to suggest solutions and targets that benefit both organizations and the rural people. Limiting them to a single livelihood activity is unrealistic as their livelihoods are determined by several engagements in diverse economic and social activities. This realization will reduce frustrations from both and increase ways on how projects can be responsive to rural frame of activities. In this paper, therefore, the mobile phones were looked at as tools to help people expedite livelihoods in totality.

In analysing the findings from the study respondents, some of the social benefits of mobile phone use including communication-calling friends and relatives, learning about farming, elevation of the social status, employment, mobile money transactions, socializing and meeting friends, and increased productivity. Figure 2 shows an NVivo node analysis extract about the positive benefits of mobile phones' use in rural communities. Each benefit is further expounded below.

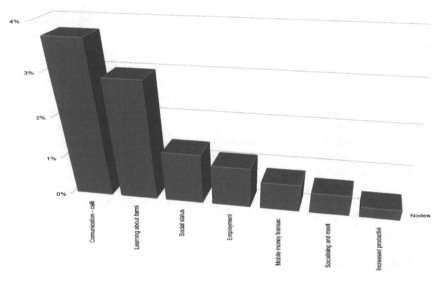

Fig. 2. Benefits/uses of mobile phones' NVivo node extract. *Source* Field data

Mobile phones, as pointed out in previous literature, have to a bigger extent facilitated communication. All farmers attested to the fact that they wanted to stay connected with their relatives and friends. This staying in touch was not only limited to people away or abroad, but also to those in the same locales. Even when the phones helped in contacting one another for meetings, it was evident that many people did not use their airtime to contact neighbours. Other local channels of communication were still relevant and majority would meet up in mobile markets, at shops in the evening and at special occasions like weddings, burials and at funeral rites to pass on information. Most calls targeted people outside these communities, especially relatives and friends who worked in town centres, countryside and abroad. This staying in touch was linked to know the health status of the family, general household well-being, as well as managing farm produce.

The phones were considered to have increased farm productivity, through facilitating marketing channels. Farmers would call to find out and negotiate prices of different food items and farm inputs, whilst selling their products to local and external markets. This, in turn, lessened travel costs, thereby increasing farmers' incomes since middlemen (brokers) exploitation was reduced. Financial transaction through mobile money was an essential benefit which the mobile phones extended to such rural areas. Most farmers saw a mobile phone as a bank. For example, most farmers and small savings groups in different villages saved returns on their mobile money accounts. In particular, woman saved on mobile phones to hide from male/husbands' exploitations in case of small personalized savings. The marketing and financial phone capabilities were so significant and like one smallholder in Kabale exclaimed, '*if you take away my phone, you take away my life*'. Another from Kabale town said, '*I feel disarmed without my mobile phone. My phone is my business*'. This shows how some farmers personalized their mobile phones and the magnitude of relevance to their day-to-day activities.

Such personalization was related to how the phones were viewed as tools to enhance and uplift ones' social status. First, for some farmers, it was a 'social fit' to own a mobile phone, thus a social good if one needed to earn respect among peers. Even when one did not have people to call or connect with promptly, their satisfaction was to own a personalized mobile phone. Thus, owning a mobile meant power which in turn earned ones' respect in the community. On the same note, the CKWs who were given smartphones felt empowered within their villages as the mobile phones came along with solar panels, chargers and the solar bulbs which improved their households' well-being. In a focused group discussion, Kato, a CKW narrated that '*...being given a smart phone in 2009 was no simple business to my life. I felt respected and am telling you; this phone has changed the way people look at me. I am considered a respectable person with knowledge on my phone, an educator, and a role model farmer. In this village, other organizations now consult me about mobilizing people and the community looks at me as a key resource. Thanks to Grameen for this opportunity*'. All CKWs felt the mobile phones and other supporting equipment impacted not only the community but also their general household well-being.

Additionally, mobile phones increased employment opportunities given the presence of many mobile phone-related business stalls in different rural towns. In all the study case sites, there were visible mobile phone shops ranging from selling airtime- and phone-related accessories, to availability of mobile phones' technicians. The mushrooming mobile money outlets also availed employment to youths who offered guidance to illiterate farmers to complete financial transactions. A jolting benefit was that mobile phones also acted as safety nets—through acting as security in case of no money. For example, one would use a mobile phone as an asset to get urgent money (in the form of a loan) from moneylenders and friends. More so, mobile phones were used as the main source of light at night for households with no solar lighting. Because the villages did not have electricity connections, mobiles were seen to support retail shops activities and other night transactions. Thus, mobile phones' benefits were incredible in such a rural community.

With the presence of 2G and 3G network coverage, mobile phones supported access to weather and market information. CKWs and some youth farmers with smartphones accessed Internet which allowed access to the weather and market information online from agricultural agencies. This benefit was, however, limited to the educated farmers since it required knowing what site to visit to access the updates. Also, Internet access meant buying mobile data packages to access actionable information which hindered majority. But other farmers with no access to smartphones and data would benefit from such updates given the social networks that existed within communities. Through their different groups, CKWs would alert farmers about the available updates for immediate action. More so, some CKW groups had a marketing strategy where farmers pooled food items and sold at a higher bargaining power which improved profits.

Access to learning on mobile phones was an intriguing social benefit mentioned by most smallholder farmers. As noted by Bhatnagar [28], ICTs have supported literacy practices, especially among the poor rural women in most developing countries. This was noticeable in specific households where mobile phones helped some farmers exercise reading and writing skills, thereby supporting literacy practices. Interestingly, as noted by one farmer, '... *even when the mobile phone is in English, I can write and send messages in my local language. Before I used to fear sending messages. But with time, I managed to learn by myself through doing it*'. This depicts some confidence building and an aspect of learning to read and write. Quite notably, in this study, the three mobiles for development projects all aimed at extending actionable information to smallholder farmers; thus, by implication, the mobile phones farmers used were learning resources.

For mobile phones as learning resources, both smartphones and small end phones had different functionalities and capabilities they afforded to farmer communities. To begin with the small end phones, farming-related information was shared through local dialect, which eased translation. Each farmer who was registered in the L3F—Lifelong learning for farmers programme was sent both audio and text messages on matters pertaining to farming. This information exchange based on farmers needs and thus, it was in line with the farmers' enterprise. While the approach of using the traditional phones seemed an option for project sustainability (that is, using the locally available farmers' traditional phones), the content shared was rather limited. Farmers who got access to this information shared with other farmers which supported horizontal learning—an aspect where the farmers shared information at the same level, highly integrated into their day-to-day activities. The Grameen and USAID CC projects gave farmers smartphones fully equipped with farming information. The smartphones were connected on a satellite and each local farmer—termed Community Knowledge Worker (CKW)—had to work with a network of other 50 farmers in order to engage with the actionable information on the mobile phones.

The most vantage point is that 'as technology becomes cheaper and more powerful, mobile phones provide an affordable platform to access data. Speech interfaces can reduce the required skills, literacy being one of the biggest impediments to access existing data' [29]. The availability and willingness of 'significant others' within the social network support use of mobile phones [29]. For example, even when some people did not know how to use mobile phones, the presence of a relative, friend, neighbour or child was a reinforcing factor for adoption and use of the technology. The use of locally available phones owned by farmers was also a sustainable option in ensuring how mobile phones can facilitate inclusive learning for all. This is especially relevant when thinking of advancing the use of locally available assets to work for the poor. Thus, the need 'to identify what people have rather than what they do not have' [30] can be a facilitating factor to advance sustainability of most technology-related initiatives in developing regions. Such a realization strengthens farmers own available resources, rather than substituting, blocking or undermining them [20, 30].

The point to take from this discussion therefore is that both smartphones and small end phones had implications to practice. While it looks justifiably right to agitate for locally available small end phones farmer communities owned, their use for learning was rather limited. On a small end phone, you hardly shared content of more than 30 characters, hence, limiting learning affordances. Besides, in rural areas, many people are still illiterate which justifies the use of significant others to support information–knowledge sharing processes. For L3F, the challenge was to envisage contextual ways of reaching out to the illiterate farmers through voice messaging, which to the project key formants was also costly and unsustainable. The cost of content production was high given the diversity of tribes in Uganda with over 65 languages, which makes it so hard to categorize a region based on a single language. As the key informant from L3F quipped, '...This is one of the biggest problems we are still facing as L3F. You realise that we still need to reach out to a particular category; but because of many languages to translate, not all farmers are reached'.

Therefore, 'rather than assuming Information and Communication Technology will always be cost effective and yield a better outcome, a more nuanced understanding of underlying institutional environment and constraints is warranted' [31]. Such institutional barriers can encompass structural and contextual barriers from within which the smallholder farmers interacted with the mobile phones. For example, cognitive barriers as suggested by [32] also affected the use of mobile phones. Many rural people lacked literacy skills, with limited/no English language command to help them navigate through the mobile applications with ease. This meant that such communities had to reply to those with basic grounding in English. Thus, responding to contextual challenges is the aspect we need to surmount if we are to take mobiles to the next-level discussion that can be inclusive of rural community groups like smallholder farmers.

On the other hand, irrespective of the contextual challenges, the smartphones given to farmers also raised sustainability implications. For example, even when the farmers (CKWs) identified that they had some English training, the approach of facilitating them through staying on the network and giving monthly volunteer allowance raised sustainability questions. At present, most Grameen CKW and USAID CC projects have closed and because of no facilitation in the form of airtime and monthly allowances, very few CKWs still reach out to other farmers. From a focused group interaction with the CKWs, most of them felt uncertain about the sustainability of these projects. One super CKW at Mitooma parish echoed '...we are over 100 CKWs trained and given smart phones with relevant actionable farming information. But, where will the cadre *of generated knowledge workers go when all the CKW projects close down'*? The CKWs started feeling this gap in fear of the next-level interactions with the different farmer groups. Nonetheless, the few CKWs who continued to reach out to other farmers are CKWs whose groups had income-generating activities in the form of marketing their produce. Thus, a central livelihood activity was a network node that tied these farmers together irrespective of Grameen or USAID support.

The livelihood enterprises were also connected to social bonds and ties a group had, and thus social capital was central in supporting farmer interactions on mobile phones. Mobile technology affordances have supported the socialization of farmers which in a way has widened social networking among those involved in the mobiles for development network. This supported collaborations among peers which in turn enhanced joint and coordinated problem-solving. This type of collaborative problem-solving supported by mobile devices increased farmers' motivation, engagements and interactions in their community of practice. Like Putman notes, organizational structures that build on horizontal linkages will increase trust and cooperative relations necessary to strengthen social capital, better than organizations that use vertical hierarchical linkages [33]. On the other hand, there is a likelihood of social capital breeding negative bonds. In this study, while mobile phones supported social interactions within farmer communities, their continuous usage also exhibited some negative bonds as presented in the next section.

4.2 Negative Social Implications of Mobile Phones

As technologies become part of everyday practice, they posit implications which in turn make not all people embrace them. Indeed, 'If you are pointing to the wrong direction, technology can help you to get there' [31]. Technologies have come with disruptions to society, which if well not thought about; soon or very later, we might be crashed by it [34]. One thing about technology is that it is never neutral. 'People will either love it or loathe it' [35]. People will always have different perceptions about a technological intervention, and this explains the need to understand people's perceptions of mobile phones' use in a wider livelihood discussion. Majority will have different expectations about such technologies and strive to relate them to their experiences and just like [35] notes, some technological interventions can change peoples' mindset where they can become open to more external influence than environmental influence.

The negative social implications depict how mobile phones have been a distraction to society. To begin with, mobile phones have facilitated and increased burglaries and theft in different areas, supported murder activities, vandalism, thereby threatening peoples' safety within communities. A significant number of respondents echoed how mobile phones had affected marital relationships making attribution to increased marriage breakdowns in their societies. Dissenting narratives from few people about why they did not possess nor use mobile phones was also relevant for this study. *'if you want to ask how mobile phones are associated with increased crime rates, go ask any police officer in that nearby police station to tell about the mobile phones related cases registered everyday'* said a respondent in Katerera town. While the study main intention was to focus on only smallholders, when it came to what people perceived as dangers from mobile phones, I was prompted to meet the local area village police post as an aspect of triangulation. Mobile phones have declined the stocks of social capital as people are increasingly becoming suspicious about the nature of calls they receive. A married smallholder, for example, had reservations buying his wife a mobile phone. To him, the mobile phone raised a lot of suspicions and he was not only comfortable with leaving his wife with a mobile phone but also the wife receiving calls from other strangers not known to him. This observation cut across as most marrieds felt uncomfortable with who was calling who. Actually, like one observed, '...*what I have seen in my friend's household, I do not want it to come to my house. The phones have caused marital challenges and that is why even at church today, the young couples are advised by the church leaders not to peep or read the others' messages* ...' said a senior farmer. This shows how mobile phones' interactions have caused suspicion even among family members, a justification as to why people (even married) have resorted to creating passwords in their phones—all aimed at limiting access of use.

More so, in rural communities, many people are tied to their cultures and religions. Once technologies like these come up, not all will embrace such innovations as some religious and cult groups did not embrace mobile phone usage. An interesting observation from this finding is that while mobile phones are thought to help rural communities in addressing development challenges, some individuals and groups do not embrace their use given their cultural and religious beliefs, expressed in their increased rigidity towards technology integration in day-to-day activities. In fact, many felt, such newer technologies are coming to destroy humanity. A religion named 'Abazukufu' (literary translated as 'the awakened') does not believe in any computerized related services. To this religion, 'technologies like mobile phones are evil and highly satanic with codes of 666'. Such groups therefore dissociated themselves with all activities on the mobile phones as they did not possess any. In informal interviews with 10 participants from this group, only two youths used mobile phones, radios and televisions. Majority relied on traditional means of communication to meet one another and to stay in touch.

Apart from this sect, others felt they need not use the mobile phones since they have caused trouble to many lives of those who use. The other non-use was also associated with gender where men stopped their wives from owning mobile phones. Although some farmers did not own individual mobile phones, they embraced their use through social networking. On a controversial note, others cited some health implications about the use of mobile phones. Like one respondent echoed, '*mobile phones have an effect on our health. Women and men put them in the bras and in under wears respectively for fear of being stolen. You can imagine breast-feeding mothers keep their phones next to their breasts and even breast feed while phones are next to the heads of their little ones*'. This disgruntled old farmer cautioned how the poor use of mobile phones had health implications and that people need to be sensitized about good use of mobile phones. This 'my device, anywhere and at any time' is coming with implications to human health. There are numerous ongoing viral messages citing dangers of mobile phones ranging from crashing, bursting, triggering road accidents, to causing cancers.

Similarly, the increased penetration of mobile phones' use presupposes mobile phone masts (base stations) that come with radiation. Thus, 'people close to mobile phone masts frequently report symptoms of electromagnetic hypersensitivity such as dizziness, headaches, skin conditions, allergies, and many others…' [36]. In Uganda, the mobile phones and many telecommunication masts might be the cause of many unknown cancer cases today. In both rural and urban town centres, the mobile phone masts are living with people, and every day, we take up the share of these ultrarays. Unlike in European states where masts are far away located from residential places, in Uganda, because of lack of space, dire need for bigger moneys and technology investments, we live and stay close to masts day by day. Now, who is there to help people understand such health implications? We do not know, we do not bother, neither does government advise about such health implications. Yet, mobile phones have an implication to the general environment and the totality of human well-being [37]. Therefore, we ought to be conscious about such bad health implications which in turn questions livelihood security in the long term.

Digital technologies such as mobile phones can have radical and disruptive effects for society as well as individual organization [38]. They present both capabilities and consequences to society. While they offer possibilities, there visible forces and consequences of society digitalization. While society, and partly media, have over publicized both the good and bad from digital technologies like mobile phones, we need to be upfront to these challenges and harness the good that comes along with technologies. The same society is still responsible for how to support the young generation, especially the digital natives in trying to bridge the bad and good divide. Therefore, there is a need for a balance between caution and encouragement [39]. Not forgetting the fact that we enter a digital culture, where even in rural areas where technologies were hard to reach, mobile penetration is high. Thus, the essence is to appreciate both positives from digitization, but also embrace and act on the negatives with respect to 'how to appropriately use' technologies in everyday practice. In this way, we shall maximally utilize the inherent benefits from mobile phone usage for livelihood security.

5 Conclusion

Understanding how mobile technologies can be effectively integrated into development-related activities requires appreciating that these technologies keep changing. Such evolving technological changes create an imagination gap among those trying to implement and use technologies [40]. I am not meaning to imply, however, that the new and increasing technological developments in ICT are bad, but rather they create a gap which always has to be fixed to keep up with the current technological pace. Ideally, this creates disruptions in systems and puts to task institutions trying to rely on ICT innovations as a way of life, but also, it constantly checks the capacities needed to keep up the pace.

Social implications about the usefulness of mobile phones' discussion are important to understand how mobile phones can support the livelihoods of many smallholder farmer communities. Actually, as emphasized in most mobile technology literature, the mobile phone is the fastest and highly diffused technology in most developing states like Uganda. Most interventions in education, health, agriculture, marketing, business and finance are looking for better possibilities to extend the reach of such services through affordable mobile phones' applications. Thus, understanding what kinds of mobile phones available and what people make use of mobile phones is important to take 'mobiles for livelihoods' research to the next-level discussions.

The view about mobile technologies as tools to support livelihoods is to realize that phones are actor tools that carry content to use; they do not plough or harvest. Like Qvortrup cited in Hetland notes, 'It is too imprecise to analyse information technology as if they only presented [all technologies in general]. Because, essentially, information technologies are tools for interactions and organization. You cannot use them to till the soil or to hammer, but you can use them to plan the tilling, to control and administer the hammering' [41]. In line with this affirmation, ICTs are not used to manipulate nature, but rather to manipulate cognitive and interactive processes essential to contribute to human well-being (Ibid.). In information technologies, information has 'an active role in shaping context [and is] not only embedded within a social structure, but creates that structure itself' [41]. Thus, to view information as an actor, and knowledge as power, we need to recognize the constructive force that it affords participation in contributing to the actors' own reality.

Therefore, in technological interventions for livelihoods, everything matters. It is not only one aspect, but rather a joint effort encompassing different actors can sustainably support livelihoods of many in developing contexts. An ecosystem approach that analyses livelihoods in totality is important, and for ICTs to be effectively utilized, they ought to be integrated within existing norms of practices, and more so, work hand-in-hand within available cultural practices. Therefore, everything is important, and we should not look at things in 'either or position', but rather, in 'both and both position'.

Amidst all the rhetoric successes of most Information Communication Technology for Development (ICTD) projects especially in Africa, very few are yet to be sustainable [42, 43]. In some villages, different farmer groups had concerns about the end of the mobiles for development projects, even others citing the need for other ICTD actors to continue with the ongoing projects. This revelation points to the lack of business models in most technological solutions that work for poorer communities. Besides, even when all the projects had some sustainability elements, it looked obvious that some activities would not continue given the way they were structured. Several interactions with communities where the projects had operated showed how initiated project activities would be rendered futile in the long run. Visible business model strategies like use of local farmers and call centres—where farmers had to incur a cost —were available but no farmer was willing to pay in search for more knowledge. This observation raises practical implications for future adherence of how ICTs like mobile phones can sustainably support development-related initiates.

To summarize this discussion, therefore, rural people use mobile phones in varying ways among which communication, mobile money transactions and learning for livelihoods were significant social benefits. Mobile phones used for learning in rural locales is rarely a researched dimension; yet, smallholders too need new knowledge and skills to secure better livelihoods. Notably, for mobile phones to really impact on equity and access to learning for all, extra support is vital to help such farmers' benefit from growing opportunities for mobile learning. Adversely, just like any technology, mobile phones also generated disruptions to society. The presence of social negative implications and dissenting voices about mobile phone usage among rural populace is worth consideration, as this has implications for sustainability of mobiles for livelihoods interventions. Perhaps, strategies on better use of mobile phones need to be harnessed in mobile for development discussions to enable communities gain better insights about mobile phone usage for livelihoods security.

Acknowledgements. The work reported in this paper was financed by DELP project; funded by NORAD and University of Agder, Faculty of social sciences. Special thanks to Makerere University who are in this research collaboration and partnership.

References

1. Omolewa, M. (2008). Rethinking open and distance learning for development in Africa.
2. CoL Lifelong Learning for Farmers Improves Household Food Security. 2013.
3. Gwali, S. (2014). Building community-based adaptation and resilience to climate change in Uganda.
4. Furuholt, B., & Matotay, E. (2011). The developmental contribution from mobile phones across the agricultural value chain in rural Africa. *The Electronic Journal of Information Systems in Developing Countries, 48*(1), 1–16.
5. Castells, M. (2011). *The rise of the network society: The information age: Economy, society, and culture* (Vol. 1). Wiley.
6. Manske, J. (2014). Innovations out of Africa. The emergence, challenges and potential of the Kenyan Tech Ecosystem, in Reports & Publications. Germany: Vodafone Institute for Society and Communications.
7. Aker, J. C., & Mbiti, I. M. Mobile phones and economic development in Africa. Center for Global Development Working Paper, 2010(211).
8. Alzouma, G. (2005). Myths of digital technology in Africa Leapfrogging development? *Global Media and Communication, 1*(3), 339–356.
9. Bolton Palumbo, L. (2014). Mobile phones in Africa: Opportunities and challenges for academic librarians. *New Library World, 115*(3/4), 179–192.
10. Crossan, A., McKelvey, N., & Curran, K. (2018). Mobile technologies impact on economic development in Sub-Saharan Africa. In *Encyclopedia of Information Science and Technology* (4th ed., pp. 6216–6222). IGI Global.
11. Zuckerman, E. (2010). Decentralizing the mobile phone: A second ICT4D revolution? *Information Technologies & International Development, 6*(SE), 99–103.
12. Heeks, R. (2015). *Digital Development Report in UN Economic and Social Council Commissions on science and technology for development.* Geneva: United Nations Economic and Social Council.
13. Mohamed, A., & Avgoustos, T. (2014). Increasing access through mobile learning, C.w.o. Learning, Editor. Common wealth of Learning, United Kingdom.
14. Mohammed, A., & Josep, P.-B. (2014). What is the future of mobile learning in education? *RUSC. Universities and Knowledge Society Journal, 11*(1), 142–151.
15. GSMA. (2018). The Mobile Economy Sub-Saharan Africa. GSM Association: United Kingdom.
16. Naismith, L., et al. (2004). Literature review in mobile technologies and learning.
17. NITA. (2018). National Information Technology (NITA) Survey 2017/18 Report, I.C. Technology, Editor. Kampala, Uganda: NITA.

18. Scoones, I. (2009). Livelihoods perspectives and rural development. *The Journal of Peasant Studies, 36*(1), 171–196.
19. Chambers, R., & Conway, G. (1992). Sustainable rural livelihoods: Practical concepts for the 21st century. Institute of Development Studies (UK).
20. Ellis, F. (2000). *Rural livelihoods and diversity in developing countries.* Oxford University Press.
21. Nakakaawa, C., et al. Collaborative resource management and rural livelihoods around protected areas: A case study of Mount Elgon National Park, Uganda. *Forest Policy and Economics, 2015*(0).
22. Beyer. (2018). Smallholder farming: Small land, large impact. The smallholder effect, in Beyer Crop Science, Beyer, Editor.
23. FAO. (2012). Smallholders and family farmers. Sustainability pathways in sustainability pathways F.a.A. Organization.
24. ASFG. (2018). Supporting smallholder farmers in Africa: A framework for an enabling environment, A.S.F.G. (ASFG).
25. Kasekende, L. (2016). Agriculture development strategy must focus on smallholder farmers, in The observer. The Observer: Kampla Uganda.
26. Lybbert, T. J., & Sumner, D. A. (2012). Agricultural technologies for climate change in developing countries: Policy options for innovation and technology diffusion. *Food Policy, 37*(1), 114–123.
27. Ngwira, A. R. (2014). Conservation agriculture systems for smallholder farmers in Malawi: An analysis of agronomic and economic benefits and constraints to adoption. Noragric, Department of International Environment and Development Studies, Norwegian University of Life Sciences.
28. Bhatnagar, S. (2000). Social implications of information and communication technology in developing countries: Lessons from Asian success stories. *The Electronic Journal of Information Systems in Developing Countries, 1*(1), 1–9.
29. Knoche, H., Rao, P. S., & Huang, J. (2010). Voices in the field: A mobile phone based application to improve marginal farmers livelihoods. In *Proceedings of SIMPE Workshop.*
30. Moser, C. O. (1998). The asset vulnerability framework: Reassessing urban poverty reduction strategies. *World Development, 26*(1), 1–19.
31. World Bank. (2016). *World development report 2016: Digital dividends.* Washington, DC: The World Bank.
32. Haseloff, A. M. (2005). Cybercafés and their potential as community development tools in India. *The Journal of Community Informatics, 1*(3).
33. Putnam, R. D., Leonardi, R., & Nanetti, R. Y. (1994). *Making democracy work: Civic traditions in modern Italy.*
34. Herselman, M., & Britton, K. (2002). Analysing the role of ICT in bridging the digital divide amongst learners. *South African Journal of Education, 22*(4), 270–274.
35. Sharples, M. (2006). Big issues in mobile learning: Report of a workshop by the kaleidoscope network of excellence mobile learning initiative: LSRI. University of Nottingham.
36. Goldsworthy, A. (2008). Why mobile phone masts can be more dangerous than the phones. Imperial College London.
37. Goldsworthy, A. (2012). Cell phone radiation and harmful effects: Just how much more proof do you need?
38. Kelly, N., Bennett, J. M., & Starasts, A. (2017). Networked learning for agricultural extension: A framework for analysis and two cases. *The Journal of Agricultural Education and Extension, 23*(5), 399–414.

39. Palfrey, J. G., & Gasser, U. (2011). Born digital: Understanding the first generation of digital natives. ReadHowYouWant.com.
40. Somekh, B. (2007). *Pedagogy and learning with ICT: Researching the art of innovation.* New York: Routledge.
41. Hetland, P. (1991). *Technology transfer to developing countries in integration, telecommunication and development: Strategies for rural telecommunications* (pp. 91–108). Norway: Norwegian Research Council.
42. Selwyn, N. (2013). *Education in a digital world: Global perspectives on technology and education.* Routledge.
43. Nampijja, D. (2010). The role of ICT in community rural development: The case of Buwama multi-media community centre Mpigi district, Uganda.

Chemical Assessment Framework and Ontology

Baboucar Diatta[1(✉)], Adrien Basse[1], Massamba Seck[1], and Samuel Ouya[2]

[1] University Alioune Diop of Bambey, Bambey, Senegal
{baboucar.diatta, adrien.basse, Massamba.seck}@uadb.edu.sn
[2] Department Computer Engineering, University Cheikh Anta Diop, Dakar, Senegal
Samuel.ouya@gmail.com

Abstract. Virtual chemical laboratory teaching helps develop and acquire skills. To build such laboratories, several research ontologies have proposed to describe chemical elements as well as their interaction. As part of the learning process, integrating such ontologies would require adding evaluation items. This article proposes an ontology that will help the learner to self-evaluate on laboratory material, safety notices and the stages through which chemistry practical work is carried out. In the literature, many methodologies address the ontology development issue. To achieve our ontology, we mainly adopt a six-step methodology: Determine the scope of the ontology; consider reusing existing ontologies; list significant terms; define classes and class hierarchy; define properties and facets of classes; and create instances. This paper presents the construction of our ontology, which can be used in many scopes of application. We plan to develop a web and mobile assessment platform to help learners to assess themselves and prepare the report of an experiment.

Keywords: Lab work · Chemical assessment · Ontology

1 Introduction

The teaching of Science, Technology, Engineering, and Mathematics (STEM) disciplines are becoming more and more important because of the increasing demand for jobs in these fields [1]. This teaching is based on theoretical knowledge as well as on the practice of experiments [2]. Defines the experimental activities as essential in science as they participate in shaping minds to rigorous, scientific method and critique. Laboratories are the places of experimentation of theoretical knowledge and the acquisition of basic skills difficult to reach by other means; [3, 4] also emphasized the importance of laboratory experiments in terms of stems; [5, 6] have demonstrated that these experiences have a real impact on the practical knowledge of learners. However, availability and access to labs are not obvious. This is why many structures have integrated virtual labs into their environment. These virtual labs make it possible to improve the performance of learners and to reduce the time of presence in classical

© Springer Nature Switzerland AG 2019
M. E. Auer and T. Tsiatsos (eds.), *Mobile Technologies and Applications for the Internet of Things*, Advances in Intelligent Systems and Computing 909,
https://doi.org/10.1007/978-3-030-11434-3_41

laboratories. Such virtual environments, accessible from a mobile or web application, will allow learners to acquire or consolidate their knowledge and skills.

Chemistry is one of the disciplines where the use of virtual labs is very present. Virtual labs in chemistry can allow learners to realize and/or follow experiments before going to the physical lab but also to evaluate themselves. The evaluation, with the analysis, was recognized as an important strategy to help learners develop knowledge and skills in their learning activities [7].

In this respect, we propose a web and mobile platform to help learners to assess themselves on the material and the products used during the experiments in chemistry, on the safety instructions to be adopted during the experiments and once in Laboratory. The learner will also be able to prepare the report of an experiment through the platform. With this platform, learners are therefore better prepared for laboratory experiments in their understanding of the safety rules to be respected, the products and the material to be used.

To achieve this platform, we have represented with the help of an ontology the steps of an experiment in chemistry, exercises on the equipment, the products used and the safety instructions. Ontology is a structured set of terms and concepts that represent the meaning of a field of information or elements of an area of knowledge [8]. Their implementation is considered as a better solution for organizing and visualizing didactic knowledge, and for this knowledge to be shared and reused by different educational applications [9]. They are, therefore, very suitable for supporting self-evaluations in online teaching.

In Sect. 2, we will review the literature on the ontology of evaluations in general and those in chemistry. Section 3 describes the construction of our ontology. Section 4 presents the web and mobile self-assessment platform. Section 5 is the conclusion.

2   Related Work

In this part, corresponding to a review of the literature, we will first address the ontology in the assessment of knowledge in general, then the ontology used in the area of chemistry. Evaluation has always been an important step in the learning process. The objective of self-evaluation activities is to improve the learning outcomes of students in order to stimulate their higher order thinking and to increase their autonomy [10].

Therefore, the development of good evaluation and feedback techniques is essential to the development of tutoring systems [11]. In [12], the authors have proposed a system that helps learners to create laboratory reports in experimental activities, which they can submit to teachers for evaluation. This evaluation gives a grade on the reports but does not provide learners with teachers' feedback. The system proposed in [13] allows teachers to give textual feedback on student Java-written programs. In [14], the authors establish an evaluation system that aims to provide a solution to improve the mathematics learning courses at the university. Students were involved in the process of designing the system of evaluation based on a mobile tablet for mathematics and deployment in real life on a mathematics course. A system for testing, examining, and evaluating the Internet learning is presented in [15]. This system is intended for use in learning modules and remote labs dealing with the design of complex digital control systems. A versatile and comprehensive mobile system, based on Android, is provided in [16] to support different types of learning and evaluation content. Another adaptive ontology for the evaluation of student learning applied to mathematics has been proposed in [17].

In the field of chemistry, most ontologies focus on the structure, characteristics of the chemical elements or on the annotation of chemical substances applied to biology. In [18], a first step toward the design of an OWL-DL ontology of functional groups for the classification of chemical compounds has been proposed. The work proposed in [18] represents a preliminary step toward the description, reasoning, and questioning of the structure and function of the molecules [19] analyses the different categories of structural classes in chemistry, presenting a list of models for the characteristics found in the class definitions. Then, compare these class definition models to tools that allow the automation of hierarchical construction within the ontology. In [20], the chemical ontology applied to biology, chebi (chemical entities of biological Interest) provides a classification of chemical entities such as atoms, molecules, and ions. The Chebi ontology contains about 15,000 chemical entities and classifies them according to common structural characteristics. Chebi is widely used as a database of chemical entities that can be queried by functional annotations in the ontology of roles. Other work such as [18] is a first step toward the design of an ontology for classifying chemical compounds into functional groups. The work presented in [18] represents a preliminary step toward the description, reasoning, and questioning of the structure and function of the molecules.

In this article, we do not focus on the structure of chemical elements such as atoms, molecules, and ions. We present a system of self-evaluation based on the ontology in order to assess knowledge of chemical materials, safety instructions, and reporting.

3 Chemical Assessment Ontology

In this paper, we have developed an ontology that allows learners to evaluate their knowledge on products and materials used, as well as safety instructions for chemistry labs. We have included, in our ontology, the preparation of a report of a chemical labwork. After defining the ontology objectives, we present the most representative methodologies used in ontology. The several methodologies that have been developed in different construction steps include [21–25]. We choose Noy and McGuinness [23] method to build our ontology because it is better suited to the context of this work. From the six-step of Noy and McGuinness methodology, we choose the three following step:

- Enumerate important terms of the ontology: the definition of important terms has been possible thanks to the checking of documents such as a book of the first year chemistry lab work and scientific articles such as [18, 26]. All those terms have been validated by an expert in the field that is responsible for the first year chemistry course of our university. This phase highlights key terms used in the proposed system.

- Define classes and class hierarchy as well as properties of classes and relations: Organization of terms by using classes, relations, and properties according to hierarchical organization principles. Free, open source ontology editor and framework protected is used from this step.
- Create instances: Populated with individual instance class of the ontology defined in step 2.

Our goal is to build our ontology by following the abovementioned steps.

3.1   Enumerate Important Terms

In this part of paper, we will proceed to define the key concepts used. This will make it possible to describe the different components of a system that will allow a self-assessment of the products and materials used, as well as safety instructions for chemical laboratories. It will describe also the preparation of a chemical laboratory report. Our first source for collecting knowledge is a book of the first year chemistry lab work at our university. We also take into account scientific articles such as [18, 26]. A part of terms used in this step is shown in Fig. 1.

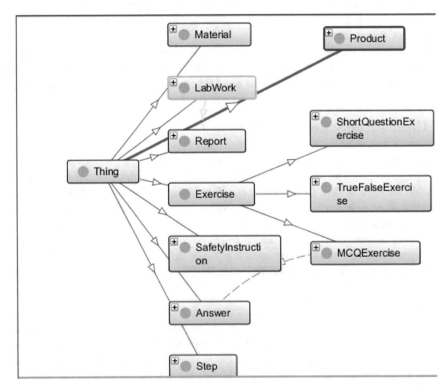

Fig. 1. Part of important terms

3.2   Classes, Properties, and Relations

This section illustrates a brief description of some classes, class hierarchy (see Fig. 2), data properties defined for each class, objet properties, and relations among elements. Figure 3 represents a part of properties and provides a brief description of data and objet properties.

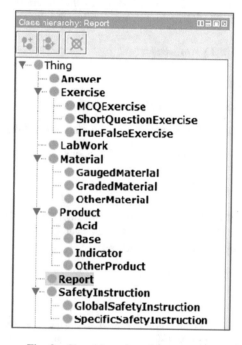

Fig. 2. Class hierarchy of the ontology

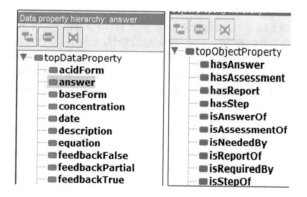

Fig. 3. Part of properties

- *Labwork*: Define lab work.
- *Exercise*: This class defines exercises used for a labwork. This class contains three subclasses: *Short Question Exercise, True False Exercise,* and *MCQ Exercise*.
- *Answer*: Define answers of exercises defined in the class Exercise.
- *Report*: This class defines the report of lab work.
- *Safety Instruction*: Describe the safety instructions. It contains two subclasses: *Global Safety Instruction* and *Specific Safety Instruction*. *Global Safety Instruction* contains safety instruction for all activities in the laboratory and *Specific Safety Instruction* regroups the instructions associated with specific material or product.

Part of Data and Object properties

- *Question* represents exercise question;

- *Name* represents a name of each element of laboratory;

- *Description* gives a small description of each element of laboratory;

- *feedback Partial/true/false* represent the feedbacks.

3.3   Instances

We have created many instances of lab work with their properties using protégé.[1] The instances that we have created include labworks, exercises, and reports in the book of the first year chemistry lab work of our university.

4   Use Case

In this section, we describe the application of ontology that we use. Our goal is to illustrate how to exploit this ontology in order to achieve a better organization of chemical assessment and report. To build web and mobile application for storing and querying RDF data, we used as SPARQL end-point Fuseki[2] and as programming language PHP through Zend framework.[3] Fuseki is a web server that implements SPARQL Protocol to expose triple stores over HTTP. We also used Bootstrap[4] and Jquery.[5] *Bootstrap* is one of the most popular HTML, CSS, and JavaScript frameworks with mobile-first fluid grid system for developing *responsive*, mobile-first websites.

Our application provides a user-friendly interface to retrieve available labworks with products and materials used, to help learners to assess themselves and to prepare a report.

---

[1] https://protege.stanford.edu/.

[2] https://jena.apache.org/documentation/fuseki2/.

[3] https://framework.zend.com/.

[4] https://getbootstrap.com/.

[5] https://jquery.com/.

Figure 4 shows a list of some exercises. For each exercise, we show the name, the description. Other properties are available through a button.

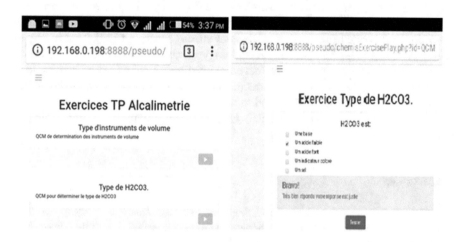

Fig. 4. Left part of exercises and right solution of exercise

The system provides information such as title of lab work report and a list of questions for lab work. Figure 5 provides a report of lab work named "Acidimetrie" with list of questions.

Fig. 5. Report of "Alcalimetrie" lab work

5 Conclusion

Self-evaluation is an important strategy that allows learners to measure the value or the quality of their work. It helps to refine the construction of knowledge and the development of skills. In this study, a self-assessment system has been proposed to enable learners so as to verify their knowledge of basic chemistry experiments. This system uses web and mobile technologies to get information from an ontology. The proposed ontology provides, among other things, a description of the products and materials used in chemistry experiments as well as the rules of safety to be observed in the laboratory. Our ontology also represents the activity of self-evaluation and incorporates reporting activities on laboratory experiments. Our system will have a positive impact on both how learners will behave in chemistry labs and how they will prepare their reports.

The monitoring of learners' learning activity is an essential component of quality education and is one of the main preachers of effective teaching [27]. That is why our future work will focus on enriching the ontology to include for example the possibility for a teacher to check if the students have prepared the exercises before coming to the lab and correcting the reports. The correction of the exercises allows the teacher to have an idea on the level of the learners and to prepare in advance a strategy to improve this level. The teacher also earns valuable time in solving complex issues during lab sessions.

References

1. DPE—Department for Professional Employee. (2013). *Fact sheet 2013, the STEM Workforce: An occupational overview*. Retrieved December 21, 2014, from http://dpeaflcio.org/wp-content/uploads/The-STEM-workforce-2013.pdf.
2. Richoux, H., & Beaufils, D. (2005). Conception de travaux pratiques par les enseignants: analyse de quelques exemples de physique en termes de transposition didactique.
3. Elawady, Y. H., & Tolba, A. S. (2010, December). A general framework for remote laboratory access: A standarization point of view. In *2010 IEEE International Symposium on Signal Processing and Information Technology (ISSPIT)* (pp. 485–490). IEEE.
4. Hashemian, R., & Riddley, J. (2007, June). FPGA e-Lab, a technique to remote access a laboratory to design and test. In *IEEE International Conference on Microelectronic Systems Education, 2007, MSE'07* (pp. 139–140). IEEE.
5. Kondabathini, V., Boutamina, S., & Vinjarapu, S. K. D. (2011, July). A theme to unite the resources of different remote laboratories. In *2011 IEEE International Conference on Technology for Education (T4E)* (pp. 51–55). IEEE.
6. Hanson, B., Culmer, P., Gallagher, J., Page, K., Read, E., Weightman, A., et al. (2009). ReLOAD: Real laboratories operated at a distance. *IEEE Transactions on Learning Technologies, 2*(4), 331–341.
7. Lai, C. L., & Hwang, G. J. (2015). An interactive peer-assessment criteria development approach to improving students' art design performance using handheld devices. *Computers & Education, 85*, 149–159.
8. Béranger, J. (2015). *Les systèmes d'information en santé et l'éthique: D'Hippocrate à e-ppocr@te*. ISTE Editions.
9. Dalipi, F., Idrizi, F., Rufati, E., & Asani, F. (2014, May). On integration of ontologies into e-learning systems. In *2014 Sixth International Conference on Computational Intelligence, Communication Systems and Networks (CICSyN)* (pp. 149–152). IEEE.
10. McMahon, T. (2010). Combining peer-assessment with negotiated learning activities on a day-release undergraduate-level certificate course (ECTS level 3). *Assessment & Evaluation in Higher Education, 35*(2), 223–239.
11. Aleven, V., Ashley, K., Lynch, C., & Pinkwart, N. (2008, June). Intelligent tutoring systems for Ill-defined domains: Assessment and feedback in Ill-defined domains. In *The 9th International Conference on Intelligent Tutoring Systems* (pp. 23–27).

12. d'Ham, C., Girault, I., Marzin, P., & Wajeman, C. LabBook un environnement collaboratif support à l'investigation scientifique pour les travaux pratiques. In *6e Conférence sur les Environnements Informatiques pour l'Apprentissage Humain* (p. 17).

13. Ahren, T. C. (2005, October). Using online annotation software to provide timely feedback in an introductory programming course. In *35th Annual Frontiers in Education* (pp. T2H-1). IEEE.

14. Isabwe, G. M. N., & Reichert, F. (2012, July). Developing a formative assessment system for mathematics using mobile technology: A student centred approach. In *2012 International Conference on Education and e-Learning Innovations (ICEELI)* (pp. 1–6). IEEE.

15. Henke, K., Debes, K., Wuttke, H. D., & Katzmann, A. (2014). Mobile assessment tools. *International Journal of Recent Contributions from Engineering, Science & IT (iJES), 2*(3), 9–14.

16. Dalir, M., Rölke, H., & Buchal, B. (2012). Android-based mobile assessment system. In *Proceedings of the 20th International Conference on Computers in Education (ICCE 2012)* (pp. 370–377). Singapore: National Institute of Education, Nanyang Technological University.

17. Lee, C. S., Wang, M. H., Chen, I. H., Lin, S. W., & Hung, P. H. (2013, December). Adaptive fuzzy ontology for student assessment. In *2013 9th International Conference on Information, Communications and Signal Processing (ICICS)* (pp. 1–5). IEEE.

18. Villanueva-Rosales, N., & Dumontier, M. (2007, June). Describing chemical functional groups in OWL-DL for the classification of chemical compounds. In *OWLED* (Vol. 258).

19. Hastings, J., Magka, D., Batchelor, C., Duan, L., Stevens, R., Ennis, M., et al. (2012). Structure-based classification and ontology in chemistry. *Journal of Cheminformatics, 4*(1), 8.

20. De Matos, P., Alcántara, R., Dekker, A., Ennis, M., Hastings, J., Haug, K., ... Steinbeck, C. (2009). Chemical entities of biological interest: An update. *Nucleic Acids Research, 38* (suppl_1), D249–D254.

21. Bachimont, B., Isaac, A., & Troncy, R. (2002, October). Semantic commitment for designing ontologies: A proposal. In *International Conference on Knowledge Engineering and Knowledge Management* (pp. 114–121). Berlin, Heidelberg: Springer.

22. Fernández-López, M., Gómez-Pérez, A., & Juristo, N. (1997). Methontology: From ontological art towards ontological engineering.

23. Noy, N. F., & McGuinness, D. L. (2001). Ontology development 101: A guide to creating your first ontology.

24. Uschold, M., & King, M. (1995). Towards a methodology for building ontologies.

25. Uschold, M., & Gruninger, M. (1996). Ontologies: Principles, methods and applications. *The Knowledge Engineering Review, 11*(2), 93–136.

26. Feldman, H. J., Dumontier, M., Ling, S., Haider, N., & Hogue, C. W. (2005). CO: A chemical ontology for identification of functional groups and semantic comparison of small molecules. *FEBS Letters, 579*(21), 4685–4691.

27. Cotton, K. J. (1988). *Monitoring student learning in the classroom*. School Improvement Research Series Close-Up# 4.

Lectures in Rotodynamic Pumps—From Design and Simulations to Testing

Djordje S. Cantrak[1(✉)], Novica Z. Jankovic[1], Milos S. Nedeljkovic[1],
Milan S. Matijevic[2], and Dejan B. Ilic[1]

[1] Faculty of Mechanical Engineering, Hydraulic Machinery and Energy Systems
Department, University of Belgrade, Belgrade, Serbia
{djcantrak, njankovic, mnedeljkovic, dilic}@mas.bg.ac.rs
[2] Faculty of Engineering, Department for Applied Mechanics and Automatic
Control, University of Kragujevac, Kragujevac, Serbia
matijevic@kg.ac.rs

Abstract. Teaching in hydraulic machines at the University of Belgrade, Faculty of Mechanical Engineering, Chair for Hydraulic Machinery and Energy Systems has a long tradition starting back in the nineteenth century. Numerous experimental test rigs are designed and novel measurement techniques are applied. Students have an ability to design, perform computational fluid dynamics (CFD), manufacture, and test pump units with the supervision of lecturers. This practical work, with concept "do it yourself", attracts students to develop their abilities in 3D geometry modeling, computational fluid dynamics, computer-aided manufacturing (CAM), and application of novel acquisition and data processing softwares. Parallel to this, they develop a piping system, which will be coupled with the assembled centrifugal pump and run the first test. Afterward, they are stimulated to connect the acquisition system with LabView software. They are supported to develop their own application for pump and hydraulic components testing and to perform the tests. The final stage is to integrate this installation in Go-Lab Sharing platform repository or at the Hydraulic machinery and Energy Systems Department website. The purpose of this educational approach is to have well educated students, not only with good theoretical background, but also with practical skills and engineering way of thinking. They have to be able to develop the operational pump hydraulic system, and not only a component.

Keywords: Dynamic learning experience · Rotodynamic pumps · Engineering education · Acquisition

1 Introduction

Students learn basics of fluid mechanics, centrifugal pump units, and hydraulic system design, as well as fluid flow measurements in the courses presented in curricula at the University of Belgrade, Faculty of Mechanical Engineering, Chair for Hydraulic Machinery and Energy Systems (HMES) [1–9]. Theoretical background is, afterward, enriched with engineering tasks through numerical problems and laboratory exercises. Some of experimental test rigs are already prepared for them and they could follow

© Springer Nature Switzerland AG 2019
M. E. Auer and T. Tsiatsos (eds.), *Mobile Technologies and Applications
for the Internet of Things*, Advances in Intelligent Systems and Computing 909,
https://doi.org/10.1007/978-3-030-11434-3_42

instructions, but the more advanced stage is to create the new installation starting from the scratch to realization [9–13]. It is proved in practice that facing the problem and providing the solution is the best way of learning.

Students generate 3D computer-aided design (CAD) model of the pump and system [14]. They, afterward, perform computational fluid dynamics (CFD) calculations. After these actions, students, with the supervision of lecturers, use computer-aided manufacturing (CAM) software to improve design and manufacture prototypes on computer numerical control (CNC) machines or by available 3D printers [15–17]. Parallel to this, they develop and manufacture piping system, with support of the laboratory staff. It will be coupled with the assembled centrifugal pump, connected with differential pressure transmitters and the first test could be performed. Afterward, they are stimulated to connect the acquisition system with the existing LabView software [10–13], which could be modified. They are supported to develop their own application for pump and hydraulic components testing and to perform the tests. The final stage is to integrate this installation in Go-Lab Sharing Platform repositories or at the HMES Department website. All these works could apply for the student competition and the best awarded project. This means that developed demonstration-educational installations could be available via the Internet [10–13]. Therefore, newer generations could be able to upgrade existing ones.

Repeating of laboratory exercises is another advantage of this education approach. It is possible to demonstrate and measure the work of a pump for raising the fluid to a geodesic height or only to circulate the water in the tank. Students are educated how to fill and start the centrifugal pump and employ energy efficiency measures. The influence of measured pump rotation speed on work, i.e., comparison of pump regulation by valve or by frequency regulation, could be demonstrated also. Students are, also, educated how to determine pump hydraulic characteristics, how to use volume method for measurement and/or calibration of flow rate measurement devices, to measure the valve hydraulic characteristic for various openings, as well as to measure pressure drop on the double elbow, as well on the test rig inlet and pipe straight section.

2 Design of the Educational Installation

Educational installation, its functionality, and possible experiments are described in the following chapters. Students, under lecturers' supervision, have prepared the drawings of the installation, calculated hydraulic losses and determined pump duty points. 3D CAD model was generated afterward. They are now able to choose adequate elements of the installation and to assemble it with the support of the laboratory staff.

2.1 Functionality and Elements of the Educational Installation

Technical task for the installation was to design a simple and not expensive one, where students would be able to:

1. follow the pump starting procedure,
2. circulate water in the tank,

3. calibrate the various differential pressure or ultrasonic flow meters by use of the volumetric method,
4. determine pump head characteristics for various rotation speeds on the basis of the international standard ISO 9906 [18],
5. regulate in-line pump duty point with three rotation numbers manually,
6. regulate in-line pump duty point manually with valve on the pressure side,
7. determine hydraulic losses, i.e. pressure drop on the regulating valve for its various positions,
8. determine hydraulic losses on other hydraulic elements in the test rig (intake element, double elbow, and friction losses on the basis of the Darcy equation in the straight pipe section) and
9. demonstrate energy efficiency issues, i.e., pump energy consumption, by comparison of pump regulation with regulating (throttling) valve versus pump rotation speed.

CAD model of this installation is presented in Fig. 1.

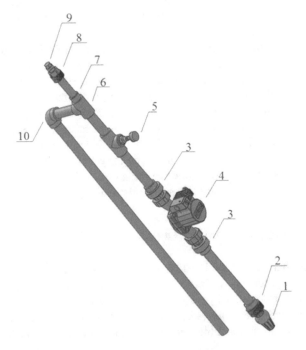

Fig. 1. CAD model of the demonstrational installation [14].

The main elements of the installation, presented in Fig. 1, are: 1—suction strainer (inlet in the installation), 2—pipe fitting, 3—coupling, 4—in-line pump, 5—regulating valve, 6—T-joint, 7—reduction, 8—fitting, 9—cap, and 10—elbow.

All hydraulic elements are modeled, except pump 3D model which is downloaded from the manufacturer's (WILO) internet page. They are presented in Fig. 2: 1—suction strainer, 2—T-joint, 3—coupling, 4—pump, 5 and 6—fittings, 7—elbow (90°), 8—fitting (reduction), 9—cap, and 10—regulating valve.

The purpose of a suction strainer is to filter ahead of the pump what prevents large particles from entering the pump. On the other hand, strainer increases hydraulic losses what is especially problem on the pump suction side, because it reduces NPSHA (net positive suction head available). NPSHA could be defined as the absolute pressure at the pump inlet, while NPSHR (NPSH required) is the minimum pressure required at the pump suction port to keep the pump from cavitating [1–4, 12]. NPSHA is a function of the system and must be calculated, while NPSHR is a function of the pump and must be provided by the pump manufacturer. In order to have the pump system operating without cavitation must be satisfied that NPSHA > NSPHR. Due to these issues with cavitation, students are instructed to position regulating valve at the pump pressure side (Fig. 1). Specific issue is determination of the valve cavitation characteristics [19] what is not demonstrated here.

Fig. 2. Hydraulic elements of the demonstrational installation [14].

2.2 Hydraulic Calculation of the Installation

In the courses of Fluid Mechanics, the students get acquainted with hydraulic calculation of the installation and three core equations: continuity, momentum and energy equations. Calculation of the liquid flow through pipes following assumptions are introduced [5–8]:

- the flow is incompressible,
- the flow is stationary, and
- the flow is one dimensional.

The main physical values, which are determined, are pressure (p) and average velocity (c). Two algebraic equations are used for this. Hydraulic calculation of the system is based on the continuity and Bernoulli equations [1, 2, 4–8]. Here is presented a simple piping system (Fig. 1).

Energy loss due to friction in turbulent flows is defined with Darcy formula. It contains friction coefficient (λ) which depends, in general, on the Reynolds number and relative roughness [1, 2, 4–8].

Weisbach equation treats local hydraulic losses (ξ). In the system presented in Fig. 1, the following local piping hydraulic losses occur: test rig inlet, regulating valve, T-joint, elbow and mechanical energy loss due to a sudden flow expansion. The last one is described with the Borda–Carnot equation and has in the case of the tank at the test rig outlet value 1 [1, 2, 4–8]. Other sudden flow expansion and contraction coefficients are neglected due to small variations in inner pipe diameter in the installation.

Coefficient of all losses (m) in the system is determined in the following way:

$$m = \frac{8}{d^4 \pi^2} \left(\Sigma \xi + \Sigma \lambda \frac{1}{d} \right), \tag{1}$$

where d is an average inner pipe diameter and l is a pipe length [1, 4, 7]. Values of the local hydraulic losses and friction are determined on the basis of many literature references [1, 2, 4, 7].

In this way is obtained the hydraulic characteristic (Y_A [J/kg]) of the simple pump system:

$$Y_A = gH_{geo} + mQ^2, \tag{2}$$

where Q [m³/s] is a volume flow rate in the installation. Geodesic height could be omitted depending on the connection of the piping system with the water tank. If it only circulates water, geodesic height does not exist.

In pump hydraulic diagram is usually presented as pump head H [m], and it could be determined on the basis of the following relation:

$$H = Y/g, \tag{3}$$

where g is Earth's gravitational acceleration, which is determined for our laboratory.

The pump head (H) is equal to the system head (H_A) under steady-state conditions and the duty point is generated.

2.3 In-built Rotodynamic Pump

In-line pump is presented in Fig. 4. It is a wet rotor pump for hot-water circulation in domestic central heating systems of the closed circuit pressurized systems, or more generalized, it is used in hot-water heating systems of all kinds, industrial circulation systems, cold water, and air-conditioning systems.

It has three preselectable rotation speeds and pretty easy and safe installation with threaded connection. It has terminal box and simplified electrical installation. It consumes following maximum powers by manufacturer's specification, respectively: 18, 30, and 45 W. It is obvious that energy consumption is really low, so it is adequate for long pump laboratory exercises. It should be installed with a horizontal shaft (Fig. 3). Maximum load pressure should not exceed 10 bar what is important if the test rig is filled directly with the tap water hosing.

Pump is presented in Fig. 3.

Fig. 3. In-built pump: 1—volute casing, 2—pump, and 3—impeller.

In-built pump has a plastic pump impeller with eight vanes. A new pump impeller without a front shroud is designed on the basis of the existing impeller (Fig. 4). Impeller is designed in CATIA software and has backward curved vanes. This pump has profiled inlet. Blade thickness is approximately 1.7 mm and it is designed to be constant.

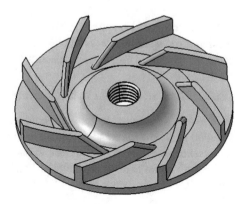

Fig. 4. Designed pump impeller without a front shroud [17].

"Vane thickness is mainly governed by the centrifugal force stresses and the manufacturing method … The minimum vane thickness is approx. 3 mm for cast iron, 4 mm for cast steel, and in special cases (e.g. inserted or welded-on sheet steel vanes) it is possible to produce even thinner vanes" [20]. Due to this fact, first test of this impeller was performed in water by use of the drill Black & Decker, model KR55CRE and no fractures occurred after applying maximum rotation number of 2800 rpm, which is declared by the manufacturer [17].

Impeller is digitally fabricated by use of the 3D printing machine—Printrbot Simple with a limited working space in the box of 150 mm side. It was used the standard black PLA/PHA ColorFabb 3D printing filament with 1.75 mm diameter. Printing parameters are provided in [17]. Printing time was approximately 36 min, with 23 layers, 22,395 lines, and 2277 mm needed filament. Therefore, this construction was pretty time and material consuming [17]. In addition, no surface treatment was applied to the fabricated pump impeller, but this could be done (Fig. 5).

Fig. 5. Digitally fabricated pump impeller.

Impeller's geometry is accommodated, so it could be installed on the in-line pump housing instead of the existing one, but some technical issues with shaft should be resolved. Hopefully, the idea of varying the impellers and testing pumps on the designed and manufactured test rig could be soon realized.

Laboratory for HMES has also capabilities to offer students to manufacture their 3D models on the CNC machines in aluminum, steel, or plastics [15, 16].

2.4   Built Complete Educational Installation and Experimental Results

After 3D modeling, calculations, collecting hydraulic elements and assembling, the complete educational installation is now built and presented in Fig. 6, where: 1—pump test rig, 2—differential pressure transmitter by Endress Hauser, model Deltabar M PMD55, range 0–1 bar, 3—differential pressure transmitter by Endress Hauser, model Deltabar M PMD55, range 0–0.1 bar, 4—water tank with barrier, and 5—pump test rig carrier.

Differential pressure transmitters are connected with transparent hoses to the built educational installation. In order to better control air bubbles and eliminate them from the installation and measuring sections.

Five measuring positions exist on the installation. The first differential pressure transmitter (Fig. 6, pos. 2) measures pressure difference after and in front of the pump, while the second one (Fig. 6, pos. 3) measures differential pressure before the T-joint and after the elbow. This "double elbow" is now treated as the volume flow meter and calibrated with the help of the calibrated water tank (Fig. 6, pos. 4). It is demonstrated, in this way, an idea that each hydraulic element with pressure drop could be flow meter.

Fig. 6. Built educational installation.

Calibration is performed in the following way. Pump transports water from the right part of the tank to the calibrated left part. Geodesic height should be kept constant during this procedure and this is a technical issue which could be resolved by using an additional reservoir. In this case, the pump system outlet is kept in the air above the calibrated part of the reservoir, presented in Fig. 6, on the specified height. The water level in the additional, i.e., suction reservoir must be kept constant. Pump is running constantly and this could be controlled by, for example, stroboscope on the rotor shaft. Regulating valve is suddenly open and the filling time is measured. Water level in the left part is measured precisely. This procedure follows standing-start-and-finish gravimetric method [21], but without the fast-acting valve. This is overcome with at least 30 s of tank filling what is satisfied even for the pump highest flow rate in the test installation. In addition, water temperature and atmospheric pressure could be recorded.

There is also a possibility to connect differential pressure transmitter to the measurement position following the test rig inlet, so the pressure drop on this hydraulic element could be determined or at the beginning and end of the pipe straight section where the pressure drop, on the basis of the Darcy equation, could be determined, etc. Therefore, possibilities are numerous.

The pump head could be determined in the following way [1, 2, 4, 18]:

$$H = (p_{II} - p_I)/(\rho g) + z_{II} - z_I, \tag{4}$$

where indexes I and II denote inlet and outlet pumps measuring sections, respectively, ρ is water density and z is a geodesic height. Due to small pressure difference, water density is considered to be constant. Difference of the kinetic energy does not exist due to the fact that pipe diameters are the same in these measuring sections. It should be, also, emphasized that in the case when water circulates, pump should take and deliver fluid in the same part of the tank in order to eliminate duty point variations.

If the power meter is connected, the measured electrical power input (P) could be used for determination of the pump unit efficiency (η) in the following manner [1, 2, 4, 18]:

$$\eta = \rho QHg/P. \tag{5}$$

After one measurement, opening of the regulating valve is now changed. This should be done in approximately five points. In this way head (H), volume flow rate (Q), and electric power (P) pump unit characteristics are obtained for one rotation speed. The whole procedure is repeated for two other rotation speeds. Students are acquainted with the pump testing procedure in this way and they could generate the report.

3   LabVIEW Application for the Educational Installation

LabVIEW application is developed for the designed hydraulic system. Students can follow the procedures described in the previous chapter and record signals from these two differential pressure transmitters (Fig. 6). These two transmitters are weird and connected with the eight channel input module National Instruments NI-9203. USB CompactDAQ chassis cDAQ-9174 is connected to a desktop computer with the installed LabVIEW software, version 2017. The developed LabVIEW application is shown in Fig. 7. Here is used application presented in [12].

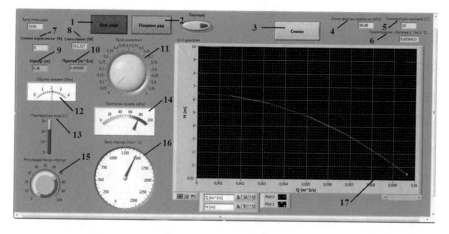

Fig. 7. LabVIEW application front panel [12].

Two transmitters are connected, one for pump head measurements, while the second one for pump flow rate. Front panel is shown in Fig. 7, where: 1—end, 2—start, 3—record, 4—atmospheric pressure, 5—ambient temperature, 6—acceleration due to gravity, 7—efficiency, 8—power, 9—head, 10—flow rate, 11—valve position, 12—torque, 13—water temperature, 14—pump maximum pressure in kPa (or in some other units), 15—pump speed in percents, 16—pump speed in rpm, and 17—simulated pump characteristic curve.

In Fig. 8 are presented experimentally obtained results on the test rig. These curves denote pressure rise on the pump in the function of the volume flow rate. Pressure difference is measured by differential pressure transmitter (Fig. 6, pos. 2), while volume flow rate by volumetric method. This could be, also, performed by using ultrasonic flow meter in the pipe straight section or by other differential pressure flow measuring devices.

Pump rotation speeds were $n_1 = 1660$ rpm, $n_2 = 2230$ rpm, and $n_3 = 2450$ rpm. They were measured by stroboscope and they varied less than 1.5%. The last curve, for the lowest speed, was determined on the basis of only three points due to the hydraulic system characteristic. This could be presented also in Q-H diagram, on the basis of the Eq. (4) and additionally measured data.

In this LabVIEW application could be determined what could be exported for students. Students, afterward, demonstrate their knowledge by conducting the experiments, discussing and presenting the results in an appropriate form.

Preparation for uploading this, or similar, application on the Go-Lab Sharing Platform repositories and/or HMES web Internet page is in progress [22].

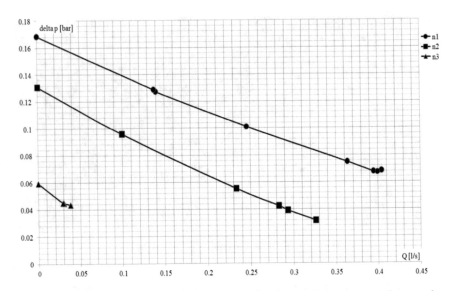

Fig. 8. Measured differential pressures on pump for three various pump rotation speeds.

4 Conclusions

This paper presents an integrated system of learning by doing in the process of facing the problems. Developed hydraulic pumps and systems, equipped with measurement and regulating devices, as well as acquisition software are demonstration-educational test rigs. This paper is a continuity of achievements presented in [10–17]. The main idea is to improve students' practical work and their overall knowledge in various hydraulic installations and problems. Better educational results are to be expected with this novel approach. Knowledge dissemination among higher number of participants is to be expected with integration of these installations in the internet repositories.

Acknowledgements. This work has been partly funded by the SCOPES project IZ74Z0_160454/1 "Enabling Web-based Remote Laboratory Community and Infrastructure" of Swiss National Science Foundation and partly by Project TR 35046 Ministry of Education, Science and Technological Development Republic of Serbia what is gratefully acknowledged.

References

1. Protić, Z., & Nedeljković, M. (2006). *Pumps and fans—Problems, solutions, theory* (in Serbian) (5th ed.). Belgrade, Serbia: Faculty of Mechanical Engineering, University of Belgrade.
2. Protić, Z. (1979). *Hydraulic machines* (in Serbian) (1st ed.). Lectures. Belgrade, Serbia: Faculty of Mechanical Engineering, University of Belgrade. Unpublished.
3. Nedeljković, M. (2012). *Design of the pumps, fans and turbocompressors* (in Serbian). Lectures. Belgrade, Serbia: Faculty of Mechanical Engineering, University of Belgrade. Unpublished.

4. Genić, S., Stamenić, M., Živković, B., Čantrak, Đ., Nikolić, A., & Brdarević, L. (2017). *Manual for energy managers' training in the industry energetics* (in Serbian). Belgrade, Serbia: Faculty of Mechanical Engineering, University of Belgrade. Chapter: 13.
5. Čantrak, S. M. (2012). *Hydrodynamics* (in Serbian) (V Rev. ed.). Belgrade, Serbia: Faculty of Mechanical Engineering, University of Belgrade.
6. Čantrak, S., Marjanović, P., Benišek, M., Pavlović, M., & Crnojević, C. (2001). *Fluid mechanics, theory and practice* (in Serbian) (VII ed.). Belgrade, Serbia: Faculty of Mechanical Engineering, University of Belgrade.
7. Čantrak, S. (2004). Applied fluid mechanics (in Serbian). In *Thermotechnical engineer* (Vol. 1, pp. 110–227). Belgrade, Serbia: Interklima-grafika, SMEITS.
8. Čantrak, S., Lečić, M., & Ćoćić, A. (2009). *Fluid mechanics B* (in Serbian). Handout. Belgrade, Serbia: Faculty of Mechanical Engineering, University of Belgrade.
9. Ilić, D., & Čantrak, Đ. (2017). *Laboratory practicum for fluid flow measurements* (in Serbian). Belgrade, Serbia: Faculty of Mechanical Engineering, University of Belgrade.
10. Nedeljkovic, M., Jankovic, N., Cantrak, D., Ilic, D., & Matijevic, M. (2018). Remote engineering education set-up of hydraulic pump and system. In *Proceedings, 15th International Conference on Remote Engineering and Virtual Instrumentation (REV 2018)*, March 21–23 (pp. 57–64), University of Applied Sciences, Duesseldorf, Germany.
11. Nedeljkovic, M., Cantrak, D., Jankovic, N., Ilic, D., & Matijevic, M. (2018). Virtual instrumentation used in engineering education set-up of hydraulic pump and system. In *Proceedings, 15th International Conference on Remote Engineering and Virtual Instrumentation (REV 2018)*, March 21–23 (pp. 341–348), University of Applied Sciences, Duesseldorf, Germany.
12. Nedeljkovic, M. S., Cantrak, D. S., Jankovic, N. Z., Ilic, D. B., & Matijevic, M. S. (2018). Virtual instruments and experiments in engineering education lab setup with hydraulic pump. In *Proceedings, 2018 IEEE Global Engineering Education Conference (EDUCON)*, April 17–20 (pp. 1145–1152), Santa Cruz de Tenerife, Canary Islands, Spain.
13. Nedeljkovic, M. S., Jankovic, N. Z., Cantrak, D. S., Ilic, D. B., & Matijevic, M. S. (2018). Engineering education lab setup ready for remote operation—Pump system hydraulic performance. In *Proceedings, 2018 IEEE Global Engineering Education Conference (EDUCON)*, April 17–20 (pp. 1175–1182), Santa Cruz de Tenerife, Canary Islands, Spain.
14. Jeremić, Đ. (2016). *Project of the installation for testing in-line pumps with small volume flow rate*. B.Sc. thesis, Faculty of Mechanical Engineering, University of Belgrade, Belgrade, Serbia.
15. Gađanski, I. I., & Čantrak, Đ. S. (2016). Kickstarting the fab lab ecosystem in Serbia—SciFabLab and FABelgrade conference, EFEA congress. In *Multidisciplinary Engineering Design Optimization—MEDO 2016, IEEE Conference, Special Session "FabLabs in Science and Education"*, September 14–16 (p. 24), Belgrade, Serbia.
16. Gadjanski, I., Čantrak, Đ., Matijević, M., & Prodanović, R. (2015). Stimulating innovations from university through the use of digital fabrication—Case study of the SciFabLab at Faculty of Mechanical Engineering, University of Belgrade. In *WBCInno2015 International Conference*, Novi Sad, Serbia.
17. Čantrak, Đ. S., Janković, N. Z., Ilić, D. B., & Lečić, M. R. (2016). Centrifugal pumps' impellers design and digital fabrication, EFEA congress. In *Multidisciplinary Engineering Design Optimization—MEDO 2016, Special Session "FabLabs in Science and Education"*, September 14–16 (p. 27), Belgrade, Serbia.
18. ISO 9906:2012 Rotodynamic pumps—Hydraulic performance acceptance tests—Grades 1, 2 and 3.

19. Chinyaev, I. R., Fominykh, A. V., & Pochivalov, E. A. (2016). Method for determining of the valve cavitation characteristics. In *International Conference on Industrial Engineering*. Procedia Engineering (Vol. 150, pp. 260–265).
20. KSB Lexikon. https://www.ksb.com/centrifugal-pump-lexicon.
21. Paton, R. (2005). Calibration and standards in flow measurements. Reproduced from the Handbook of measuring system design. Wiley.
22. Salzmann, C., Govaerts, S., Halimi, W., & Gillet, D. (2015). The smart device specification for remote labs. *International Journal of Online Engineering, 11*(4), 20–29.

Correction to: Learning Analytics for Motivating Self-regulated Learning and Fostering the Improvement of Digital MOOC Resources

D. F. O. Onah, E. L. L. Pang, J. E. Sinclair, and J. Uhomoibhi

Correction to:
Chapter "Learning Analytics for Motivating Self-regulated Learning and Fostering the Improvement of Digital MOOC Resources" in: M. E. Auer and T. Tsiatsos (eds.), *Mobile Technologies and Applications for the Internet of Things*, **Advances in Intelligent Systems and Computing 909, https://doi.org/10.1007/978-3-030-11434-3_3**

The original version of the book was inadvertently published with incorrect second author's name as "E. E. L. Pang", which has been corrected to "E. L. L. Pang" in chapter "Learning Analytics for Motivating Self-regulated Learning and Fostering the Improvement of Digital MOOC Resources".

The erratum chapter and the book have been updated with the change.

The updated version of the chapter can be found at
https://doi.org/10.1007/978-3-030-11434-3_3

Printed in the United States
By Bookmasters